NEUROMETHODS

Series Editor
Wolfgang Walz
University of Saskatchewan
Saskatoon, SK, Canada

For further volumes:
http://www.springer.com/series/7657

Analysis of Post-Translational Modifications and Proteolysis in Neuroscience

Edited by

Jennifer Elizabeth Grant

Department of Biology
University of Wisconsin-Stout
Menomonie, Wisconsin, USA

Hong Li

Department of Microbiology, Biochemistry and Molecular Genetics
Rutgers University—New Jersey Medical
School Cancer Center Newark
Newark, NJ, USA

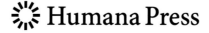

Editors
Jennifer Elizabeth Grant
Department of Biology
University of Wisconsin-Stout
Menomonie, Wisconsin, USA

Hong Li
Department of Microbiology, Biochemistry
 and Molecular Genetics
Rutgers University—New Jersey Medical
 School Cancer Center Newark
Newark, NJ, USA

ISSN 0893-2336   ISSN 1940-6045   (electronic)
Neuromethods
ISBN 978-1-4939-3470-6   ISBN 978-1-4939-3472-0   (eBook)
DOI 10.1007/978-1-4939-3472-0

Library of Congress Control Number: 2016946421

Printed on acid-free paper

This Humana Press imprint is published by Springer Nature
The registered company is Springer Science+Business Media LLC New York

Preface to the Series

Experimental life sciences have two basic foundations: concepts and tools. The *Neuromethods* series focuses on the tools and techniques unique to the investigation of the nervous system and excitable cells. It will not, however, shortchange the concept side of things as care has been taken to integrate these tools within the context of the concepts and questions under investigation. In this way, the series is unique in that it not only collects protocols but also includes theoretical background information and critiques which led to the methods and their development. Thus it gives the reader a better understanding of the origin of the techniques and their potential future development. The *Neuromethods* publishing program strikes a balance between recent and exciting developments like those concerning new animal models of disease, imaging, in vivo methods, and more established techniques, including, for example, immunocytochemistry and electrophysiological technologies. New trainees in neurosciences still need a sound footing in these older methods in order to apply a critical approach to their results.

Under the guidance of its founders, Alan Boulton and Glen Baker, the *Neuromethods* series has been a success since its first volume published through Humana Press in 1985. The series continues to flourish through many changes over the years. It is now published under the umbrella of Springer Protocols. While methods involving brain research have changed a lot since the series started, the publishing environment and technology have changed even more radically. Neuromethods has the distinct layout and style of the Springer Protocols program, designed specifically for readability and ease of reference in a laboratory setting.

The careful application of methods is potentially the most important step in the process of scientific inquiry. In the past, new methodologies led the way in developing new disciplines in the biological and medical sciences. For example, Physiology emerged out of Anatomy in the nineteenth century by harnessing new methods based on the newly discovered phenomenon of electricity. Nowadays, the relationships between disciplines and methods are more complex. Methods are now widely shared between disciplines and research areas. New developments in electronic publishing make it possible for scientists that encounter new methods to quickly find sources of information electronically. The design of individual volumes and chapters in this series takes this new access technology into account. Springer Protocols makes it possible to download single protocols separately. In addition, Springer makes its print-on-demand technology available globally. A print copy can therefore be acquired quickly and for a competitive price anywhere in the world.

Saskatoon, Canada *Wolfgang Walz*

Preface

Analysis of Post-translational Modifications and Proteolysis in Neuroscience is designed to provide detailed experimental procedures for proteomics studies of post-translational modifications (PTMs) to answer basic and translational research questions in neuroscience. Our main focus is on recent studies that apply state-of-the-art mass spectrometry techniques for either qualitative identifications of PTMs or the determination of their dynamics in neuronal cells and tissues. That said, we have also included additional cutting-edge approaches that may benefit future neuroscience studies. In many cases, the chapters presented here include novel and sophisticated examples of targeted proteomics, where much creativity is employed in the sample work-up.

Given that space is held at a premium in journal articles, some of the technical intricacies are not always described precisely within the methods and protocols sections, complicating the efforts of others to either reproduce the results or apply similar methods to study PTMs in neuronal cells and tissues. We hope this collection of articles will fill this gap and provide sufficient details that will help readers to successfully conduct similar experiments in their labs.

Applications utilizing proteomics to answer neuroscience questions have made tremendous progress. These approaches are becoming more effective due to significant improvement in the sensitivity of mass spectrometers and the sophistication of affiliated bioinformatics software. The recent availability of the Orbitrap and Triple-TOF mass spectrometers has revolutionized protein detection sensitivities to low attomole levels, while the development of electron transfer dissociation (ETD) has allowed preservation of PTMs that often degraded during MS/MS fragmentation. On the digital end, development of MaxQuant and other software packages for rapid analysis of either unlabeled or stable isotope-labeled phosphopeptides has greatly enabled our ability to understand protein-signaling networks in cells and tissues. Furthermore, state-of-the-art software such as Skyline has made targeted quantification of selected proteins in a large number of biological samples much more feasible.

Yet, despite this progress, neuroproteomicists must still confront multiple layers of complexity associated with neurological processes and structures. In both the central and peripheral nervous systems, many cells types form intricate networks, highly intertwined, to accomplish the complex biological tasks such as the ones associated with memory and thought formation, judgment, and neuromuscular activity. In this complex milieu it is very difficult to dissect cell-specific functions, where contributions of the neighboring cells often obfuscate analysis. An additional complication is that proteomes are more complex compared to genomes, where a diverse array of dynamic processes regulate PTMs, including proteolysis. Such post-translational activities are crucial for maintaining protein function, enzymatic activities, protein subcellular localization, protein-protein interactions, and even protein turnover. On a systems level, abnormally modified and processed proteins have been implicated to be accumulated in brains of patients suffering from many neurodegenerative diseases including Alzheimer's disease, Parkinson's disease, Huntington's disease, and prion diseases among others. In addition, proteins with rare PTMs, including citrullination, have been shown to be associated with increased autoimmunity in multiple sclerosis and other degenerative neuronal diseases. Therefore, it is essential to be able to accurately identify

PTMs and regulated proteolysis in neuronal proteins and quantify their dynamic changes during normal signaling events, and during disease progression.

Each individual chapter *in Analysis of Post-translational Modifications and Proteolysis in Neuroscience* is designed to provide detailed information regarding the analysis of protein processing and PTMs, as well as notes that draw attention to key steps that require extra care for the successful execution of the experiment. The methods described here are contributed by experts in neuroproteomics, most of whom published their original research on proteins derived from neuronal cells and tissues in peer-reviewed journals.

Post-translational Modifications and Processing in Neuroscience begins by considering PTM analysis from the systems level, where Gu et al. describe their technology to analyze a bevy of PTMs using an immunoaffinity LC-MS/MS proteomics approach. Their approach, termed PTMScan Direct, is focused on detecting peptide epitopes representing proteins found at key nodes within major signaling cascades, especially kinases. These authors describe methods for working with both cell culture and frozen mouse brain tissue and utilize examples from studies of dopamine signaling.

From there, a bead-based method for large-scale capture and analysis of glycoproteins, using boronic acid's ability to covalently bond with *cis*-diols, is presented by Xiao et al. This covalent bond can be reversed, leading to an effective method for large-scale glycoprotein analysis when combined with enzymatic deglycosylation in ^{18}O water, followed by mass spectrometric analysis.

Incorporation of D-amino acids is a unique modification that is often overlooked and technically challenging to detect, yet may have significant consequences in neuroscience. Jia et al. employ ion mobility spectrometry (IMS) to identify the site-specific location of D-amino acids within peptides. First, fragment ions are generated by collision-induced dissociation (CID), whereupon the presence of D-amino acid epimers is detected by analyzing potential arrival time differences between fragment ion epimers that become apparent during IMS.

Protein S-nitrosylation (SNO) or S-nitrosation is often delicately balanced with S-oxidative events and both are intimately associated with intracellular signaling. Wang and Thatcher discuss the use of the d-SSwitch method to simultaneously detect and quantify S-nitrosylation and S-oxidation events. In this instance, they use this method to elucidate the status of these PTMs in neuroblastoma cells, employing Fourier Transform Inductively Coupled Resonance (LTQ-FT-ICR) for MS and MS/MS detections. Showcasing nitrosylation impacts on a mouse model of Alzheimer's disease, Zaręba-Kozioł et al. utilize Biotin Switch as well as SNO site identification technique to identify SNO targets present within synaptosomes.

Turning additional attention to oxidation of cysteine, Jain et al. describe the use of a recombinant form of the nucleus-directed thioredoxin 1 (Trx1) mutant to trap protein targets of its activity, namely proteins targeted for disulfide reduction or denitrosylation. The authors illustrate the use of this method in neuroblastoma cells, utilizing an LTQ-Orbitrap Velos mass spectrometer for identification and quantification of affinity-captured peptides.

An additional contribution to understanding oxidative stress, Silva et al. present their method for large-scale proteomic analysis of lysine ubiquination in mouse HT22 cells, which also allows for quantification of these events. Taking advantage of the K63-TUBE recombinant system, this technique utilizes a SILAC approach to quantify ubiquitination, relying on an LTQ Orbitrap Velos mass spectrometer for detection.

Among the more prevalent PTMs, phosphorylation is crucial to a myriad number of regulatory and cell signaling events. On the one hand, phosphorylation of serine and threonine plays a key role in modulating the activity of enzymes and represents a large and diverse overall kinome. Tyrosine phosphorylation, on the other hand, occurs more rarely, functioning in select cell signaling cascades. Both PTMs provide technical challenges. In this volume, Huang et al. present a method for global phosphoproteomics profiling utilizing both hydrophobic interaction chromatography (HILIC) and titanium dioxide (TiO_2) phosphopeptide enrichment, and focusing on hybrid mouse neuroblastoma/rat glioma NG108 cells. LC-MS/MS analysis was performed with a hybrid quadrupole/Orbitrap mass spectrometer operated in tandem with HPLC.

Taking advantage of a robust epitope formed by phosphotyrosine, Chalkely and Bradshaw describe how they couple immunoprecipitation method with LC-MS/MS, in order to decipher the phosphorylation of tyrosines. Because CID utilizes beams of ions, as opposed to resonant excitation, it is more likely to generate the pTyr immonium ion at m/z 216, providing confirmation of the presence of Tyr during MS/MS analysis. In a third article, Mendoza et al. provide insight into the analysis of phosphorylation of Tau protein, a major component of neurofibrillary tangles associated with Alzheimer's disease, using an $LC-MS^3$ method. In this instance, an LTQ-Orbitrap mass spectrometer was used to acquire mass spectra, allowing for use of data-dependent neutral loss analysis of the MS^3 data. Rounding out this survey of effective methods of studying phosphorylation, Li et al. describe procedures for identifying substrates of the PINK and CAMKII kinases using CID, ETD, and higher-energy C-trap dissociation (HCD) techniques, with emphasis on use of the Orbitrap mass spectrometer.

While PTMs play diverse roles within the neuronal cell, there are also proteolytically generated bioactive peptides that shape endocrine and cellular signaling and functions. In their first description of a method to analyze proteolytic PTMs, Kockmann et al. provide expertise on how to identify candidate protease substrates using the iTRAQ-TAILS approach, utilizing a Hybrid Quadrupole-Orbitrap Mass Spectrometer. In a separate chapter, the Fricker laboratory reports on the use of isotopic labels built on a trimethylammonium butyrate base to analyze protein proteolysis in neuronal tissue, a method that allows quantification of up to five different samples within one LC-MS/MS run. For these studies, electrospray ionization Quadrupole Time of Flight (ESI q-TOF) instruments were used for detection.

Myelin, an important neuronal protein in its own right, is subjected to a variety of modifications. Proteolytic degradations can impair axon function and are recognized as a biomarker for traumatic brain injury. Complicating matters, myelin exists in any of several isoforms, presenting a severe challenge to PTM analysis. Ottens, in his contribution, provides an isotope dilution method for quantifying specific isoforms of myelin, illustrating strategies for monitoring proteolytic degradation of myelin in brain tissues. Detection involves multiple reaction monitoring (MRM) mass spectrometry. He provides a list of precursor/product ion m/z values, helpful for dynamic data acquisition (DDA) for studying the myelin isoforms, and then offers advice on detecting the MRM transitions and validating the method.

We've included one DIGE application in this volume, a testament to the power of DIGE quantitative gel electrophoresis in detecting PTMs. With this method, free sulfhydryls are first labeled with N-ethyl-maleimide. As illustrated for a palmitoylated protein, the palmitoylation is then removed chemically, freeing a sulfhydryl. This sulfhydryl is subsequently labeled with a fluorescent Cy dye as a marker for palmitoylation. Protocols for analyzing phosphorylation, ubiquination, and nitrosylation are also included. A strength of this 2D gel-based method is that specific PTM-modified protein bands/spots can be excised from the gel and sequenced using MS/MS methods.

Detyrosination and polyglutamylation events are not to be overlooked and have serious consequences for neuronal issue. Recognizing the drawbacks of antibody approaches, Mori et al. provide mass spectrometric protocols for the measurement of detyrosination and polyglutamylation of tubulin within neuronal cells. The techniques presented utilize 2-D gel electrophoresis in conjunction with strategic endoprotease digestion to enrich samples for C-terminal tubulin peptides. In this case, an AXIMA-QIT mass spectrometer was employed.

Attachment of poly-(R)-3-hydroxybutyrate (PHB) to a protein is a novel PTM whose importance in ion transport process, among others, is emerging. PHB modification presents several challenges, including disintegration of this PTM during MS analysis, and the fact that PHB adducts are polymers, resulting in variable mass additions according to the number of PHB molecules present at the site of attachment. In this application, Zakharian et al. illustrate the use of an Orbitrap mass spectrometric analysis of PHB modifications of the mammalian ion channel TRMP8 using LC-MS/MS.

This volume provides insightful protocols in the study of PTMs using proteomics techniques to understand neurological processes, provided by experts in their field. Even as much progress has been made understanding PTMs, and their impact on cell function and disease, this is an exciting time in their study as techniques are expanded to provide more global information more effectively. We, the editors, can safely anticipate more progress as sophisticated experimental regimens are more consistently brought to bear to solve challenges in neuroscience.

Menomonie, Wisconsin, USA   *Jennifer Elizabeth Grant*
Newark, NJ, USA   *Hong Li*

Contents

Contributors

OSCAR ALZATE • *Center for Translational Research in Aging and Longevity, Texas A&M Health Science Center, College Station, TX, USA*

MICHAEL A. BEMBEN • *Receptor Biology Section, National Institute of Neurological Disorders and Stroke (NINDS), National Institutes of Health (NIH), Bethesda, MD, USA; Department of Biology, Johns Hopkins University, Baltimore, MD, USA*

RALPH A. BRADSHAW • *Department of Pharmaceutical Chemistry, University of California, San Francisco, CA, USA*

NATHALIE CARTE • *Novartis Institutes for BioMedical Research, Basel, Switzerland*

ROBERT J. CHALKLEY • *Department of Pharmaceutical Chemistry, University of California, San Francisco, CA, USA*

WEIXUAN CHEN • *School of Chemistry and Biochemistry, Georgia Institute of Technology, Atlanta, GA, USA; Petit Institute for Bioengineering and Bioscience, Georgia Institute of Technology, Atlanta, GA, USA*

WEI CHEN • *Center for Advanced Proteomics Research, Rutgers University—New Jersey Medical School Cancer Center, Newark, NJ, USA; Department of Biochemistry and Molecular Biology, Rutgers University—New Jersey Medical School Cancer Center, Newark, NJ, USA*

CHUANLONG CUI • *Center for Advanced Proteomics Research, Rutgers University-New Jersey Medical School Cancer Center, Newark, NJ, USA; Department of Biochemistry and Molecular Biology, Rutgers University-New Jersey Medical School Cancer Center, Newark, NJ, USA*

HUACHENG DAI • *Center for Advanced Proteomics Research and Department of Microbiology, Biochemistry and Molecular Genetics, Rutgers University-New Jersey Medical School Cancer Center, Newark, NJ, USA*

GEORGIA DOLIOS • *Department of Genetics and Genomic Sciences, Icahn School of Medicine at Mount Sinai, New York, NY, USA*

LLOYD D. FRICKER • *Department of Molecular Pharmacology, Albert Einstein College of Medicine, Bronx, NY, USA*

HONGBO GU • *Cell Signaling Technology, Inc., Danvers, MA, USA*

ERIC D. HAMLETT • *Medical University of South Carolina, Charleston, SC, USA*

FANG-KE HUANG • *Department of Biochemistry and Molecular Pharmacology and Kimmel Center for Molecular Medicine at the Skirball Institute, New York University School of Medicine, New York, NY, USA*

KOJI IKEGAMI • *Department of Cell Biology and Anatomy, Hamamatsu University School of Medicine, Hamamatsu, Japan*

MOHIT RAJA JAIN • *Center for Advanced Proteomics Research and Department of Microbiology, Biochemistry and Molecular Genetics, Rutgers University-New Jersey Medical School Cancer Center, Newark, NJ, USA*

CHENXI JIA • *School of Pharmacy, University of Wisconsin–Madison, Madison, WI, USA; National Center for Protein Sciences-Beijing, and Beijing Proteome Research Center, Beijing, China*

LESLEY A. KANE • *Biochemistry Section, Surgical Neurology Branch, National Institute of Neurological Disorders and Stroke (NINDS), National Institutes of Health (NIH), Bethesda, MD, USA*

ULRICH AUF DEM KELLER • *ETH Zurich, Department of Biology, Institute of Molecular Health Sciences, Zurich, Switzerland*

TOBIAS KOCKMANN • *ETH Zurich, Department of Biology, Institute of Molecular Health Sciences, Zurich, Switzerland*

ALU KONNO • *Department of Cell Biology and Anatomy, Hamamatsu University School of Medicine, Hamamatsu, Japan*

MACIEJ LALOWSKI • *Medicum, Institute of Biomedicine, Biochemistry/Developmental Biology, Meilahti Clinical Proteomics Core Facility, University of Helsinki, Helsinki, Finland; Folkhälsan Institute of Genetics, Helsinki, Finland*

YAN LI • *Protein/Peptide Sequencing Facility, National Institute of Neurological Disorders and Stroke (NINDS), National Institutes of Health (NIH), Bethesda, MD, USA*

LINGJUN LI • *School of Pharmacy, University of Wisconsin–Madison, Madison, WI, USA; Department of Chemistry, University of Wisconsin–Madison, Madison, WI, USA; School of Life Sciences, Tianjin University, Tianjin, China*

QING LI • *Center for Advanced Proteomics Research and Department of Microbiology, Biochemistry and Molecular Genetics, Rutgers University-New Jersey Medical School Cancer Center, Newark, NJ, USA*

HONG LI • *Department of Microbiology, Biochemistry and Molecular Genetics, Rutgers University—New Jersey Medical School Cancer Center, Newark, NJ, USA*

CHRISTOPHER B. LIETZ • *Department of Chemistry, University of Wisconsin–Madison, Madison, WI, USA*

TONG LIU • *Center for Advanced Proteomics Research, Rutgers University—New Jersey Medical School Cancer Center, Newark, NJ, USA; Department of Biochemistry and Molecular Biology, Rutgers University—New Jersey Medical School Cancer Center, Newark, NJ, USA*

SANDHYA MANOHAR • *Department of Biology, Center for Genomics and Systems Biology, New York University, New York, NY, USA*

SAMU MELKKO • *Novartis Institutes for BioMedical Research, Basel, Switzerland*

JHOANA MENDOZA • *Biomarker Discovery and Development Laboratory, Sanford-Burnham Medical Research Institute, Orlando, FL, USA*

YASUKO MORI • *Department of Cell Biology and Anatomy, Hamamatsu University School of Medicine, Hamamatsu, Japan*

THOMAS A. NEUBERT • *Department of Biochemistry and Molecular Pharmacology and Kimmel Center for Molecular Medicine at the Skirball Institute, New York University School of Medicine, New York, NY, USA*

CRISTINA OSORIO • *Department of Biological Sciences, University of Texas at El Paso, El Paso, TX, USA*

ANDREW K. OTTENS • *Department of Anatomy and Neurobiology, Virginia Commonwealth University School of Medicine, Richmond, VA, USA*

STACEY PAN • *Center for Advanced Proteomics Research, Rutgers University—New Jersey Medical School Cancer Center, Newark, NJ, USA; Department of Biochemistry and Molecular Biology, Rutgers University—New Jersey Medical School Cancer Center, Newark, NJ, USA*

SUJEEWA C. PIYANKARAGE • *Department of Medicinal Chemistry and Pharmacognosy, University of Illinois at Chicago, Chicago, IL, USA*

MITSUTOSHI SETOU • *Department of Cell Biology and Anatomy, Hamamatsu University School of Medicine, Hamamatsu, Japan*

JEFFREY C. SILVA • *Cell Signaling Technology, Inc., Danvers, MA, USA; Lighthouse Proteomics LLC, Beverly, MA, USA*

GUSTAVO MONTEIRO SILVA • *Department of Biology, Center for Genomics and Systems Biology, New York University, New York, NY, USA*

MATTHEW P. STOKES • *Cell Signaling Technology, Inc., Danvers, MA, USA*

GEORGE X. TANG • *School of Chemistry and Biochemistry, Georgia Institute of Technology, Atlanta, GA, USA; Petit Institute for Bioengineering and Bioscience, Georgia Institute of Technology, Atlanta, GA, USA*

GREGORY R.J. THATCHER • *Department of Medicinal Chemistry and Pharmacognosy, University of Illinois at Chicago, Chicago, IL, USA*

CHRISTINE VOGEL • *Department of Biology, Center for Genomics and Systems Biology, New York University, New York, NY, USA*

KATHERINE W. ROCHE • *Receptor Biology Section, National Institute of Neurological Disorders and Stroke (NINDS), National Institutes of Health (NIH), Bethesda, MD, USA*

YUE-TING WANG • *UICentre (Drug Discovery at UIC), University of Illinois at Chicago, Chicago, IL, USA*

RONG WANG • *Department of Genetics and Genomic Sciences, Icahn School of Medicine at Mount Sinai, New York, NY, USA*

WEI WEI • *Department of Biology, Center for Genomics and Systems Biology, New York University, New York, NY, USA*

CHANGGONG WU • *Center for Advanced Proteomics Research and Department of Microbiology, Biochemistry and Molecular Genetics, Rutgers University-New Jersey Medical School Cancer Center, Newark, NJ, USA*

RONGHU WU • *School of Chemistry and Biochemistry, Georgia Institute of Technology, Atlanta, GA, USA; Petit Institute for Bioengineering and Bioscience, Georgia Institute of Technology, Atlanta, GA, USA*

ALEKSANDRA WYSŁOUCH-CIESZYŃSKA • *Institute of Biochemistry and Biophysics, Polish Academy of Sciences, Warsaw, Poland*

HAOPENG XIAO • *School of Chemistry and Biochemistry, Georgia Institute of Technology, Atlanta, GA, USA; Petit Institute for Bioengineering and Bioscience, Georgia Institute of Technology, Atlanta, GA, USA*

QING YU • *School of Pharmacy, University of Wisconsin–Madison, Madison, WI, USA*

ELEONORA ZAKHARIAN • *Department of Cancer Biology and Pharmacology, University of Illinois College of Medicine, Peoria, IL, USA*

MONIKA ZARĘBA-KOZIOŁ • *Institute of Biochemistry and Biophysics, Polish Academy of Sciences, Warsaw, Poland*

GUOAN ZHANG • *Department of Biochemistry and Molecular Pharmacology and Kimmel Center for Molecular Medicine at the Skirball Institute, New York University School of Medicine, New York, NY, USA*

Neuromethods (2016) 114: 1–29
DOI 10.1007/7657_2015_99
© Springer Science+Business Media New York 2015
Published online: 08 November 2015

Proteomic Analysis of Posttranslational Modifications in Neurobiology

Hongbo Gu, Matthew P. Stokes, and Jeffrey C. Silva

Abstract

Advancements in mass spectrometry instrumentation and proteomics software have made LC-MS technology the primary tool for discovering protein biomarkers in many different areas of biological research, including neurobiology. Coupling novel approaches for sample preparation such as antibody enrichment and other affinity techniques has facilitated proteomic efforts to characterize proteins on a global scale, and has led to a better understanding of the complex dynamic changes in protein expression and posttranslational modifications (PTMs). Together these techniques have proven effective in the study of the functional role of complex signaling and regulatory networks in many model systems. Here we describe a general approach for immunoaffinity purification (IAP) and LC-MS/MS-directed identification and quantification of a variety of critical PTMs such as phosphorylation, methylation, acetylation, succinylation, ubiquitination, and cleaved caspase proteolytic processing. Immunoaffinity LC-MS has been routinely used for characterizing changes in response to a multitude of biological perturbations or treatment conditions, and is applicable for numerous model systems and species whose genome sequences are available.

Keywords: Posttranslational modification, Immunoaffinity LC-MS/MS, Phosphopeptide profiling, Acetylation, Succinylation, Ubiquitination, Methylation

Abbreviations

MeCN	Acetonitrile
IAP	Immunoaffinity purification
IMAC	Immobilized metal affinity chromatography
LC	Liquid chromatography
LC-MS/MS	Tandem mass spectrometry
MS	Mass spectrometry
PTM	Posttranslational modification
TFA	Trifluoroacetic acid

1   Introduction

Developmental neuroscience focuses on growth, death, and morphological changes of neuronal cells and involves many complex signaling networks. These topics include the differentiation of neuronal stem cells along the neural crest, glial or neuronal lineages, and the growth and maturation of neurons and neuronal connectivity [1–4]. These networks are modulated by posttranslational modifications of key proteins that drive each of these critical pathways. Posttranslational modifications are key regulators of protein function, and they play fundamental roles in cell development, but due to the nature of their low abundance, their identification and the analysis of the cellular processes that they regulate are challenging. Proteomic methods that can efficiently assay protein/peptide PTMs, and quantitatively monitor their alterations in the context of signaling pathways, are urgently needed for the study of disease mechanisms and advancement of basic and translational research and would be a valuable tool for drug development programs.

Genetic methods have long been available to profile many genes or whole genomes simultaneously, such as comparative genomic hybridization arrays, single-nucleotide polymorphism analysis, or whole-genome sequencing. However, these methods have the disadvantage that many changes observed at the genetic level do not necessarily affect progression of the disease. Quantitative proteomic methods represent a more direct measure of changes that affect various disease states, providing information complementary to that obtained through genomic methods [5–8].

Historically, the study of protein activity in complex diseases and cellular signaling pathways has either focused on a few proteins known to be critical to the model system of interest or has employed proteomic methods that provide rich data sets that randomly sample the entire proteome with limited coverage of the signaling pathways known to be involved in the underlying pathology. The detailed study of one or a few specific proteins has the advantage of focusing on known pathway components but suffers from an inability to sample many data points from complex systems. Proteomic analyses using liquid chromatography-tandem mass spectrometry (LC-MS/MS) have allowed simultaneous profiling of many thousands of proteins and PTMs but some peptide enrichment methods, such as IMAC, can suffer from a lack of sensitivity for low-abundance proteins and their associated PTMs [9, 10]. The metal affinity methods tend to enrich for the more abundant phosphoserine and phosphothreonine peptides present in a sample, whereas critical signaling may occur through phosphotyrosine-modified proteins that are present at low levels. For example, in a yeast protein extract, the IMAC method was used to identify 216 phosphopeptides but only 3 phosphotyrosine phosphorylation sites [11]. In a separate

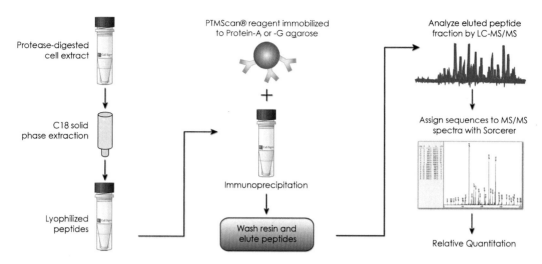

Fig. 1 PTMScan™ method. An immunoaffinity LC-MS/MS method for profiling a variety of different PTMs by utilizing the appropriate motif antibody or PTM-specific antibody of interest (serine, threonine, and tyrosine phosphorylation, ubiquitination, neddylation, ISGylation, acetylation, succinylation, methylation, and/or cleaved caspase)

study, IMAC was used to perform a phosphopeptide analysis of capacitated human sperm cells, whereby only a small fraction of the isolated phosphopeptides were phosphotyrosine [12]. The use of antibodies to immunoaffinity enrich for specific PTMs can overcome these limitations [13, 14].

Using antibodies to enrich for specific PTMs allows studies to be focused on a particular class of related PTM peptides, such as those sharing a consensus phosphorylation motif or against a particular posttranslational modification type [15–20]. Employing these antibody-based strategies has yielded many insights into signaling pathways and key regulators of disease. Current methods allow for simultaneous identification and quantification of thousands of PTM peptides across serine, threonine, and tyrosine phosphorylation, as well as ubiquitination, neddylation, ISGylation, acetylation, succinylation, methylation, and cleaved caspase substrates (*see* Fig. 1) using antibodies engineered for peptide enrichment (http://www.cellsignal.com/common/content/content.jsp?id= motif-antibody-kits) [21–24]. This particular approach can be applied to any model system (prokaryotic, eukaryotic, or viral) since the motif antibody reagents were designed to recognize peptides with a particular PTM or a PTM in the context of a conserved sequence motif. In a typical experiment, cells, tissues, or other biological materials are prepared under denaturing conditions to produce a protein lysate. The protein lysate is subsequently digested to peptides with a standard protease such as trypsin (or endoproteinase LysC) and the resulting peptides are purified by reverse-phase chromatography. The purified peptides are subjected to immunoaffinity purification (IAP) using antibody bound to Protein A (or Protein G) beads. Bound peptides are eluted off the beads, desalted over C18, and analyzed under standard LC-MS conditions. The relative abundance of the each identified PTM-modified peptide ion is compared across multiple samples using either a label-free quantitative method that reports the integrated peak area (or height) in the MS1 channel (corresponding to the intact peptide ion) or labeling methods such as SILAC, reductive amination, or isobaric tagging [24–33]. This method provides quantitative analysis of hundreds to thousands of PTM-modified peptides in a single experiment.

One of the key factors contributing to the success of the immunoaffinity LC-MS method for PTM profiling was the development of motif antibodies [34]. In a typical antibody development program, one peptide or protein antigen is used to elicit an immune response for the antigen-specific antibody. However, with motif antibodies (and PTM antibodies), degenerate peptide libraries are utilized for immunization, in which one or a few key residues are fixed and present in all peptides in the library, while other amino acids in the conserved peptide sequence are varied. The concept of a phosphorylation motif antibody is illustrated by a phospho-Akt substrate antibody, which recognizes phosphopeptides with Arg in the -3 and -5 positions, relative to the site of phosphorylation (RXXRX[pS/pT]). This sequence corresponds to the conserved substrate sequence motif recognized and phosphorylated by the kinase, Akt [15]. This strategy has allowed the development of antibodies that recognize phosphotyrosine, a series of antibodies that recognize consensus kinase substrate motifs, and other posttranslational modifications such as lysine-acetylation, lysine-succinylation, lysine- and arginine-methylation, and lysine-ubiquitination.

The combination of IAP enrichment and LC-MS/MS analysis is not limited to analysis of PTM classes, but can be applied as well to the study of specific signaling pathways [25, 35]. This approach, termed PTMScan Direct, focuses on a defined set of peptides from proteins to monitor specific nodes in key signaling pathways or derived from a single protein type, such as kinases (http://www.cellsignal.com/common/content/content.jsp?id=proteomics-direct). To date, PTMScan Direct antibody mixtures have been developed for IAP enrichment and LC-MS/MS analysis in the following signaling areas: (1) tyrosine kinases, (2) serine/threonine kinases, (3) Akt/PI3K signaling, (4) cell cycle and DNA damage, (5) apoptosis and autophagy, and (6) a multi-pathway reagent that monitors key regulators in approximately 13 critical signaling pathways, including Akt signaling, MAP kinase signaling, cell cycle regulation, apoptosis, and TGF signaling (*see* Fig. 2 and Table 1).

The combination of IAP enrichment and LC-MS/MS analysis has been used to conduct many types of proteomic studies, as illustrated above. The immunoaffinity LC-MS method can be used to probe a variety of PTM-related studies, such as the following examples listed below:

- Discovering candidate biomarkers linked to activation or repression of a specific class of PTMs.
- Performing a comprehensive proteomic survey of thousands of PTM sites associated with a new model organism or disease model, integrating the data into known signaling pathways or helping to define novel signaling networks.

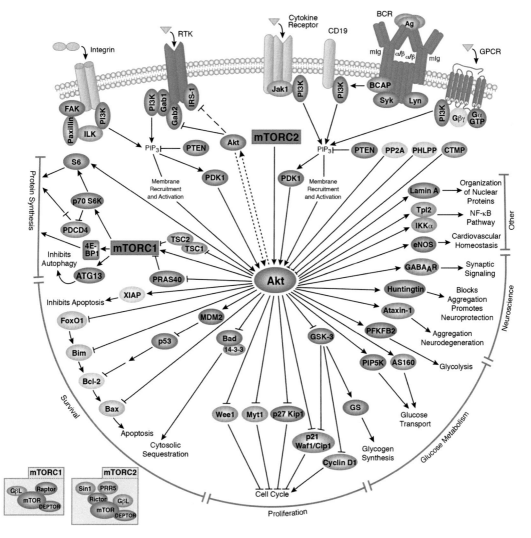

Fig. 2 Two representative signaling pathways targeted by PTMScan Direct Multi-Pathway. Two of the nine critical signaling pathways monitored by PTMScan Direct Multi-Pathway, Akt/PI3K signaling (*left*) and AMPK signaling (*right*). Phosphopeptides to the proteins highlighted in *purple* are accessible using the immunoaffinity enrichment reagent

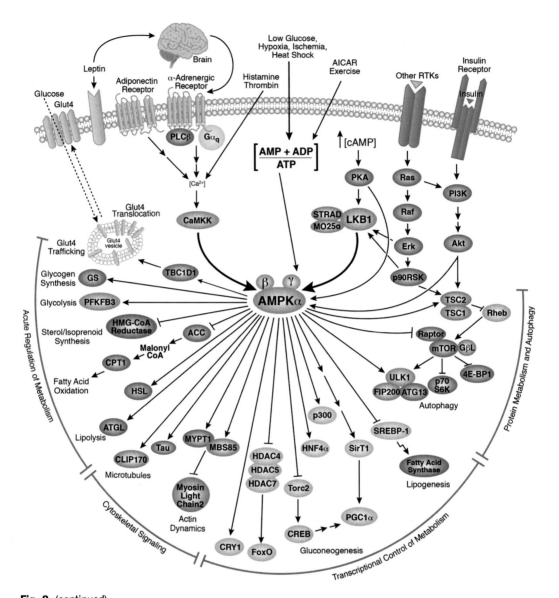

Fig. 2 (continued)

- Detecting substrates of novel signaling proteins (kinases, phosphatases, ubiquitin ligases, deubiquitinases, acetyl transferases, methyl transferases, or succinyl transferases).

- Profiling global effects of a candidate therapeutic on a specific type of PTM, identifying nodes of interest for further study.

- Exploring how cross talk among various PTMs (phosphorylation, acetylation, succinylation, methylation, ubiquitination) is related to a particular biological response or involved in cell development or differentiation.

Table 1
Targeted coverage of signaling pathways or kinase classes using PTMScan

PTMScan direct reagent type	PTM sites recognized	Proteins covered
Multi-pathway	1006	409
PI3K/Akt signaling	296	105
Apoptosis/autophagy signaling	175	100
Cell cycle/DNA damage signaling	168	263
Serine/threonine kinases	385	130
Tyrosine kinases	671	120

A total of six different PTMScan Direct reagents are available for pathway-specific PTM profiling. For each immunoaffinity enrichment reagent, the total number of validated PTM sites recognized and the total number of corresponding proteins are indicated

- Analyzing downstream effects of targeted gene silencing (siRNA, CRISPR) on signaling and activation of alternative compensatory pathways.

- Identifying protein-protein interaction binding partners along with their corresponding PTMs.

In addition to the LC-MS/MS methods described in this chapter, there are hundreds of activation-state and total protein antibodies available for the study of molecular, cellular, and developmental neuroscience as well as neurodegenerative diseases. These reagents have been validated for a number of different applications such as western blot, immunofluorescence, immunohistochemistry, flow cytometry, and chromatin immunoprecipitation. Among these include antibodies directed against neurotransmitter receptors such as glutamate, GABA, and serotonin. These neurotransmitters cross the synaptic gap between individual neurons, resulting in intercellular signaling and the ability to rapidly communicate nervous system signals throughout the body. At the molecular level, reagents are available to monitor protein participants of neuronal signaling events and brain function. Such targets include Akt, CDK3, p35, Tau, and GSK-3, which are involved in the production of the amyloid plaque and neurofibrillary tangles characteristic of Alzheimer's disease (Fig. 3—Alzheimer's Pathway) as well as those critical nodes involved in dopamine signaling in Parkinson's disease (Fig. 4—Parkinson's Pathway).

The following method will describe in detail the protocol for performing IAP and liquid chromatography-tandem mass spectrometry (IAP LC-MS/MS) in nine sections: (a) general solutions and reagents, (b) cell lysis, (c) reduction and alkylation of proteins,

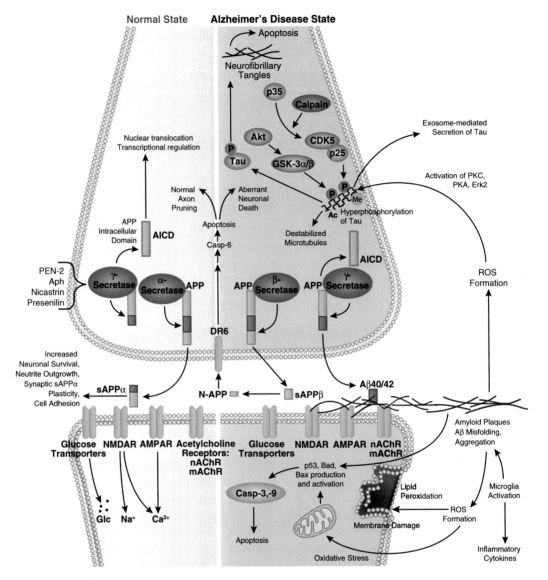

Fig. 3 Amyloid plaque and neurofibrillary tangle formation in Alzheimer's disease. Antibody reagents to total and site-specific PTMs are available for many proteins involved in amyloid plaque and neurofibrillary tangle formation (http:/www.cellsignal.com/common/content/content.jsp?id=pathways-park)

(d) protein lysate digestion, (e) Sep-Pak C18 purification of lysate peptides, (f) IAP of PTM peptides, (g) concentration and purification of peptide for LC-MS analysis, (h) LC-MS/MS analysis of peptides, and (i) sample data (http://www.phosphosite.org/staticSupp.do) review from mouse brain tissue.

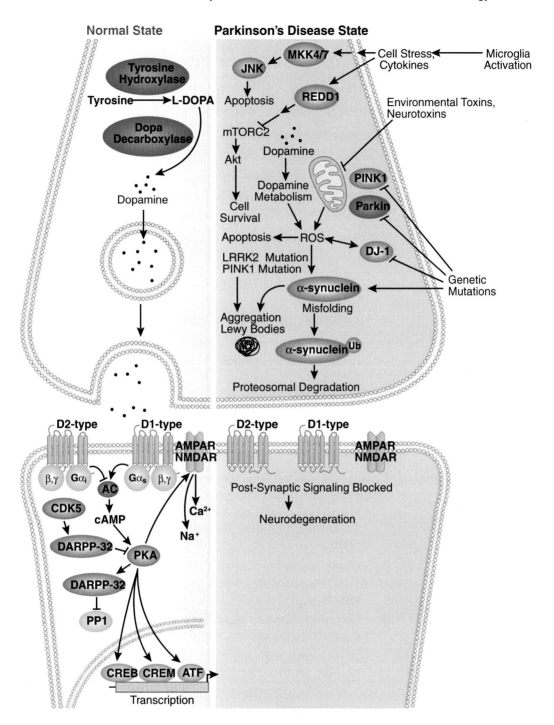

Fig. 4 Dopamine signaling in Parkinson's disease. Antibody reagents to total and site-specific PTMs are available for many proteins involved in dopamine signaling (http:/media.cellsignal.com/www/pdfs/science/pathways/Dopamine_Parkinsons.pdf)

2   Immunoaffinity LC-MS Protocol for PTM Peptides Using PTMSCAN™ Antibodies

**2.1   General
Solutions
and Reagents**

1. HEPES (Sigma, H-4034)
2. Sodium pyrophosphate (Sigma, S-6422)
3. β-Glycerophosphate (Sigma, G-9891)
4. Urea, Sequanal Grade (Thermo Fisher Scientific, 29700)
5. Sodium orthovanadate (Sigma, S-6508)
6. Iodoacetamide (Sigma, I-6125)
7. Dithiothreitol (DTT) (American Bioanalytical, AB-00490)
8. Trypsin-TPCK (Worthington, LS-003744)
9. Trypsin (Promega, V5113)
10. Lysyl Endopeptidase, LysC (Wako, 129-02541)
11. Trifluoroacetic acid, Sequanal Grade (Thermo Fisher Scientific, 28903)
12. Acetonitrile (Thermo Fisher Scientific, 51101)
13. Sep-Pak® Classic C18 columns, 0.7 mL (Waters, WAT051910)
14. Burdick and Jackson Water (Honeywell, AH365-4)

2.1.1   Note

Prepare solutions for cell lysis, Sep-Pak purification, and IAP enrichment with Milli-Q or equivalent grade water. Prepare solutions for subsequent steps with HPLC-grade water (Burdick and Jackson water).

2.1.2   Stock Solutions

1. 200 mM HEPES/NaOH, pH 8.0: Dissolve 23.8 g HEPES in approximately 450 mL water, adjust pH with 5 M NaOH to 8.0, and bring to a final volume of 500 mL. Filter through a 0.22 μM filter (as used for cell culture), and use for up to 6 months.

2. Sodium pyrophosphate: Make 50× stock (125 mM, MW = 446): 1.1 g/20 mL. Store at 4 °C, and use for up to 1 month.

3. β-Glycerophosphate: Make 1000× stock (1 M, MW = 216): 2.2 g/10 mL. Divide into 100 μL aliquots and store at −20 °C.

4. Sodium orthovanadate (Na_3VO_4): Make 100× stock (100 mM, MW = 184): 1.84 g/100 mL. Sodium orthovanadate must be depolymerized (activated) according to the following protocol:

   (a) For a 100 mL solution, fill up with water to approximately 90 mL. Adjust the pH to 10.0 using 1 M NaOH with stirring. At this pH, the solution will be yellow.

   (b) Boil the solution until it turns colorless and cool to room temperature (put on ice for cooling).

(c) Readjust the pH to 10.0 and repeat **step 2** until the solution remains colorless and the pH stabilizes at 10.0 (usually it takes two rounds). Adjust the final volume with water.

(d) Store the activated sodium orthovanadate in 1 mL aliquots at -20 °C. Thaw one aliquot for each experiment; do not refreeze thawed vial.

5. DTT: Make 1.25 M stock (MW = 154): 19.25 g/100 mL. Divide into 200 μL aliquots, and store at -20 °C for up to 1 year. Thaw one aliquot for each experiment.

6. Trypsin-TPCK: Store dry powder for up to 2 years at -80 °C. Parafilm cap of trypsin container (Worthington) to avoid collecting moisture, which can lead to degradation of the reagent. Prepare 1 mg/mL stock in 1 mM HCl. Divide into 1 mL aliquots, and store at -80 °C for up to 1 year.

7. Lysyl Endopeptidase (LysC): Store dry powder up to 2 years at -80 °C. Parafilm cap of LysC container to avoid collecting moisture, which can lead to degradation of the reagent. Prepare 5 mg/mL stock in 20 mM HEPES pH 8.0. Divide into single-use aliquots, and store at -80 °C for up to 1 year.

2.2 Cell Lysis

2.2.1 Solutions and Reagents

1. **NOTE**: Prepare solutions with Milli-Q or equivalent grade water.

1. Urea lysis buffer: 20 mM HEPES pH 8.0, 9 M urea, 1 mM sodium orthovanadate, 2.5 mM sodium pyrophosphate, 1 mM β-glycerophosphate.

2. **NOTE**: The urea lysis buffer should be prepared prior to each experiment. Aliquots of the urea lysis buffer can be stored in the -80 °C freezer for up to 6 months.

3. **NOTE**: Dissolving urea is an endothermic reaction. Urea lysis buffer preparation can be facilitated by placing a stir bar in the beaker and by using a warm (not hot) water bath on a stir plate. 9 M urea is used so that upon lysis, the final concentration is approximately 8 M. The urea lysis buffer should be used at room temperature. Placing the urea lysis buffer on ice will cause the urea to precipitate out of solution.

2. DTT solution, 1.25 M (see stock solutions for preparation).

3. Iodoacetamide solution: Dissolve 95 mg of iodoacetamide (formula weight = 184.96 mg/mmol) in water to a final volume of 5 mL. After weighing the powder, store in the dark and add water only immediately before use. The iodoacetamide solution should be prepared fresh prior to each experiment.

2.2.2 Preparation of Cell Lysate from Suspension Cells

1. Grow approximately $1–2 \times 10^8$ cells for each experimental condition (enough cells to produce approximately 10–20 mg of soluble protein).

1. NOTE: Cells should be washed with $1\times$ PBS before lysis to remove any media containing protein contaminants. Elevated levels of media-related proteins will interfere with the total protein determination.

2. Harvest cells by centrifugation at 130 rcf (g), for 5 min at room temperature. Carefully remove supernatant, wash cells with 20 mL of cold PBS, centrifuge and remove PBS wash, and add 10 mL urea lysis buffer (room temperature) to the cell pellet. Pipet the slurry up and down a few times (do not cool lysate on ice as this may cause precipitation of the urea).

2. NOTE: If desired, the PTMScan™ protocol may be interrupted at this stage. The harvested cells can be frozen and stored at $-80\,°C$ for several weeks.

3. Using a microtip, sonicate at 15 W output with three bursts of 15 s each. Cool on ice for 1 min between each burst. Clear the lysate by centrifugation at 20,000 rcf (g) for 15 min at 15 °C or room temperature and transfer the protein extract (supernatant) into a new tube.

3. NOTE: Centrifugation is performed at 15 °C or room temperature to prevent urea from precipitating out of solution. Centrifugation should be performed in an appropriate container rated for at least 20,000 rcf (g).

2.3 Preparation of Cell Lysate fromAdherent Cells

1. Grow $1–2 \times 10^8$ cells for each experimental condition (enough cells to produce approximately 10–20 mg of soluble protein). The cell number corresponds to approximately three to ten 150 mm culture dishes (depending on the cell type), grown to between 70 and 80 % confluence.

1. NOTE: Cells should be washed with $1\times$ PBS before lysis to remove any media containing protein contaminants. Elevated levels of media-related proteins will interfere with the total protein determination.

2. Take all 150 mm culture dishes for one sample, remove media from the first dish by decanting, and let stand in a tilted position for 30 s so the remaining medium flows to the bottom edge. Remove the remainder of the medium at the bottom edge with a P-1000 micropipettor. Wash each dish with 5 mL of cold PBS. Remove PBS as described above.

3. Add 10 mL of urea lysis buffer (at room temperature) to the first dish, scrape the cells into the buffer, and let the dish stand in tilted position after scraping the buffer to the bottom edge

of the tilted dish. Remove the medium from the second dish as above. Transfer the lysis buffer from the first dish to the second dish using a 10 mL pipette, then tilt the first dish with the lid on for 30 s, and remove the remaining buffer from the dish and collect. Scrape cells from the second dish and repeat the process until the cells from all the dishes have been scraped into the lysis buffer. Collect all lysate in a 50 mL conical tube.

2. NOTE: DO NOT place urea lysis buffer or culture dishes on ice during harvesting. Harvest cells using urea lysis buffer at room temperature. During lysis, the buffer becomes viscous due to DNA released from the cells.

4. The yield will be approximately 9–12 mL lysate after harvesting all the culture plates.

3. NOTE: If desired, the PTMScan protocol may be stopped at this stage. The cell lysate can be frozen and stored at −80 °C for several weeks.

5. Using a microtip, sonicate at 15 W output with three bursts of 15 s each. Cool on ice for 1 min between each burst. Clear the lysate by centrifugation at 20,000 rcf (g) for 15 min at 15 °C or room temperature and transfer the protein extract (supernatant) into a new tube.

4. NOTE: Lysate sonication fragments DNA and reduces sample viscosity. Ensure that the sonicator tip is submerged in the lysate. If the sonicator tip is not submerged properly, it may induce foaming and degradation of your sample (refer to the manufacturer's instruction manual for the sonication apparatus).

2.4 Preparation of Cell Lysate from Frozen Tissue

1. NOTE: Where possible, with xenograft or general tissue samples, the tissue representing each experimental condition should be pooled from at least three different animals. This is to average the biological variability for each condition.

2. NOTE: For xenograft tissue, the tissue weight from each animal should be no greater than 150 mg to ensure the healthy tissue, without any signs of necrosis influencing the experiment.

1. Harvest tissue. Separate approximately 50 mg of tissue (wet weight) for each experimental condition and reserve the material for other supplemental analyses (e.g., western blots, IHC staining). Flash freeze the 50 mg samples directly in liquid nitrogen and transfer them into labeled cryo-vials. Store the tissue samples at −86 °C.

2. Cut the remaining tissue (300–450 mg, wet weight) into small pieces and place into a round-bottom centrifuge tube.

3. Add 1 mL of freshly prepared urea lysis buffer for each 100 mg of wet tissue.

4. Homogenize the tissue sample using a Polytron set at maximum speed: 2 × 20-s pulses. Chill on ice for 1 min between each pulse.

5. Sonicate the tissue homogenate. First cool on ice for about 1 min, and then sonicate using a microtip set to 15 W output with three bursts of 30 s each. Cool on ice for 1 min between each burst.

6. Centrifuge the cell homogenate/lysate to clear cell debris. Centrifuge at 20,000 rcf (g) at 4 °C for 15 min, and then transfer the supernatant (this is your protein sample) to a 50 mL screw-cap bottle.

3. NOTE: Reserve 100 µL of each sample in a microfuge tube for protein concentration determination and any subsequent western blot analysis.

4. NOTE: If the cell lysate is left on ice for a prolonged period of time, the urea may precipitate. If a precipitate forms, remove from ice and warm slightly by hand until urea is in solution.

2.5 Reduction and Alkylation of Proteins

1. Add 1/278 volume of 1.25 M DTT to the cleared cell supernatant (e.g., 36 µL of 1.25 M DTT for 10 mL of protein extract), mix well, and place the tube into a 55 °C incubator for 30 min.

2. Cool the solution on ice briefly until it has reached room temperature (tube should feel neither warm nor ice-cold by hand).

3. Add 1/10 volume of iodoacetamide solution to the cleared cell supernatant, mix well, and incubate for 15 min at room temperature in the dark.

2.6 Protein Lysate Digestion

1. NOTE: Alternative proteases such as GluC, chymotrypsin, and others can be used in addition to the protease treatments outlined above to expand the coverage of modified peptides from each antibody reagent. When considering the use of additional protease treatments it should be compatible with the respective motif antibody by not cleaving residues within the designated sequence motif. Protease treatments that generate larger proteolytic peptides may not be ideal if the resulting peptides do not ionize well in the mass spectrometer (Table 2).

1. Dilute threefold with 20 mM HEPES pH 8.0 to a final concentration of 2 M urea and 20 mM HEPES, pH 8.0. For example, for 10 mL of lysate add 30 mL 20 mM HEPES pH 8.0.

Table 2
Protease digest reference for PTM antibodies

Catalogue No	PTMScan kit description	Motif	Protease
5563	Phospho-Akt Substrate motif Kit	RXRXX(S*/T*)	LysC[a]
5561	Phospho-Akt Substrate motif Kit	RXX(S*/T*)	LysC[a]
5564	Phospho-AMPK Substrate motif Kit	LXRXX(S*/T*)	LysC[a]
5565	Phospho-PKA Substrate motif Kit	RRX(S*/T*)	LysC[a]
8803	Phospho-Tyrosine Kit	(Y*)	Trypsin
5567	Phospho-T*PP Motif Kit	T*PP	Trypsin
5566	Phospho-ST*P Motif Kit	ST*P	Trypsin
4652	Phospho-MAPK/CDK Substrate Motif Kit	PXS*P and S*PX (K/R)	Trypsin
13416	Acetyl-Lysine Kit	K-acetyl	Trypsin
12810	Cleaved Caspase Substrate Kit	DE(T/S/A)D	Trypsin
13474	Di-methyl-Arginine (asymmetric) Kit	R-(methyl)$_2$	Trypsin
13563	Di-methyl-Arginine (symmetric) Kit	R-(methyl)$_2$	Trypsin
12235	Mono-methyl-Arginine Kit	R-methyl	Trypsin
5562	Ubiquitin Remnant Motif Kit	K-epsilon-GG	Trypsin

[a]For LysC-digested material, there is a second digestion performed after the StageTip purification of enriched peptides (see the protocol after StageTip Purification). Refer to the following link for an updated version of available PTMScan Kits: http://www.cellsignal.com/services/ptmscan_kits.html

2.6.1 Trypsin Digestion

1. Add 1/100 volume of 1 mg/mL trypsin-TPCK (Worthington) stock in 1 mM HCl and digest overnight at room temperature with mixing.

2. Analyze the lysate before and after digest by SDS-PAGE to check for complete digestion.

3. Continue through the Sep-Pak, IAP, and StageTip protocols prior to LC-MS analysis of enriched peptides.

2.6.2 LysC Digestion

1. Prepare 5 mg/mL stock solution of LysC in 20 mM HEPES pH 8.0. Aliquot for single use and store at −80 °C.

2. Add LysC solution to peptides at 1:250 (w:w). For 20 mg sample, use 20 mg ÷ 250 = 80 μg × 1 μL/5 μg = 16 μL LysC and digest overnight at room temperature.

3. Analyze the lysate before and after digest by SDS-PAGE to check for complete digestion.

4. Continue through the Sep-Pak®, IAP, and StageTip protocols before conducting the SECONDARY DIGESTION with trypsin (*see* end of Section 2.12).

2.7 Sep-Pak® C18 Purification of Lysate Peptides

1. NOTE: Purification of peptides is performed at room temperature on 0.7 mL Sep-Pak columns from Waters Corporation, WAT051910.

2. NOTE: Sep-Pak® C18 purification utilizes reversed-phase (hydrophobic) solid-phase extraction. Peptides and lipids bind to the chromatographic material. Large molecules such as DNA, RNA, and most protein, as well as hydrophilic molecules such as many small metabolites are separated from peptides using this technique. Peptides are eluted from the column with 40 % acetonitrile (MeCN) and separated from lipids and proteins, which elute at approximately 60 % MeCN and above.

3. NOTE: About 20 mg of protease-digested peptides can be purified from one Sep-Pak column. Purify peptides immediately after proteolytic digestion.

2.7.1 Solutions and Reagents

1. NOTE: Prepare solutions with Milli-Q® or equivalent grade water. Organic solvents (trifluoroacetic acid, acetonitrile) should be of the highest grade. All percentage specifications for solutions are vol/vol.

1. 20 % trifluoroacetic acid (TFA): Add 10 mL TFA to water to a total volume of 50 mL.

2. Solvent A (0.1 % TFA): Add 5 mL of 20 % TFA to 995 mL water.

3. Solvent B (0.1 % TFA, 40 % acetonitrile): Add 400 mL of acetonitrile (MeCN) and 5 mL of 20 % TFA to 500 mL of water, and adjust final volume to 1 L with water.

2.8 Acidification of Digested Cell Lysate

NOTE: Before loading the peptides from the protein digest on the column, the digest must be acidified with TFA for efficient peptide binding. The acidification step helps remove fatty acids from the digested peptide mixture.

1. Add 1/20 volume of 20 % TFA to the digest for a final concentration of 1 % TFA. Check the pH by spotting a small amount of peptide sample on a pH strip (the pH should be under 3). After acidification, allow precipitate to form by letting stand for 15 min on ice.

2. Centrifuge the acidified peptide solution for 15 min at 1780 rcf (*g*) at room temperature to remove the precipitate. Transfer peptide-containing supernatant into a new 50 mL conical tube without dislodging the precipitated material.

2.9 Peptide Purification

1. NOTE: Application of all solutions should be performed by gravity flow.

1. Connect a 10 cc reservoir (remove 10 cc plunger) to the SHORT END of the Sep-Pak column.

2. Pre-wet the column with 5 mL 100 % MeCN.

2. NOTE: Each time solution is applied to the column air bubbles form in the junction where the 10 cc reservoir meets the narrow inlet of the column. These must be removed with a gel-loader tip placed on a P-200 micropipettor; otherwise the solution will not flow through the column efficiently. Always check for appropriate flow.

3. Wash sequentially with 1, 3, and 6 mL of Solvent A (0.1 % TFA).

4. Load acidified and cleared digest (from Section 2.2).

3. NOTE: In rare cases, if the flow rates decrease dramatically upon (or after) loading of sample, the purification procedure can be accelerated by gently applying pressure to the column using the 10 cc plunger after cleaning it with organic solvent. Again make sure to remove air bubbles from the narrow inlet of the column before doing so. Do not apply vacuum (as advised against by the manufacturer).

5. Wash sequentially with 1, 5, and 6 mL of Solvent A (0.1 % TFA).

6. Wash with 2 mL of 5 % MeCN and 0.1 % TFA.

7. Place columns above new 15 or 50 mL polypropylene tubes to collect eluate. Elute peptides with a sequential wash of 3 × 2 mL of Solvent B (0.1 % TFA, 40 % acetonitrile).

8. Freeze the eluate on dry ice (or −80 °C freezer) for 2 h to overnight and lyophilize frozen peptide solution for a minimum of 2 days to assure that TFA has been removed from the peptide sample.

4. NOTE: The lyophilization should be performed in a standard lyophilization apparatus. DO NOT USE a SPEED-VAC apparatus at this stage of the protocol.

5. NOTE: The lysate digest may have a much higher volume than the 10 cc reservoir will hold (up to 50–60 mL from adherent cells) and therefore the peptides must be applied in several fractions. If available a 60 cc syringe may be used in place of a 10 cc syringe to allow all sample to be loaded into the syringe at once.

6. NOTE: Lyophilization: The digested peptides are stable at −80 °C for several months (seal the closed tube with parafilm for storage). The PTMScan procedure can be interrupted before or after lyophilization. Once the lyophilized peptide is dissolved in IAP buffer (*see* next step), continue to the end of the procedure.

2.10 Immunoaffinity Purification of PTM Peptides

2.10.1 Solutions and Reagents

1. NOTE: Prepare solutions with Milli-Q or equivalent grade water. Trifluoroacetic acid should be of the highest grade. All percentage specifications for solutions are vol/vol.

2. NOTE: Dilute 10× IAP buffer with water to 1× buffer before use. Store 1× buffer for up to 1 month at 4 °C.

1. Centrifuge the tube containing lyophilized peptide in order to collect all material to be dissolved. Add 1.4 mL IAP buffer. Resuspend pellets mechanically by pipetting repeatedly with a P-1000 micropipettor taking care not to introduce excessive bubbles into the solution. Transfer solution to a 1.7 mL Eppendorf tube.

3. NOTE: After dissolving the peptide, check the pH of the peptide solution by spotting a small volume on pH indicator paper (the pH should be close to neutral, or no lower than 6.0). In the rare case that the pH is more acidic (due to insufficient removal of TFA from the peptide under suboptimal conditions of lyophilization), titrate the peptide solution with 1 M Tris base solution that has not been adjusted for pH. 5–10 µL is usually sufficient to neutralize the solution.

2. Clear solution by centrifugation for 5 min at 10,000 rcf (*g*) at 4 °C in a microcentrifuge. The insoluble pellet may appear considerable. This will not pose a problem since most of the peptide will be soluble. Cool on ice.

3. Wash motif antibody-bead slurry sequentially, four times with 1 mL of 1× PBS, and resuspend as a 50 % slurry in PBS to remove the glycerol contaminating buffer.

4. Transfer the peptide solution into the microfuge tube containing motif antibody beads. Pipet sample directly on top of the beads at the bottom of the tube to ensure immediate mixing. Avoid creating bubbles upon pipetting.

5. Incubate for 2 h on a rotator at 4 °C. Before incubation, seal the microfuge tube with parafilm in order to avoid leakage.

6. Centrifuge at 2000 rcf (*g*) for 30 s and transfer the supernatant with a P-1000 micropipettor to a labeled Eppendorf tube to save for future use. Flow-through material can be used for subsequent IAPs.

4. NOTE: In order to recover the beads quantitatively, do not spin the beads at lower g-forces than what is specified in this procedure. Avoid substantially higher g-forces as well, since this may cause the bead matrix to collapse. All centrifugation steps should be performed at the recommended speeds throughout the protocol.

5. NOTE: If the cells were directly harvested from culture medium without PBS washing, some Phenol Red pH indicator will remain (it co-elutes during the Sep-Pak® C18 purification of peptides) and color the peptide solution yellow. This coloration has no effect on the immunoaffinity purification step.

6. NOTE: All subsequent wash steps are at 0–4 °C.

7. NOTE: In all wash steps, the supernatant should be removed reasonably well. Avoid removing the last few microliters, except in the last step, since this may cause inadvertent carryover of the beads.

7. Add 1 mL of IAP buffer to the beads, mix by inverting tube five times, centrifuge for 30 s, and remove supernatant with a P-1000 micropipettor.

8. Repeat **step 7** once for a total of two IAP buffer washes.

8. NOTE: All steps from this point forward should be performed with solutions prepared with Burdick and Jackson or other HPLC-grade water.

9. Add 1 mL chilled Burdick and Jackson water to the beads, mix by inverting tube five times, centrifuge for 30 s, and remove supernatant with a P-1000 micropipettor.

10. Repeat **step 9** two times for a total of three water washes. During the last water wash, the tube may need to be shaken while inverting in order to ensure efficient mixing.

9. NOTE: After the last wash step, remove supernatant with a P-1000 micropipettor as before, then centrifuge for 5 s to remove fluid from the tube walls, and carefully remove all remaining supernatant with a gel-loading tip attached to a P-200 micropipettor.

11. Add 55 μL of 0.15 % TFA to the beads, tap the bottom of the tube several times (do not vortex), and let stand at room temperature for 10 min, mixing gently every 2–3 min.

10. NOTE: In this step, the posttranslationally modified peptides of interest will be in the eluent.

12. Centrifuge for 30 s at 2000 rcf (g) in a microcentrifuge and transfer supernatant to a new 1.7 mL Eppendorf tube.

13. Add 50 μL of 0.15 % TFA to the beads, and repeat the elution/centrifugation steps. Combine both eluents in the

same 1.7 mL tube. Briefly centrifuge the eluent to pellet any remaining beads and carefully transfer eluent to a new 1.7 mL tube taking care not to transfer any beads.

2.11 Concentration and Purification of PTM Peptides for LC-MS/MS Analysis

1. **NOTE**: We recommend concentrating peptides using the following protocol by Rappsilber and co-workers [36].

2. **NOTE**: We recognize that there are many other routine methods for concentrating peptides using commercial products such as ZipTip® (*see* link provided below) and StageTips (*see* link provided below) that have been optimized for peptide desalting/concentration. Regardless of the particular method, we recommend that the method of choice be optimized for recovery and be amenable for peptide loading capacities of at least 10 μg.

2.11.1 StageTips

http://www.proxeon.com/productrange/sample_preparation_and_purification/stage_tips/index.html

2.11.2 ZipTip®

http://www.millipore.com/catalogue/item/ZTC18S096

2.11.3 Solutions and Reagents

NOTE: Prepare solutions with HPLC-grade water. Organic solvents (trifluoroacetic acid, acetonitrile) should be of the highest grade.

1. Solvent C (0.1 % trifluoroacetic acid, 50 % acetonitrile): Add 0.1 mL trifluoroacetic acid to 40 mL water, then add 50 mL acetonitrile, and adjust the final volume to 100 mL with water.

2. Solvent D (0.1 % trifluoroacetic acid): Add 0.1 mL trifluoroacetic acid to 50 mL water, and adjust the final volume to 100 mL with water.

3. Solvent E (0.1 % trifluoroacetic acid, 40 % acetonitrile): Add 0.1 mL trifluoroacetic acid to 30 mL water, then add 40 mL acetonitrile, and adjust the final volume to 100 mL with water.

NOTE: Organic solvents are volatile. Tubes containing small volumes of these solutions should be prepared immediately before use and should be kept capped as much as possible, because the organic components evaporate quickly.

2.11.4 Procedure

1. Equilibrate the StageTip by passing 50 μL of Solvent C through (once) followed by 50 μL of Solvent D two times.

2. Load sample by passing IP eluent through the StageTip. Load IAP eluent in two steps using 50 μL in each step.

3. Wash the StageTip by passing 55 μL of Solvent D through two times.

4. Elute peptides off the StageTip by passing 10 μL of Solvent E through two times, pooling the resulting eluent.

NOTE: For enriched LysC peptides, a second digest with trypsin will be performed. Therefore, we recommend eluting the LysC peptides into a 0.5 mL Eppendorf tube in preparation for the trypsin digestion protocol, described below.

5. Dry down the StageTip eluent in a vacuum concentrator (Speed-Vac) and redissolve the peptides in an appropriate solvent for LC-MS analysis such as 5 % acetonitrile and 0.1 % TFA.

2.12 Trypsin Digestion of Enriched LysC PTM Peptides

NOTE: Continued from Section 2.6 of the Protein Lysate Digestion

NOTE: Trypsin digestion of enriched LysC peptides is recommended for all basophilic motif antibodies.

1. Prepare 1 M ammonium bicarbonate stock solution.

2. Prepare digestion buffer, 50 mM ammonium bicarbonate with 5 % acetonitrile.

3. Dilute a stock solution of sequencing-grade trypsin (Promega) with digestion buffer from 0.4 μg/μL to a final concentration of 25 ng/μL.

4. Resuspend the dried, LysC-digested peptides generated from the StageTip concentration protocol above with 10 μL of trypsin solution (25 ng/μL, 250 ng total). Vortex three times to redissolve the peptides and microfuge the sample to collect peptide/trypsin solution at the bottom of the microfuge tube as the final step.

5. Incubate the solution at 37 °C for 2 h.

6. After trypsin digestion, add 1 μL of 5 % TFA to the digest solution. Vortex to mix and microfuge to collect peptide solution at the bottom of the microfuge tube.

7. Transfer the acidified peptide solution to a newly conditioned StageTip, rinse the 0.5 mL Eppendorf tube with 40 μL of 0.1 % TFA once, and apply the rinse solution to the StageTip.

8. Perform the StageTip desalting of the peptide digest and elute the peptides into an HPLC insert. Dry purified peptides under vacuum prior to LC-MS analysis (as described above).

2.13 LC-MS/MS Analysis of Peptides

1. Resuspend vacuum-dried, immunoaffinity-purified peptides in 0.125 % formic acid. 15 mg of starting material will generate sufficient peptide for two to three injections on the instrument.

2. Separate on a reverse-phase column (75 μm inner diameter × 10 cm) packed into a PicoTop emitter (~8 μm diameter tip) with Magic C18 AQ (100 Å × 5 μm). Elute PTM peptides using a 90-min gradient of acetonitrile (5–40 %) in 0.125 % formic acid delivered at 280 nL/min.

NOTE: Run samples on a high-mass-accuracy instrument (ppm accuracy in the MS1 channel) to ensure high-quality peptide identifications and accurate quantification. We use the LTQ-Orbitrap VELOS or ELITE systems (Thermo Scientific).

3. LC-MS/MS instrument parameter settings: MS run time 120 min, MS1 scan range (300.0–1500.00), top 20 MS/MS, min signal 500, isolation width 2.0, normalized coll. energy 35.0, activation-Q 0.250, activation time 20.0, lock mass 371.101237, charge state rejection enabled, charge state 1+ rejected, dynamic exclusion enabled, repeat count 1, repeat duration 35.0, exclusion list size 500, exclusion duration 40.0, exclusion mass width relative to mass, exclusion mass width 10 ppm.

4. Database search the files generated in the run versus the correct species database with a reverse decoy database included to estimate false discovery rates [37]. Search settings: mass accuracy of parent ions: 50 ppm, mass accuracy of product ions: 1 Da, up to four missed cleavages, up to four variable modifications, max charge = 5, variable modifications allowed on methionine (oxidation, +15.9949), serine, threonine, and tyrosine (phosphorylation, +79.9663). Semi-tryptic peptides are allowed (K or R residue on one side of peptide only). Results can be further narrowed by MError (usually ±3 ppm) and the presence of the intended motif (phosphorylation, caspase cleavage, validated apoptosis/autophagy peptides, etc.).

5. Quantification can be performed via a number of methods such as SILAC, reductive amination, isobaric tags, or label-free. For label-free quantification, we use Progenesis (Nonlinear Dynamics) to retrieve the integrated peak area for each observed PTM peptide in the MS1 channel. To quantify changes of PTM peptides in a profiling study, this software only requires one MS/MS event for a particular peptide across all samples, eliminating "holes" (no data) in the study due to LC-MS duty cycle limitations.

2.14 Sample Data Review: Mouse Brain Tissue

In the current study, mouse brain tissue was used as a neuronal model system for identifying posttranslational modifications. Tryptic or Lys-C peptides from mouse brain tissue was subjected to immuno-affinity purification using phospho-motif antibody mixtures (atypical phospho-motif antibody pool, basophilic phospho-motif antibody pool, proline-rich phospho-motif antibody pool, and phospho-Ser/Thr motif antibody pool), phospho-tyrosine antibody, immobilized metal affinity chromatography (Fe^{3+}-IMAC), lysine-acetylation antibody, lysine-succinylation antibody, ubiquitin-remnant (K-GG) antibody, mono-methyl-arginine antibody, mono-methyl-lysine antibody, and caspase cleavage antibody,

Table 3
Summary of PTM identifications

PTM type	Antibody/reagent	Unique PTM peptides identified	Unique PTM sites	Proteins identified
Phosphorylation	Atypical motif mix	527	397	290
Phosphorylation	Basophilic motif mix	1775	1253	628
Phosphorylation	Proline-rich motif mix	1740	1239	528
Phosphorylation	Phospho-Ser/Thr motif mix	3791	2733	1133
Phosphorylation	IMAC	8371	5786	2010
Phosphorylation	Phospho-tyrosine	1977	1177	775
Methylation	Mono-methyl Arginine	1441	787	485
Methylation	Mono-methyl Lysine	407	201	330
Acetylation	Acetyl-lysine	5128	2344	756
Succinylation	Succinyl-lysine	6778	3143	727
Ubiquitination	Ubiquitin remnant (K-e-GG)	3778	2423	1335
Caspase cleavage	Caspase cleavage	186	139	132
			Total	4692

A summary of the unique PTM peptide identifications and sites is provided for various types of PTM enrichment experiments performed with mouse brain tissue (5 mg). The list of qualitative identifications can be downloaded from PhosphoSitePlus at the following link, http://www.phosphosite.org/staticSupp.do

respectively. A total of 4692 unique proteins with at least one PTM were identified in this sample data set (http://www.phosphosite.org/staticSupp.do) (Table 3).

From the comparative analysis results of phosphorylation using phospho-motif antibody pools and Fe^{3+}-IMAC, we have observed significant complementarity between these two types of enrichment methods (Fig. 5). As an established highly efficient enrichment method for phosphopeptide, IMAC enabled highly specific enrichment (>92 % phosphopeptides) and identification of 5786 unique phosphorylation sites from a single LC-MS/MS analysis. However, the phosphoproteome contains a large number of different phosphorylation sites, and the general affinity of the Fe^{3+} ion for phosphate groups could lead to a bias towards enrichment of the more abundant phosphopeptides. In the alternative, antibody-based enrichment approach, phospho-motif antibodies are designed based on the recognition motif of substrates by a specific kinase (Table 4). For example, phospho-Akt substrate antibody recognizes phosphopeptides with the −3 and −5 positions as arginine residues (RXRXX[pS/pT]), as defined by the substrate sequence

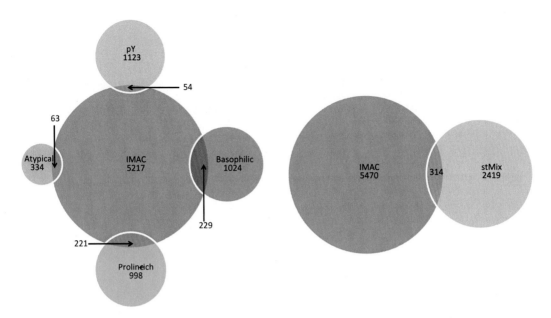

Fig. 5 Overlap of unique phosphorylation sites enriched by Fe^{3+}-IMAC and phospho-motif antibody pools. (*Left*) Phospho-motif antibodies are grouped based on the recognition motif; (*right*) s/t Mix is the pool of all phospho-Ser/Thr motif antibodies in the atypical-, basophilic-, and proline-rich groups

recognized and phosphorylated by Akt. Therefore, phospho-motif antibodies are able to enrich for phosphopeptides from a particular branch of the Kinome Tree, and are not dependent on the relative abundance of the target phosphopeptides. In addition, all motif antibodies can be used as a primary antibody reagent for western blots, providing a fast pre-screening for the pathways that are most affected by different biological interventions or perturbations. In our study, the overlap between IMAC and phospho-motif antibodies ranges from 5 to 22 %, showing the complementarity of the two methods for phosphopeptide enrichment. Therefore, parallel enrichment and analysis using IMAC and phospho-motif antibody pools are recommended if a comprehensive phosphoproteome study is desired.

The analysis of cellular perturbations by peptide-level affinity enrichment of PTMs and LC-MS/MS always generates a rich data set. To better understand the significance of a given data set, it is important to analyze the involvement of PTMs in individual pathways and the potential for cross talk between pathways. One common strategy used for this purpose is to query the list proteins and PTMs identified using pathway analysis tools such as STRING (http://string-db.org/), KEGG (http://www.genome.jp/kegg/), Cytoscape (http://www.cytoscape.org/), Ingenuity (http://www.ingenuity.com/products/ipa), or Reactome (http://www.reactome.org/), to name a few examples. Peptide-level affinity

Table 4
Summary of phospho-motif antibodies

Antibody	Recognition motif	Group
ATM/ATR substrate	(s/t)QG	Atypical
ATM/ATR substrate	(s/t)Q	Atypical
CK substrate	t(D/E)X(D/E)	Atypical
PDK1 docking motif	(F/T)(s/t)(F/Y)	Atypical
tXR	tXR	Atypical
Akt substrate	RXX(s/t)	Basophilic
Akt substrate	RXRXX(s/t)	Basophilic
AMPK substrate	LXRXX(s/t)	Basophilic
PKA substrate	(K/R)(K/R)X(s/t)	Basophilic
PKC substrate	(K/R)XsX(K/R)	Basophilic
PKD substrate	LXRXX(s/t)	Basophilic
14-3-3 binding motif	(R/K)XXsXP	Basophilic
CDK substrate	(K/R)sPX(K/R)	Proline-rich
MAPK substrate	PXsP	Proline-rich
PLK binding motif	StP	Proline-rich
tP motif	tP	Proline-rich
tPE motif	tPE	Proline-rich

Different recognition motifs can be grouped into atypical-, basophilic-, and proline-rich pools, respectively. All phospho-motif antibodies are combined together in the phospho-Ser/Thr pool. The site of phosphorylation within the recognition motif is indicated by lower case "s" or "t"

enrichment of PTM and following LC-MS/MS analysis always generate rich datasets; therefore, subsequent pathway analysis and interacting protein analysis facilitate illustration of the involvement of PTMs in individual pathway and their potential cross talks. In the current study, we analyzed the compiled PTM data set at unique protein level by KEGG pathway analysis by merging all individual PTM subsets together. Interestingly, we have identified most of the proteins in various neuronal pathways including dopaminergic synapse, GABAergic synapse, and axon guidance. Among all PTMs, ubiquitination and phosphorylation are the most highly represented in the identified proteins. Besides, large numbers of novel modification sites were identified on some crucial signaling nodes in neuronal pathways including mono-methyl-arginine in voltage-dependent calcium channel (VGCC), lysine-succinylation on GABA transaminase, and mono-methyl-lysine on PAK (Fig. 6).

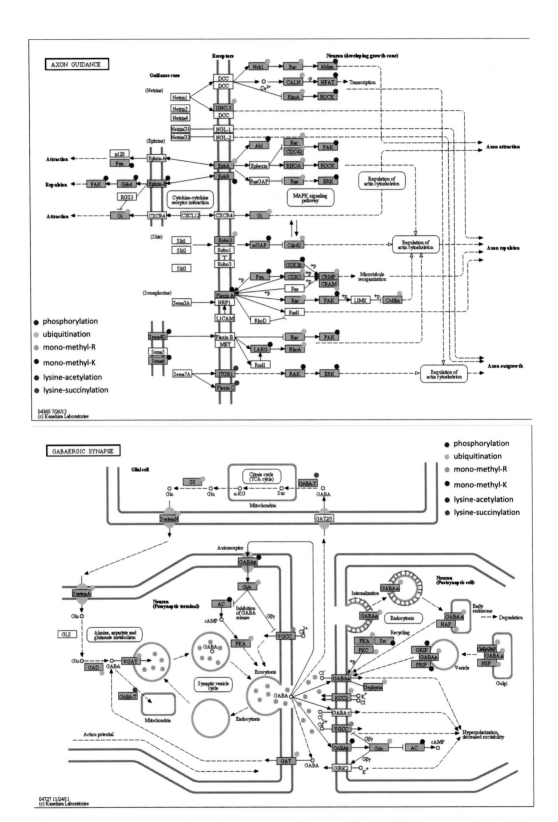

Fig. 6 KEGG pathway analysis of identified proteins (*pink rectangle*) with PTMs in mouse brain. (**a**) Axon guidance, (**b**) GABAergic synapse. Proteins are color-coded to indicate the various modification types of PTMs identified in the sample data

PhosphoSitePlus is another commonly used bioinformatics resource that can provide comprehensive information related to the biological role of associated posttranslational modifications (http://www.phosphosite.org) along with examples from a variety of sample data sets related to the characterization of many types of PTM-modified proteins (http://www.phosphosite.org/staticDownloads.do). The mouse brain PTM sample data described here can be downloaded from the following link.

The example data set described above provides a representative sampling of PTMs that are accessible from mouse brain tissue using the immunoaffinity LC-MS techniques outlined in this chapter. The qualitative results presented in this method can be accessed from PhosphoSitePlus at the following URL, http://www.phosphosite.org/staticSupp.do.

3   Conclusions

The development of the nervous system, its function, and its continued viability are initiated and maintained through a complex milieu of interacting networks consisting of critical signaling pathways and cellular interactions. These pathways can be perturbed in response to a multitude of endogenous and exogenous cellular stresses or stimuli. A shift in the balance of these carefully orchestrated signaling pathways after stress or in response to disease can have severe consequences on the development of the nervous system. Characterizing the numerous events necessary to support these complex biological processes is challenging, but technical developments in proteomics have evolved to enable us to probe these systems for a better understanding of protein expression, function, and organization in complex signaling and regulatory networks. Improvements in mass spectrometry instrumentation, the implementation of protein arrays, sample preparation techniques, the availability of antibody reagents to probe specific activated proteins, and the development of robust informatics software have made proteomics a powerful analytical platform for many areas of biology. These advances have provided sensitive and robust techniques for high-throughput technologies to enable large-scale identification and quantification of protein expression, thorough characterization of protein modifications, efficient monitoring of subcellular localization of protein targets, insight into protein-protein interactions, and protein function. Utilizing immunoaffinity methods within the proteomics workflow will have significant implications for advancing our understanding of how cellular proteomes are regulated in the nervous system in developing and mature neural networks and in healthy and disease states.

Acknowledgements

The authors would like to acknowledge David Titus, Anjli Venkateswaran, and Carrie Ann Brown for their thoughtful comments and suggestions during the editing of this chapter.

References

1. Peng J, Kim MJ, Cheng D, Duong DM, Gygi SP, Sheng M (2004) The. J Biol Chem 279:21003–21011
2. Witzmann FA, Arnold RJ, Bai F, Hrncirova P, Kimpel MW, Mechref YS, McBride WJ, Novotny MV, Pedrick NM, Ringham HN, Simon JR (2005) Proteomics 5:2177–2201
3. Morciano M, Burre J, Corvey C, Karas M, Zimmermann H, Volknandt W (2005) J Neurochem 95:1732–1745
4. Willert K, Brown JD, Danenberg E, Duncan AW, Weissman IL, Reya T, Yates JR III, Nusse R (2003) Nature 423:448–452
5. Humphery-Smith I, Cordwell SJ, Blackstock WP (1997) Electrophoresis 18:1217–1242
6. Anderson L, Seilhamer J (1997) Electrophoresis 18:533–537
7. Gygi SP, Rochon Y, Franza BR, Aebersold R (1999) Mol Cell Biol 19:1720–1730
8. Godovac-Zimmermann J, Brown LR (2001) Mass Spectrom Rev 20:1–57
9. Rush J, Moritz A, Lee KA, Guo A, Goss VL, Spek EJ, Zhang H, Zha XM, Polakiewicz RD, Comb MJ (2005) Nat Biotechnol 23:94–101
10. Thingholm TE, Jensen ON, Larsen MR (2009) Proteomics 9:1451–1468
11. Ficarro SB, McCleland ML, Stukenberg PT, Burke DJ, Ross MM, Shabanowitz J, Hunt DF, White FM (2002) Nat Biotechnol 20:301–305
12. Ficarro S, Chertihin O, Westbrook VA, White F, Jayes F, Kalab P, Marto JA, Shabanowitz J, Herr JC, Hunt DF, Visconti PE (2003) J Biol Chem 278:11579–11589
13. Goss VL, Lee KA, Moritz A, Nardone J, Spek EJ, MacNeill J, Rush J, Comb MJ, Polakiewicz RD (2006) Blood 107:4888–4897
14. Rikova K, Guo A, Zeng Q, Possemato A, Yu J, Haack H, Nardone J, Lee K, Reeves C, Li Y, Hu Y, Tan Z, Stokes M, Sullivan L, Mitchell J, Wetzel R, Macneill J, Ren JM, Yuan J, Bakalarski CE, Villen J, Kornhauser JM, Smith B, Li D, Zhou X, Gygi SP, Gu TL, Polakiewicz RD, Rush J, Comb MJ (2007) Cell 131:1190–1203
15. Andersen JN, Sathyanarayanan S, Di Bacco A, Chi A, Zhang T, Chen AH, Dolinski B, Kraus M, Roberts B, Arthur W, Klinghoffer RA, Gargano D, Li L, Feldman I, Lynch B, Rush J, Hendrickson RC, Blume-Jensen P, Paweletz CP (2010) Sci Transl Med 2:43ra55
16. Stokes MP, Rush J, Macneill J, Ren JM, Sprott K, Nardone J, Yang V, Beausoleil SA, Gygi SP, Livingstone M, Zhang H, Polakiewicz RD, Comb MJ (2007) Proc Natl Acad Sci U S A 104:19855–19860
17. Bonnette PC, Robinson BS, Silva JC, Stokes MP, Brosius AD, Baumann A, Buckbinder L (2010) J Proteomics 73:1306–1320
18. Moritz A, Li Y, Guo A, Villen J, Wang Y, MacNeill J, Kornhauser J, Sprott K, Zhou J, Possemato A, Ren JM, Hornbeck P, Cantley LC, Gygi SP, Rush J, Comb MJ (2010) Sci Signal 3:ra64
19. Gnad F, Young A, Zhou W, Lyle K, Ong CC, Stokes MP, Silva JC, Belvin M, Friedman LS, Koeppen H, Minden A, Hoeflich KP (2013) Mol Cell Proteomics 12:2070–2080
20. Giansanti P, Stokes MP, Silva JC, Scholten A, Heck AJ (2013) Mol Cell Proteomics 12:3350–3359
21. Schwer B, Eckersdorff M, Li Y, Silva JC, Fermin D, Kurtev MV, Giallourakis C, Comb MJ, Alt FW, Lombard DB (2009) Aging Cell 8:604–606
22. Lee KA, Hammerle LP, Andrews PS, Stokes MP, Mustelin T, Silva JC, Black RA, Doedens JR (2011) J Biol Chem 286:41530–41538
23. Pham VC, Pitti R, Anania VG, Bakalarski CE, Bustos D, Jhunjhunwala S, Phung QT, Yu K, Forrest WF, Kirkpatrick DS, Ashkenazi A, Lill JR (2012) J Proteome Res 11:2947–2954
24. Guo A, Gu H, Zhou J, Mulhern D, Wang Y, Lee KA, Yang V, Aguiar M, Kornhauser J, Jia X, Ren J, Beausoleil SA, Silva JC, Vemulapalli V, Bedford MT, Comb MJ (2014) Mol Cell Proteomics 13:372–387
25. Stokes MP, Farnsworth CL, Moritz A, Silva JC, Jia X, Lee KA, Guo A, Polakiewicz RD,

Comb MJ (2012) Mol Cell Proteomics 11:187–201

26. Dayon L, Hainard A, Licker V, Turck N, Kuhn K, Hochstrasser DF, Burkhard PR, Sanchez JC (2008) Anal Chem 80:2921–2931

27. Hsu JL, Huang SY, Chow NH, Chen SH (2003) Anal Chem 75:6843–6852

28. Ibarrola N, Kalume DE, Gronborg M, Iwahori A, Pandey A (2003) Anal Chem 75:6043–6049

29. Ong SE, Blagoev B, Kratchmarova I, Kristensen DB, Steen H, Pandey A, Mann M (2002) Mol Cell Proteomics 1:376–386

30. Paardekooper Overman J, Yi JS, Bonetti M, Soulsby M, Preisinger C, Stokes MP, Hui L, Silva JC, Overvoorde J, Giansanti P, Heck AJ, Kontaridis MI, den Hertog J, Bennett AM (2014) Mol Cell Biol 34:2874–2889

31. Unwin RD, Pierce A, Watson RB, Sternberg DW, Whetton AD (2005) Mol Cell Proteomics 4:924–935

32. Viner RI, Zhang T, Second T, Zabrouskov V (2009) J Proteomics 72:874–885

33. Wiese S, Reidegeld KA, Meyer HE, Warscheid B (2007) Proteomics 7:340–350

34. Zhang H, Zha X, Tan Y, Hornbeck PV, Mastrangelo AJ, Alessi DR, Polakiewicz RD, Comb MJ (2002) J Biol Chem 277:39379–39387

35. Stokes MP, Silva JC, Jia X, Lee KA, Polakiewicz RD, Comb MJ (2012) Int J Mol Sci 14:286–307

36. Rappsilber J, Ishihama Y, Mann M (2003) Anal Chem 75:663–670

37. Lundgren DH, Martinez H, Wright ME, Han DK (2009) Curr Protoc Bioinformatics Chapter 13, Unit 13 13

Neuromethods (2016) 114: 31–41
DOI 10.1007/7657_2015_94
© Springer Science+Business Media New York 2015
Published online: 08 November 2015

A Boronic Acid-Based Enrichment for Site-Specific Identification of the N-glycoproteome Using MS-Based Proteomics

Haopeng Xiao, George X. Tang, Weixuan Chen, and Ronghu Wu

Abstract

Modification of proteins by N-linked glycans plays a critically important role in biological systems, including determining protein folding and trafficking as well as regulating many biological processes. Aberrant glycosylation is well known to be related to disease, including cancer and neurodegenerative diseases. Current mass spectrometry (MS)-based proteomics provides the possibility for site-specific identification of the N-glycoproteome; however, this is extraordinarily challenging because of the low abundance of many N-glycoproteins and the heterogeneity of glycans. Effective enrichment is essential to comprehensively analyze N-glycoproteins in complex biological samples. The covalent interaction between boronic acid and *cis*-diols allows us to selectively capture glycopeptides and glycoproteins, whereas the reversible nature of the bond enables them to be released after non-glycopeptides are removed. By virtue of the universal boronic acid-diol recognition, large-scale mapping of N-glycoproteins can be achieved by combining boronic acid-based enrichment, PNGase F treatment in the presence of heavy oxygen (^{18}O) water, and MS analysis. This method can be extensively applied for the comprehensive analysis of N-glycoproteins in a wide variety of complex biological samples.

Keywords: N-glycoproteome, Mammalian cells, Boronic acid-based enrichment, PNGase F, MS-based proteomics

1 Introduction

Protein co-translational and posttranslational modifications regulate almost all aspects of protein functions and cellular activity [1–4]. Glycosylation is particularly important because it plays critical roles in protein folding and trafficking, antigenicity, molecular recognition, and cell-cell interactions [5–8]. Numerous studies have proven that aberrant glycosylation events are hallmarks of disease states, highlighting the clinical importance of protein glycosylation analysis [9–13]. However, the low abundance of glycoproteins and the heterogeneity of glycan structures make the global analysis of protein glycosylation extremely challenging [14, 15].

Lectin-based methods have most commonly been used to enrich glycoproteins [16, 17], but each lectin is inherently specific to one or several types of carbohydrates, rendering a single type of lectin or a group of lectins incapable of effectively enriching glycoproteins. While the reaction between boronic acid and *cis*-diols has previously been applied for small-scale analysis [18, 19], a universal boronic acid-based chemical enrichment method was employed to comprehensively investigate protein N-glycosylation in complex biological samples in combination with mass spectrometry (MS)-based proteomics [20]. Boronic acid was conjugated onto magnetic beads to selectively enrich glycopeptides from whole cell lysate peptides. In order to generate a common tag for MS analysis, enriched peptides were treated with peptide-N4-(*N*-acetyl-beta-glucosaminyl) asparagine amidase (PNGase F) in heavy-oxygen water ($H_2{}^{18}O$) to remove N-glycans, which converted asparagine (Asn) to heavy oxygen labeled aspartic acid (Asp) and created a mass shift of +2.9883 Da. Heavy oxygen on Asp can distinguish authentic N-glycosylation sites from those caused by accumulated deamidation. Finally, the peptides were analyzed by an on-line LC-MS/MS system. Here in this chapter, we describe the protocol to prepare peptide samples, enrich glycopeptides using boronic acid conjugated beads, and analyze them by MS (Fig. 1).

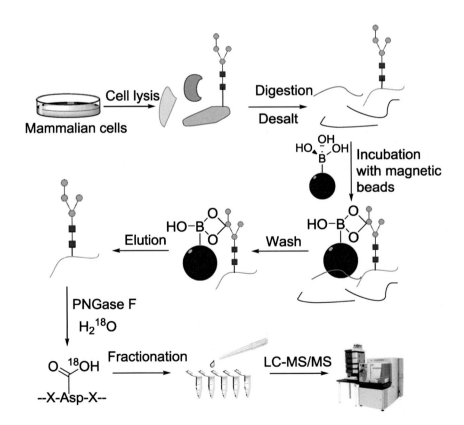

Fig. 1 Experimental procedures of the boronic acid-based enrichment strategy for the global analysis of N-glycoproteins in mammalian cells

2   Materials

2.1   Cell Culture

1. Human HeLa or HEK293 cells (or any other types of mammalian cells)
2. Culture medium: Dulbecco's Modified Eagle Medium (DMEM), low glucose (1 g/L), with 10 % fetal bovine serum (FBS).
3. Appropriate cell culture flasks and supplies

2.2   Protein Extraction

1. Phosphate-buffered saline (PBS), pH = 7.4
2. Lysis buffer: 150 mM NaCl, 50 mM 4-(2-hydroxyethyl)-1-piperazineethanesulfonic acid (HEPES) (pH = 7.4), 0.5 % sodium deoxycholate SDC, 1 % NP-40, 10 U/mL benzonase inhibitor, 0.1 pellet/mL protease inhibitor EDTA-free (Roche)
3. Ice bath
4. Refrigerated benchtop centrifuge (Thermo)
5. Appropriate experimental supplies, e.g., 15 and 50 mL centrifuge tubes, pipettes, Eppendorf tubes, etc.

2.3   Protein Reduction, Alkylation, Precipitation, and In-Solution Digestion

1. 1 M dithiothreitol (DTT)
2. Iodoacetamide (IAA, powder)
3. Heat block set to 56 °C
4. Refrigerated benchtop centrifuge (Thermo)
5. Incubating shaker (Troemner)
6. Methanol
7. Chloroform
8. Water from a Milli-Q water purification system (EMD Millipore)
9. Digestion buffer: 1.6 M urea, 50 mM HEPES pH = 8.6, 5 % acetonitrile (ACN)
10. Lys-C, mass spectrometry grade (Wako)
11. Trypsin, mass spectrometry grade (Promega)

2.4   Peptide Desalting

1. 10 % trifluoroacetic acid (TFA) in water
2. tC18 Sep-Pak cartridge (Waters)
3. Activation buffer: (1) ACN; (2) 50 % ACN, 0.5 % Acetic acid in water
4. Equilibration buffer: 0.1 % TFA in water
5. 0.5 % Acetic acid in water
6. Elution buffer: (1) 50 % ACN, 0.5 % Acetic acid in water; (2) 75 % ACN, 0.5 % Acetic acid in water
7. −80 °C freezer (Thermo)
8. Speed-vacuum (Labconco)

2.5 Boronic Acid Enrichment

1. Boronic acid (BA) conjugated magnetic beads stored in ethanol (BA concentration = 6 mM)

2. Binding buffer: 200 mM ammonium acetate buffer (pH = 10.0)

3. Elution buffer: ACN:H$_2$O:TFA = 50:49:1

4. Incubating shaker

5. Magnetic rack

6. −80 °C freezer

7. A speed-vacuum sample dry system

2.6 PNGase F Treatment

1. PNGase F (Sigma)

2. Heavy-oxygen water (H$_2$^{18}O) (Cambridge Isotope Laboratories, Inc.)

3. Incubating shaker

2.7 Glycopeptide Fractionation

1. Materials for peptide desalting (*see* Section 2.4)

2. HPLC (Agilent)

3. 4.6 × 250 mm 5 μm particle reversed phase column (Waters)

4. Buffer A: 10 mM ammonium acetate pH = 10.0 in water

5. Buffer B: 10 mM ammonium acetate pH = 10.0 in 90 % ACN, 10 % water

2.8 Stage Tip

1. Stage tips, C18 material

2. Activation buffer: (1) Methanol; (2) 80 % ACN, 0.5 % acetic acid

3. Equilibrium buffer: 1 % formic acid (FA)

4. Elution buffer: 50 % ACN, 0.5 % acetic acid

5. Benchtop centrifuge (Thermo)

2.9 LC-MS/MS Analysis

1. WPS-3000TPLRS autosampler (UltiMate 3000 thermostatted Rapid Separation Pulled Loop Wellplate Sampler, Dionex)

2. Microcapillary column packed with C18 beads (Magic C18AQ, 5 μm, 200 Å, 100 μm × 16 cm)

3. Buffer A: 97.5 % H$_2$O, 2.375 % ACN, 0.125 % FA

4. Buffer B: 97.5 % ACN, 2.375 % H$_2$O, 0.125 % FA

5. Hybrid dual-cell quadrupole linear ion trap-orbitrap mass spectrometer (LTQ Orbitrap Elite, Thermo)

6. Software: Xcalibur 3.0.63

2.10 Database Searches, Data Filtering, and Glycosylation Site Localization	1. SEQUEST algorithm (version 28) 2. UniProt Human (*Homo sapiens*) database (88,591 protein entries, downloaded in February 2014)

3 Methods

3.1 Cell Culture	1. Place cell seeds in the culture flask at a ratio of 1:5 2. Change medium every 2 days. When cells reaches ~80 % confluency, harvest cells by scraping in PBS pH = 7.4 or trypsinize cells and passage to new flasks
3.2 Protein Extraction	1. Wash cells twice with PBS (pH = 7.4) 2. Freshly prepare lysis buffer (*see* Section 2.2). For every 8.4×10^6 cells, add 1 mL lysis buffer 3. Incubate the lysis buffer and cells on an end-over-end rotor for 45 min at 4 °C 4. Centrifuge the resulting cellular extract at 20k × *g* for 10 min at 4 °C 5. Carefully transfer the supernatant to a new tube
3.3 Protein Reduction, Alkylation, Precipitation, and In-Solution Digestion	1. Add 5 µL of 1 M DTT for every 1 mL of supernatant collected in Section 3.2, incubate in heat block for 30 min at 56 °C, then cool to room temperature 2. Add IAA to a final concentration of 14 mM in the solution and incubate for 25 min at room temperature in the dark (**Note 1**) 3. To every one starting volume of sample add four volumes of methanol, then vortex 4. Add one volume of chloroform and three volumes of water, then vortex well 5. Centrifuge sample at 4500 × *g* for 10 min at 4 °C 6. Carefully pipette out the top liquid layer (**Note 2**) 7. Add four volumes of methanol and vortex well 8. Centrifuge sample at 4500 × *g* for 10 min at 4 °C 9. Carefully pipette out all the liquid and dry the protein pellet in air 10. Dissolve the protein pellet in digestion buffer (*see* Section 2.3) 11. Add Lys-C at a 1/100–1/200 enzyme:protein ratio. Incubate at 31 °C overnight 12. Add trypsin at a 1/100–1/200 enzyme:protein ratio. Incubate at 37 °C for 4 h (**Note 3**)

3.4 Peptide Desalting

1. Quench the digestion by acidification with 10 % TFA to 0.4 % (vol/vol). Verify that the pH is <2.0; otherwise add more 10 % TFA

2. Peptides are then desalted using a tC18 Sep-Pak cartridge. Here we use a 500 mg cartridge as an example. The maximum capacity is 5 % cartridge weight

3. Wash and condition the cartridge with 10 mL ACN and then with 5 mL of 50 % ACN and 0.5 % acetic acid

4. Equilibrate with 10 mL of 0.1 % TFA

5. Load sample

6. Wash with 10 mL of 0.1 % TFA

7. Wash with 1 mL of 0.5 % acetic acid to remove TFA

8. Elute peptides with (1) 3.5 mL of 50 % ACN, 0.5 % acetic acid in water; (2) 1.5 mL of 75 % ACN, 0.5 % acetic acid in water

9. Freeze the eluate in a −80 °C freezer for 15 min

10. Lyophilize the sample using a speed-vacuum sample dry system. The resulting purified peptides should appear as a white fluffy powder

3.5 Boronic Acid Enrichment

1. For every 5 mg of peptides, pick up 250 μL of BA magnetic bead slurry to work with (**Note 4**)

2. Wash beads three times with 2 mL of 200 mM ammonium acetate buffer pH = 10.0 (binding buffer)

3. Resuspend BA beads in 500 μL of binding buffer, then transfer the slurry into the Eppendorf tube containing all peptides

4. Incubate the tube at 37 °C for 1 h with appropriate shaking

5. Wash beads five times with 500 μL binding buffer

6. Elute glycosylated peptides with 1.5 mL solution containing $ACN:H_2O:TFA = 50:49:1$ (elution buffer), incubate the reaction at 37 °C for 30 min with appropriate shaking

7. Transfer the eluate into a new tube. Wash the beads twice with 200 μL elution buffer, then combine the resulting solutions with the eluate

8. Freeze the final solution in −80 °C freezer for 15 min then lyophilize the sample overnight (**Note 5**)

3.6 PNGase F Treatment

1. Dissolve PNGase F in heavy-oxygen water to a concentration of 1 U/μL

2. Dissolve lyophilized glycopeptide (entirely dry) in 100 μL of heavy-oxygen water

3. Add 5 μL of 1 U/μL PNGase F

4. Incubate the reaction at 37 °C for 3 h with appropriate shaking

Table 1
The HPLC gradient for glycopeptide fractionation

Time (min)	Buffer A (%)	Buffer B (%)	Flow (mL/min)
0.0	100.0	0.0	0.7
3.0	100.0	0.0	0.7
4.0	95.0	5.0	0.7
7.5	86.5	13.5	0.7
30.0	68.0	32.0	0.7
34.5	66.0	34.0	0.7
44.5	45.0	55.0	0.7
47.0	0.0	100.0	0.7
51.0	0.0	100.0	0.7
55.0	100.0	0.0	0.7
80.0	100.0	0.0	0.7

3.7 Glycopeptide Fractionation

1. Quench the reaction by acidification with 10 % TFA to 0.4 % (vol/vol). Verify the pH is <2.0; otherwise add more 10 % TFA

2. Desalt glycopeptides using a 50 mg cartridge (*see* Section 3.4)

3. Fractionate glycopeptides using HPLC with a 4.6 × 250 mm 5 μm particle reversed phase column. Set up the HPLC gradient method as follows (Table 1)

4. Wash the column for 30 min with buffer B (**Note 6**)

5. Equilibrate the column for 40 min with buffer A

6. Load the glycopeptides to the column in 300 μL buffer A (**Note 7**)

7. Starting from minute 10, collect 10 fractions in total with a 4-min interval between fractions

8. Freeze the final solution in a −80 °C freezer for 15 min, then lyophilize

3.8 Stage Tip

1. Prepare ten stage tips for ten glycopeptide fractions

2. For each stage tip, add 50 μL methanol, then spin at 2500 rpm. Remove the flow-through (**Note 8**)

3. Add 40 μL 80 % ACN and 0.5 % acetic acid to each stage tip, then spin at 2500 rpm. Remove the flow-through

4. Equilibrate the stage tip with 40 μL 1 % FA. Spin at 3.0 × k rpm, then remove the flow-through

5. Dissolve the glycopeptide sample in 50 μL 1 % FA (**Note 9**)

Table 2
Some parameters set for MS analysis

Resolution	60,000
AGC target	1,000,000
Analyzer	FTMS
Polarity	Positive
Data type	Centroid
Activation	CID
Isolation width (m/z)	1
Normalized collision energy	35
Activation time	10
Mass range (m/z)	300–1500

6. Load the sample to the stage tip. Spin at 2000 rpm, then remove the flow-through

7. Desalt with 50 μL 1 % FA. Spin at 3000 rpm, then remove the flow-through

8. Elute purified glycopeptides with 20 μL 50 % ACN and 0.5 % acetic acid. Collect the eluate in a mass spectrometry insert

9. Freeze the final solution in −80 °C freezer for 2 min, then dry them in a Speed-Vacuum system

3.9 LC-MS/MS Analysis

1. Dissolve the dried glycopeptides in 10 μL 5 % ACN and 4 % FA. Sonicate the samples in a water bath for 15 s and vortex at high speed for 15 s, then spin briefly

2. Set up acquisition methods for full MS (Table 2) and MS^2

3. Load 4 μL of each enriched glycopeptide sample from **step 1** on to the column and perform LC-MS/MS analysis

3.10 Database Searches, Data Filtering, and Glycosylation Site Localization

1. Convert the raw files into mzXML format

2. Check the precursors for MS/MS fragmentation for incorrect monoisotopic peak assignments while refining precursor ion mass measurements

3. Set the following parameters for database searching (Table 3)

4. Search all MS/MS spectra using SEQUEST algorithm, matching mass spectra against UniProt Human (*Homo sapiens*) database protein entries

5. Perform linear discriminant analysis (LDA) [21] to distinguish correct and incorrect peptide identifications using parameters such as Xcorr, ΔCn, and precursor mass error

Table 3
The parameters for database searching

Precursor mass tolerance	20 ppm
Product mass tolerance	1.0 Da
Digestion	Fully tryptic digestion
Miscleavages	Up to two
Fixed modifications	Carbamidomethylation of cysteine (+57.0214)
Variable modifications	Oxidation of Methionine (+15.9949)
	Tag for the glycosylation site on Asn (+2.9883)
False positive rate	<1 %

6. Discard peptides fewer than six amino acids in length

7. Filter the peptides to a less than 1 % false positive rate based on the number of decoy sequences in the final data set [22] (**Note 10**)

8. Use Modscore (similar to Ascore, indicating the likelihood that the best site match is correct when compared with the second best match) to evaluate the site confidence of site localizations [23]. We consider sites with a score ≥ 19 $(P < 0.01)$ to be confidently localized

9. Further data analysis

4   Notes

1. IAA must be freshly prepared before use. A higher IAA concentration or longer reaction time may induce protein N- and S-carbamidomethylation, which should be avoided

2. Proteins exist between the two layers and may be visible as a thin wafer

3. Longer incubation time may induce miscleavages

4. The composition of the slurry is beads:solution = 1:1

5. Peptides must be completely dried to allow full incorporation of O^{18} during the PNGase F treatment

6. Isopropanol could be used for stringent wash

7. We recommend "sandwich injection"—pick up 100 µL of buffer A, 300 µl of sample solution, and 100 µL of buffer A accordingly, then load to the column

8. Make sure there is not any solution left above the packing

9. If the peptides cannot be dissolved completely, sonicate the mixture for 30 s, then spin at $17,000 \times g$ for 5 min before loading to the stage tip.

10. The data set should be restricted to glycopeptides when determining false positive rate.

Acknowledgements

This work is supported by the National Science Foundation (CAREER Award, CMI-1454501).

References

1. Witze ES, Old WM, Resing KA, Ahn NG (2007) Mapping protein post-translational modifications with mass spectrometry. Nat Methods 4(10):798–806

2. Huang H, Lin S, Garcia BA, Zhao YM (2015) Quantitative proteomic analysis of histone modifications. Chem Rev 115(6):2376–2418

3. Xiao HP, Chen WX, Tang GX, Smeekens JM, Wu RH (2015) Systematic investigation of cellular response and pleiotropic effects in atorvastatin-treated liver cells by MS-based proteomics. J Proteome Res 14(3):1600–1611

4. Alvarez-Errico D, Vento-Tormo R, Sieweke M, Ballestar E (2015) Epigenetic control of myeloid cell differentiation, identity and function. Nat Rev Immunol 15(1):7–17

5. Dwek RA (1996) Glycobiology: toward understanding the function of sugars. Chem Rev 96(2):683–720

6. Varki A (1993) Biological roles of oligosaccharides - all of the theories are correct. Glycobiology 3(2):97–130

7. Spiciarich DR, Maund SL, Peehl DM, Bertozzi CR (2014) Identifying prostate cancer biomarkers by profiling glycoproteins in human prostate tissue. Abstr Pap Am Chem S, 248

8. Stowell SR, Arthur CM, McBride R, Berger O, Razi N, Heimburg-Molinaro J, Rodrigues LC, Gourdine JP, Noll AJ, von Gunten S, Smith DF, Knirel YA, Paulson JC, Cummings RD (2014) Microbial glycan microarrays define key features of host-microbial interactions. Nat Chem Biol 10(6):470–476

9. Wang XC, Chen J, Li QK, Peskoe SB, Zhang B, Choi C, Platz EA, Zhang H (2014) Overexpression of alpha (1,6) fucosyltransferase associated with aggressive prostate cancer. Glycobiology 24(10):935–944

10. Gilgunn S, Conroy PJ, Saldova R, Rudd PM, O'Kennedy RJ (2013) Aberrant PSA glycosylation-a sweet predictor of prostate cancer. Nat Rev Urol 10(2):99–107

11. Remmers N, Anderson JM, Linde EM, DiMaio DJ, Lazenby AJ, Wandall HH, Mandel U, Clausen H, Yu F, Hollingsworth MA (2013) Aberrant expression of mucin core proteins and O-linked glycans associated with progression of pancreatic cancer. Clin Cancer Res 19(8):1981–1993

12. Ma JF, Hart GW (2013) Protein O-glcnacylation in diabetes and diabetic complications. Exp Rev Proteomic 10(4):365–380

13. Ju TZ, Otto VI, Cummings RD (2011) The Tn antigen-structural simplicity and biological complexity. Angew Chem Int Ed 50(8):1770–1791

14. Chen WX, Smeekens JM, Wu RH (2014) Comprehensive analysis of protein N-glycosylation sites by combining chemical deglycosylation with LC-MS. J Proteome Res 13(3):1466–1473

15. Zhang H, Li XJ, Martin DB, Aebersold R (2003) Identification and quantification of N-linked glycoproteins using hydrazide chemistry, stable isotope labeling and mass spectrometry. Nat Biotechnol 21(6):660–666

16. Roy B, Chattopadhyay G, Mishra D, Das T, Chakraborty S, Maiti TK (2014) On-chip lectin microarray for glycoprofiling of different gastritis types and gastric cancer. Biomicrofluidics 8(3):034107

17. Maenuma K, Yim M, Komatsu K, Hoshino M, Takahashi Y, Bovin N, Irimura T (2008) Use of a library of mutated Maackia amurensis hemagglutinin for profiling the cell lineage and differentiation. Proteomics 8(16):3274–3283

18. Xu GB, Zhang W, Wei LM, Lu HJ, Yang PY (2013) Boronic acid-functionalized detonation nanodiamond for specific enrichment of glyco-peptides in glycoproteome analysis. Analyst 138(6):1876–1885

19. Zeng ZF, Wang YD, Guo XH, Wang L, Lu N (2013) On-plate glycoproteins/glycopeptides selective enrichment and purification based on surface pattern for direct MALDI MS analysis. Analyst 138(10):3032–3037

20. Chen WX, Smeekens JM, Wu RH (2014) A universal chemical enrichment method for mapping the yeast *N*-glycoproteome by mass spectrometry (MS). Mol Cell Proteomics 13 (6):1563–1572

21. Huttlin EL, Jedrychowski MP, Elias JE, Goswami T, Rad R, Beausoleil SA, Villen J, Haas W, Sowa ME, Gygi SP (2010) A tissue-specific atlas of mouse protein phosphorylation and expression. Cell 143(7):1174–1189

22. Elias JE, Gygi SP (2007) Target-decoy search strategy for increased confidence in large-scale protein identifications by mass spectrometry. Nat Methods 4(3):207–214

23. Beausoleil SA, Villen J, Gerber SA, Rush J, Gygi SP (2006) A probability-based approach for high-throughput protein phosphorylation analysis and site localization. Nat Biotechnol 24(10):1285–1292

Neuromethods (2016) 114: 43–53
DOI 10.1007/7657_2015_82
© Springer Science+Business Media New York 2016
Published online: 19 April 2016

Site-specific Localization of D-Amino Acids in Bioactive Peptides by Ion Mobility Spectrometry

Chenxi Jia, Christopher B. Lietz, Qing Yu, and Lingjun Li

Abstract

In this study, we describe a site-specific strategy to rapidly and precisely localize D-amino acids in peptides by ion mobility spectrometry (IMS) analysis of mass spectrometry (MS)-generated epimeric fragment ions. Briefly, the D/L-peptide epimers are separated by online reversed-phase liquid chromatography (LC) and fragmented by collision induced dissociation (CID), followed by IMS analysis. The epimeric fragment ions resulting from D/L-peptide epimers exhibit arrival time differences, thus showing different mobility in IMS. The arrival time shift between the epimeric fragment ions is used as criteria to localize the D-amino acid substitution. We provide the technical details on sample preparation, LC-tandem mass spectrometry analysis, data processing, and collisional cross-section calibration.

Keywords: Neuropeptide, Peptidomics, De novo sequencing, Mass spectrometry, Crustacean, Ion mobility

1 Introduction

The isomerization of an L- to D-amino acid is a remarkable post-translational modification of peptides in RNA-based protein synthesis and has been documented in amphibians, invertebrates, and mammals [1–9]. In many cases, the D-amino acid-containing peptides (DAACPs) exhibit dramatically higher affinity and selectivity for receptor binding than their all-L counterparts and thus are essential for biological function [3]. Generally, the targeted approaches for discovery of endogenous DAACPs include two steps: screening DAACP candidates in biological samples and then localizing D-amino acid residues [10, 11]. Many new DAACPs were found by observing the differences in biological activity or chromatographic retention time between synthetic peptides and naturally occurring peptides [3, 10, 11, 12]. In addition, immunoassays based on conformational antibodies have been successfully used to screen DAACPs at the tissue and cellular levels [11, 13]. For localization of D-amino acids in DAACP candidates, the most popular approach relies on matching chromatographic retention time of the naturally occurring peptide with a panel of synthetic peptides [11]. For example, validation of a deca-DAACP presumably requires testing ten synthetic peptides, each of which contains a D-amino acid at a different position, leading to high cost and limited analytical throughput. Other techniques utilize Edman degradation [14] or acid hydrolysis [11] to release free amino acids, followed by chromatographic analysis of the free or derivatized amino acids. However, cleavage of amide bond by chemical methods induces a 3–15 % level of racemization [5]. Therefore, there is a great demand for development of a simple and low-cost method to localize D-amino acids in a wide range of DAACP candidates.

The ion mobility spectrometry (IMS)-mass spectrometry (MS) technique has been widely used to probe the gas-phase conformations of biomolecules by measuring their mobility in a buffer gas and has shown very broad applicability in the separation and identification of isomeric peptides [15–21]. In this study, we developed a novel liquid chromatography (LC)-MS/MS-IMS strategy which allows site-specific characterization of peptide epimers. Using a theoretical 7-mer peptide, Fig. 1 illustrates the workflow for the strategy discussed in this chapter. The analysis can be completed in one LC-MS/MS-IMS run. First, peptide epimers are separated by reversed-phase LC and online submitted to CID fragmentation.

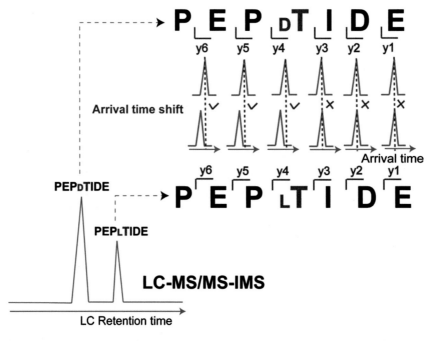

Fig. 1 Workflow of the proposed strategy for localization of D-amino acids in peptides. The analysis can be performed in one LC-MS/MS-IMS run. The two peptide epimers are separated by RPLC and respectively fragmented by CID. Their fragment ions are then submitted to IMS for arrival time measurement. By comparing the arrival time distributions between the two sets of fragment ions, the position of D-amino acid can be determined. √, Arrival time shift. ×, No shift. For illustration purpose, only y ions are listed in this workflow. Note that other fragment ions can also act as indicators for localization of D-amino acids

The resulting peptide fragment ions are then subjected to IMS for measurement of arrival time, or the ion transit time from the entrance of the mobility cell to the entrance of the mass analyzer. The epimeric ions of y_6, y_5, and y_4 derived from the two peptide epimers respectively contain the L- or D-Thr, which possibly leads to conformational differences between each epimeric y ion pair, resulting in arrival time shift during IMS analysis. In contrast, the two peptide epimers produce the same y_3, y_2, and y_1 ions containing all-L amino acids, since the D- or L-Thr has been removed from peptide chain by CID fragmentation. Thus, these y ion pairs show identical arrival times. By determining at which the arrival time shift suddenly appears, the D-amino acid can be confidently localized at the threonine. The practical utility was demonstrated by analysis of a peptide standard, [D-Trp]-melanocyte-stimulating hormone (MSH) and an endogenous large neuropeptide, crustacean hyperglycemic hormone (CHH) isolated from American lobster *Homarus americanus.*

2 Materials

1. American lobster *Homarus americanus* were purchased from Maine Lobster Direct Web site (http://www.mainelobsterdirect.com). All animals were kept in a circulating artificial seawater tank at $10 - 15\ °C$ [22].

2. The reagents and buffers for peptide digestion are 2.5 mM Dithiothreitol (DTT, Promega), 7 mM iodoacetamide (IAA, Sigma-Aldrich), 50 mM ammonium bicarbonate, and sequencing-grade trypsin (Promega).

3. The LC-MS/MS-IMS experiments were performed on a Waters nanoAcquity ultra performance LC system coupled to a Synapt G2 high-definition mass spectrometer.

4. Chromatographic separations were performed on a Waters BEH 300 Å C18 reversed-phase capillary column (150 mm × 75 μm, 1.7 μm).

5. The mobile phases used were 0.1 % formic acid in water (A); 0.1 % formic acid in ACN (B).

3 Methods

3.1 Animal Dissection and Sample Preparation

1. Animals were anesthetized in ice, and the sinus glands were dissected and collected in chilled acidified methanol and stored in −80 °C freezer prior to further sample processing [23].

2. The tissues were homogenized and extracted with 100 μL of acidified methanol (methanol:H_2O:acetic acid, 90:9:1, v:v:v) three times.

3. For trypsin digestion of CHHs, 1 μL of tissue extract was reduced and alkylated by incubation in 2.5 mM DTT for 1 h at 37 °C followed by incubation in 7 mM IAA in the dark at room temperature for 1 h, and then digested at 37 °C overnight after addition of 50 mM ammonium bicarbonate buffer with 0.5 μg of trypsin.

3.2 LC-MS/MS Coupled to Ion Mobility Spectrometry

1. The peptide sample was injected and loaded onto the Waters Symmetry C18 trap column (180 μm × 20 mm, 5 μm) using 97 % mobile phase A and 3 % mobile phase B at a flow rate of 5 μL/min for 3 min.

2. The gradient started from 3 to 10 % B during the first 5 min, increased to 45 % B in the next 65 min, and then was kept at 90 % B for 20 min.

3. For LC-MS-IMS experiment, an LC-MS survey was carried out to separate the peptide epimers. Subsequently, each epimer was analyzed by IMS.

4. For LC-MS/MS-IMS experiment, a fixed MS/MS survey was employed to select the peptide molecular ions in a traveling-wave (T-Wave) trap cell for CID fragmentation with adjusted collision energy 22–30 eV.

5. The resulting fragment ions were online submitted to T-Wave mobility cell and time-of-flight analyzer to measure the arrival time. Instrument acquisition parameters used were as follows: an inlet capillary voltage of 2.8 kV, a sampling cone setting of 35 V, and a source temperature of 70 °C.

6. The argon gas pressure in the traveling wave ion guide trap and the traveling wave ion guide transfer cell were 2.44×10^{-2} and 2.61×10^{-2} mbar, respectively.

7. The wave height, the wave velocity, and the nitrogen pressure in the traveling wave IM drift cell were 32.0 V, 800 *m/s*, and 2.96 mbar, respectively.

8. The .raw data was processed by DriftScope V2.4. The "Use Selection Tool" was used to select two separated LC peaks of the two peptide epimers, respectively. The .raw file of each peptide epimer was exported from DriftScope and further opened by Masslynx V4.1. The arrival time distributions of the two peptide epimers and their fragment ions were generated as shown in Figs. 2 and 3 [24].

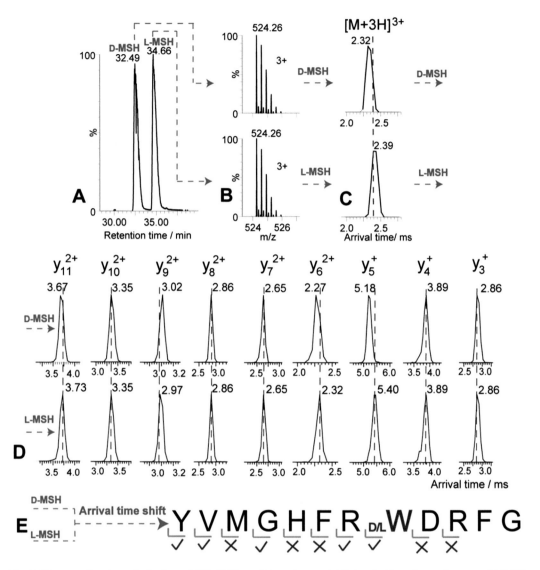

Fig. 2 Site-specific characterization of D/L-MSH peptide epimers. (**a**) Extracted ion chromatogram of LC-MS analysis of D/L-MSH peptides. (**b**) Molecular ions and (**c**) the corresponding IMS distributions of D/L-MSH peptides. (**d**) IMS distributions of fragment ions of D/L-MSH peptides. (**e**) Localization of D-amino acid residue position by comparison of arrival time shift. √, Arrival time shift. ×, No shift. Adapted with permission from ref. [24]

3.3 Collision Cross-Section Measurement

1. Gas-phase helium collision cross-section values (CCS_{He}) for all of the ions described in the manuscript were measured on the Synapt G2 travelling-wave ion mobility mass spectrometer with nitrogen buffer gas [25–27]. Although the nitrogen drift gas is more commonly employed experimentally, CCS_{He} is preferred for its easier integration into computational simulations. Using

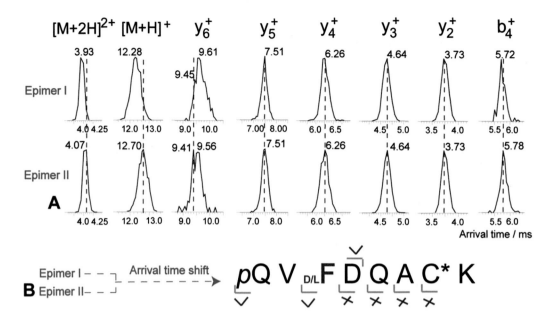

Fig. 3 Localization of D-amino acid in tryptic peptides, pQV$_{D/L}$FDQAC*K. (**a**) IMS distributions of fragment ions from tryptic peptide epimers I and II. (**b**) Localization of D-amino acid residue by comparison of arrival time shift. √, Arrival time shift. ×, No shift. It should be noted that the elution order of the two $_{D/L}$-peptide epimers cannot be determined by our current method, so we use Epimer I and II for annotation. Adapted with permission from ref. [24]

CCS$_{He}$ values for the calibrant ions allows relatively accurate calculation of unknown CCS$_{He}$ values from nitrogen gas arrival times [25].

2. Polyalanine peptides (Sigma Aldrich) were dissolved in 49.5/49.5/1 water/ACN/formic acid at a concentration of 10–100 μg/mL.

3. A separate calibration spectrum was acquired for each of the following wave velocities (m/s): 600, 700, and 800. The wave height was kept constant at 40 V. Following acquisitions, the log of the arrival times (t_D), or the portion of the arrival times spent in the mobility cell, were plotted against the log of the reduced CCS to determine the constants needed to calculate unknown Ω_{He}.

4. The equations in the following paragraph were obtained from previously published sources [25, 26]. In travelling wave ion mobility, t_D and CCS$_{He}$ (Ω_{He}) are nonlinearly related by Eq. 1:

$$\Omega_{He} = \frac{ze}{16}\left[\frac{18\pi}{k_b T}\left(\frac{1}{m}+\frac{1}{M_{He}}\right)\right]^{1/2}\frac{760}{P}\frac{T}{273.2}\frac{1}{NL}At_D^B \quad (1)$$

The variables z and e make up the charge of the analyte, T is the temperature of the drift gas, M_{He} is the mass of the helium drift gas, m is the mass of the analyte, P is the pressure inside the drift

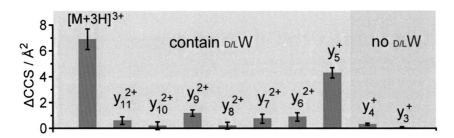

Fig. 4 CCS differences (ΔCCS, absolute values) of peptide precursor and fragment ions. *Error bars* stand for standard deviations. Adapted with permission from ref. [24]

cell, N is the number density of drift gas molecules, L is the length of the drift cell, and k_b is Boltzmann's constant. A and B are constants that arise from the nonuniformity of the travelling-wave electric field and must be empirically determined by calibration.

5. The t_D for each calibration standard was converted to corrected arrival time (t_D') by Eq. 2 to account for the m/z-dependent travel time through the Synapt G2's ion optics:

$$t_D' = t_D - \left(\frac{c\sqrt{m/z}}{1000}\right) \tag{2}$$

C is the delay constant set by the MS control software.

6. The reduced collision cross section (Ω') was normalized for mass and charge contributions and calculated by Eq. 3:

$$\Omega' = \frac{\Omega_{He}}{z\left(\frac{1}{m} + \frac{1}{M_{He}}\right)^{1/2}} \tag{3}$$

7. Plotting the natural log of Ω' versus the natural log of t_D' yielded a linear best-fit line, the slope of which is B from Eq. 1. From here, the doubly corrected arrival time (t_D'') was calculated by Eq. 4:

$$t_D'' = z\left(t_D'\right)^B\left(\frac{1}{m} + \frac{1}{M_{He}}\right)^{1/2} \tag{4}$$

A final plot was constructed with t_D'' on the x-axis and Ω_{He} on the y-axis. The equation of the best-fit line was then used to calculate the Ω_{He} of the unknown peptides and peptide fragments. The results are shown in Fig. 4 [24].

4   Notes

1. To achieve unambiguous discrimination, the peptide epimers need to be baseline-separated by LC. If the two peptide epimers are co-eluted during the reversed-phase LC separation, the LC gradient needs to be adjusted to achieve a base-line separation. Usually, increasing the elution time can result in a better separation.

2. Efficient fragmentation of peptides is essential for producing abundant sequential fragment ions, such as y and b ions. Otherwise, over-fragmentation may generate internal fragment ions, which is not useful for localization of the D-amino acids. Therefore, the collision energy needs to be optimized to ensure that the peptide can be efficiently fragmented but also avoid over-fragmentation.

3. Use of inappropriate settings for the trap ion gate may cause poor resolution of ion mobility. For the analysis of small peptides, we set the trap voltage at 20 V, to ensure that the ions do not escape from the trapping ion guide before the injection pulse to the ion mobility cell separator.

4. When extracting the arrival time distribution from .raw data, using a wide mass window may cause the contaminant ions to be included in the arrival time distribution as well. Therefore, a narrow mass window less than 0.02 Da was used.

5. It is important to observe a high signal-to-noise ratio and high mass accuracy for the polyalanine calibrants. Signal-to-noise ratios can be improved by increasing acquisition times. In our experience, acquisition times of 5–10 min are often sufficient. Under poor instrumental conditions, calibration acquisition times may need to be increased to 20 min or more.

6. For mass accuracy, it is crucial to use a lockspray correction during calibration. Our lab has developed transparent, open-source software called *pepCCScal* for CCS calibration. It is available for free by request.

7. The two peptide epimers should be fragmented at the same collision energy, so it is important to set the same collision energy.

8. Optimal wave height and wave velocity are absolutely crucial to IMS separation. We recommend always keeping the wave height at its maximum value, 40 V, and only adjusting the wave velocity. Lower wave velocities will decrease the IMS peak width, but may also decrease the temporal separation of two IMS peak apexes.

9. The sample amount for analysis is usually 5–50 ng. Overloading of samples may cause saturation of the ion mobility analysis and poor distribution of arrival time.

10. Arrival times are dependent on internal ion energies, and thus they can be affected by temperature changes in the room where analysis takes place. It is important to perform a new CCS calibration at the start of each set of experiments to minimize such errors.

11. A complete series of fragment ions is necessary for the comparison and to determine the identity of a DAACP. To maximize the duty cycle for a peptide of interest, targeted MS/MS can be performed to generate a satisfactory spectrum.

12. If the D-amino acid is at the N-terminus of the peptides, a LC-MS-IMS run without fragmentation is needed to measure the arrival times of peptide molecular ions.

13. While avoiding over-fragmentation, efficient ion transmission has to be maintained by using high enough voltage settings.

14. Although unlikely to occur on the SYNAPT G2 under settings specific to this kind of experiment, the possibility that peptide ions still retain their solution-phase structure preferences has to be ruled out since the approach is focused on gas-phase CCS, and solution-phase structures can complicate the interpretation of resulting spectra. This can be done by using standard peptides in different solvent compositions.

15. Since temperature and humidity from surrounding environment could introduce systematic errors into actual measurements, calibration has to be done right before each data acquisition to minimize such effects.

16. Due to limited resolving power of ion mobility separation on Synapt G2 $\left(\frac{\Omega}{\Delta\Omega} \approx 30 \right)$, the large peptide CHHs cannot be directly analyzed. Therefore, the large peptide was digested to produce small segments for enhanced discrimination. For analysis of small peptides with molecular weights of less than 2 kDa, the step of tryptic digestion is unnecessary.

17. This site-specific strategy is also applicable to other ion mobility instruments, such as Agilent 6560 Ion Mobility Quadrupole Time-of-Flight mass spectrometer.

Acknowledgements

This work is supported in part by the National Institutes of Health (NIH) grant (R01DK071801 to LL) and the National Science Foundation grant (CHE-1413596 to LL). LL acknowledges an H. I. Romnes Faculty Research Fellowship from UW-Madison, Tianjin 1000 Talent Plan from Tianjin China and Changjiang

Professorship from the Chinese Ministry of Education. C.L. acknowledges an NIH-supported Chemistry Biology Interface Training Program Predoctoral Fellowship (grant number T32-GM008505) and an NSF Graduate Research Fellowship (DGE-1256259). We are grateful to Prof. Jonathan V. Sweedler, Dr. Lu Bai, and Itamar Livnat (UIUC) for helpful discussions and insightful suggestions on this project.

References

1. Iida T, Santa T, Toriba A, Imai K (2001) Amino acid sequence and D/L-configuration determination methods for D-amino acid-containing peptides in living organisms. Biomed Chromatogr 15(5):319–327

2. Bai L, Romanova EV, Sweedler JV (2011) Distinguishing endogenous D-amino acid-containing neuropeptides in individual neurons using tandem mass spectrometry. Anal Chem 83(7):2794–2800

3. Bai L, Sheeley S, Sweedler JV (2009) Analysis of endogenous D-amino acid-containing peptides in metazoa. Bioanal Rev 1(1):7–24

4. Bai L, Livnat I, Romanova EV, Alexeeva V, Yau PM, Vilim FS, Weiss KR, Jing J, Sweedler JV (2013) Characterization of GdFFD, a D-amino acid-containing neuropeptide that functions as an extrinsic modulator of the Aplysia feeding circuit. J Biol Chem 288 (46):32837–32851

5. Adams CM, Zubarev RA (2005) Distinguishing and quantifying peptides and proteins containing D-amino acids by tandem mass spectrometry. Anal Chem 77(14):4571–4580

6. Hurtado PP, O'Connor PB (2012) Differentiation of isomeric amino acid residues in proteins and peptides using mass spectrometry. Mass Spectrom Rev 31(6):609–625

7. Tao WA, Cooks RG (2001) Parallel reactions for enantiomeric quantification of peptides by mass spectrometry this work was supported by the U.S. Department of energy, office of energy research. W.A.T. Acknowledges fellowship support from triangle pharmaceuticals. Angew Chem Int Ed Engl 40(4):757–760

8. Serafin SV, Maranan R, Zhang K, Morton TH (2005) Mass spectrometric differentiation of linear peptides composed of L-amino acids from isomers containing one D-amino acid residue. Anal Chem 77(17):5480–5487

9. Sachon E, Clodic G, Galanth C, Amiche M, Ollivaux C, Soyez D, Bolbach G (2009) D-amino acid detection in peptides by MALDI-TOF-TOF. Anal Chem 81(11):4389–4396

10. Soyez D, Van Herp F, Rossier J, Le Caer JP, Tensen CP, Lafont R (1994) Evidence for a conformational polymorphism of invertebrate neurohormones. D-amino acid residue in crustacean hyperglycemic peptides. J Biol Chem 269(28):18295–18298

11. Soyez D, Toullec JY, Montagne N, Ollivaux C (2011) Experimental strategies for the analysis of D-amino acid containing peptides in crustaceans: a review. J Chromatogr B Analyt Technol Biomed Life Sci 879(29):3102–3107

12. Buczek O, Yoshikami D, Bulaj G, Jimenez EC, Olivera BM (2005) Post-translational amino acid isomerization: a functionally important D-amino acid in an excitatory peptide. J Biol Chem 280(6):4247–4253

13. Serrano L, Grousset E, Charmantier G, Spanings-Pierrot C (2004) Occurrence of L- and D-crustacean hyperglycemic hormone isoforms in the eyestalk X-organ/sinus gland complex during the ontogeny of the crayfish Astacus leptodactylus. J Histochem Cytochem 52(9):1129–1140

14. Iida T, Matsunaga H, Santa T, Fukushima T, Homma H, Imai K (1998) Amino acid sequence and D/L-configuration determination of peptides utilizing liberated N-terminus phenylthiohydantoin amino acids. J Chromatogr A 813(2):267–275

15. Bohrer BC, Merenbloom SI, Koeniger SL, Hilderbrand AE, Clemmer DE (2008) Biomolecule analysis by ion mobility spectrometry. Annu Rev Anal Chem (Palo Alto Calif) 1:293–327

16. Enders JR, McLean JA (2009) Chiral and structural analysis of biomolecules using mass spectrometry and ion mobility-mass spectrometry. Chirality 21(Suppl 1):E253–E264

17. Verbeck GF, Ruotolo BT, Sawyer HA, Gillig KJ, Russell DH (2002) A fundamental introduction to ion mobility mass spectrometry applied to the analysis of biomolecules. J Biomol Tech 13(2):56–61

18. Shvartsburg AA, Tang K, Smith RD (2009) Two-dimensional ion mobility analyses of

proteins and peptides. Methods Mol Biol 492:417–445

19. Fenn LS, McLean JA (2008) Biomolecular structural separations by ion mobility-mass spectrometry. Anal Bioanal Chem 391 (3):905–909

20. Lanucara F, Holman SW, Gray CJ, Eyers CE (2014) The power of ion mobility-mass spectrometry for structural characterization and the study of conformational dynamics. Nat Chem 6(4):281–294

21. Jurneczko E, Barran PE (2011) How useful is ion mobility mass spectrometry for structural biology? The relationship between protein crystal structures and their collision cross sections in the gas phase. Analyst 136(1):20–28

22. Jia C, Hui L, Cao W, Lietz CB, Jiang X, Chen R, Catherman AD, Thomas PM, Ge Y, Kelleher NL, Li L (2012) High-definition de novo sequencing of crustacean hyperglycemic hormone (CHH)-family neuropeptides. Mol Cell Proteomics 11(12):1951–1964

23. Hui L, Cunningham R, Zhang Z, Cao W, Jia C, Li L (2011) Discovery and characterization of the Crustacean hyperglycemic hormone precursor related peptides (CPRP) and orcokinin neuropeptides in the sinus glands of the blue crab Callinectes sapidus using multiple tandem mass spectrometry techniques. J Proteome Res 10(9):4219–4229

24. Jia C, Lietz CB, Yu Q, Li L (2014) Site-specific characterization of D-amino acid containing peptide epimers by ion mobility spectrometry. Anal Chem 86(6):2972–2981

25. Bush MF, Campuzano ID, Robinson CV (2012) Ion mobility mass spectrometry of peptide ions: effects of drift gas and calibration strategies. Anal Chem 84(16):7124–7130

26. Ruotolo BT, Benesch JL, Sandercock AM, Hyung SJ, Robinson CV (2008) Ion mobility-mass spectrometry analysis of large protein complexes. Nat Protoc 3 (7):1139–1152

27. Lietz CB, Yu Q, Li L (2014) Large-scale collision cross-section profiling on a traveling wave Ion mobility mass spectrometer. J Am Soc Mass Spectrom 25(12):2009–2019

Neuromethods (2016) 114: 55–72
DOI 10.1007/7657_2015_88
© Springer Science+Business Media New York 2016
Published online: 23 February 2016

Quantitative Profiling of Reversible Cysteome Modification Under Nitrosative Stress

Yue-Ting Wang, Sujeewa C. Piyankarage, and Gregory R.J. Thatcher

Abstract

Reversible modifications of protein cysteine residues via *S*-nitrosylation and *S*-oxidation via disulfide formation are posttranslational modifications (PTM) regulating a broad range of protein activities and cellular signaling. Dysregulated protein nitrosothiol and disulfide formation have been implicated in pathogenesis of neurodegenerative disorders. Under nitrosative or nitroxidative stress, both nitrosylation and oxidation can theoretically occur at redox-sensitive cysteine residues, mediating thiol-regulated stress response. However, few detection strategies address both modifications. Nonquantitative approaches used to observe *S*-nitrosylation, regardless of unmodified and oxidized thiol forms, may lead to causal conclusions about the importance of protein nitrosothiol in NO-mediated signaling, regulation, and stress response. To observe quantitatively the modification spectrum of the cysteome, we developed a mass spectrometry-based approach, denoted as d-SSwitch, using isotopic labeling and shotgun proteomics to simultaneously identify and quantify different modification states at individual cysteine residues. Both recombinant protein and intact neuroblastoma cells were analyzed by d-SSwitch after treatment with nitrosothiol or NO. In proteins identified to be modified after nitrosothiol treatment, *S*-oxidation was always observed concomitant with *S*-nitrosylation and was quantitatively dominant. Herein, we describe the detailed procedures of d-SSwitch and important notes in practice.

Keywords: *S*-Nitrosylation, Disulfide formation, Nitrosative stress, Posttranslational modification (PTM), MS, d-SSwitch

1 Introduction

Despite the low occurrence within the human proteome [1], cysteine residues play key roles in sensing and responding to the perturbation of cellular redox homeostasis. The electronic structure of the cysteine sulfhydryl group (SH) permits multiple modifications in response to oxidative, nitrosative, or nitroxidative stimuli. These posttranslational modifications include sulfenylation to form sulfenic acid (SOH, often the initial step to form disulfides), sulfinylation to form sulfinic acid (SO_2H), sulfonylation to form sulfonic acid (SO_3H irreversible), disulfide formation (SS); and nitrosation to form protein nitrosothiol (SNO). Among all these modifications that may occur at a redox-sensitive cysteine thiol, SNO formation (widely termed *S*-nitrosylation) and SS formation have attracted interest as reversible signal transduction events. These transient PTMs have been implicated as mediators of protein activity, protein relocalization/interaction, and cell signaling as regulatory "switches," showing great similarity to the consequences of phosphorylation [2–5].

S-Nitrosylation is a product of nitrosation of cysteine sulfhydryl group with a nitrosonium (NO^+) equivalent *readily provided by endogenous nitrosothiols*, such as S-nitrosoglutathione (GSNO) and S-nitrosocysteine (CysNO), N_2O_3, or Lewis acid-catalyzed reactions of nitric oxide (NO) and NO_2^- [6–8]. Extensive studies implicating protein SNO have led to the hypothesis that this PTM is the major mechanism by which NO elicits diverse effects on cell proliferation, apoptosis, and the immune response [5, 9–11]. The disulfide bond is a reversible covalent linkage formed between the sulfur atoms from two sulfhydryl groups. It can form between protein thiols as protein disulfides, or between protein thiol and low molecular weight (LMW) thiols as mixed disulfides (e.g. S-glutathionylation by GSH). The disulfide formed after reaction with H_2S, though potentially important, has received little attention. Protein disulfides are essential structural components and can participate protein–protein interactions and/or catalytic activity [2, 12, 13]. Over the past 30 years our perspective of disulfides has dramatically evolved: the importance of disulfides in redox regulation under both normal and stress conditions is emerging [2–4, 14]. Nitrosative or nitroxidative stress is often linked to formation of reactive nitrogen species (RNS) [15–17], which often occur concomitantly with reactive oxygen species (ROS), i.e. ROS/RNS. Oxidation is the dominant modification caused by RNS such as N_2O_3, NO_2, and peroxynitrite; a process usefully termed nitroxidation [15, 16, 18]. Protein S-nitrosylation can also be nitroxidative, providing a mechanism for regulatory disulfide formation including glutathionylation [19, 20]. Aberrant nitrosative and/or nitroxidative stress likely play a role in pathogenesis of many disease states, including Alzheimer's and Parkinson's disease; although the majority of researchers focus on S-nitrosylation [21–24].

The identification of modified cysteine residue(s), and quantitatively assessment of the different modification events at each cysteine will provide information crucial to understand the regulatory role of nitrosative stress at protein levels via transient PTMs. A number of mass spectrometry (MS)-based methods have been established to identify and sometimes quantify protein nitrosothiols and disulfides in separate experiments [25]. While direct detection of SNO and SS is still technically challenging, most methods used today require selective reduction of the reversible PTM and differential labeling prior to shotgun proteomic analysis. The most commonly used approach for detecting S-nitrosylation is the modified biotin switch technique (BST) that reduces SNO with large amount of ascorbate and labels the nascent thiol with a biotin tag, followed by MS analysis in place of the original qualitative analysis by western blot [26]. Several modifications of the BST approach published after the initial report include stable isotope labeling to address limitations in quantification [27, 28]. Alternative approaches for analyzing *S*-oxidation also rely on isotopic labeling of unmodified and oxidized thiols using various isotopologues upon reduction [29]. Unfortunately very few of these methods described above can provide quantitative information of SNO, SS, and unmodified thiol for specific cysteine residues in parallel. In *S*-nitrosylation studies, complete neglect of cysteine oxidation and unmodified cysteine can be a serious flaw causing gross overestimation of protein-SNO formation. Given the complexity in nitrosative stress-induced modification profiles, it is sensible to develop methods allowing simultaneous quantitation of cysteome inventory (SNO + SS + SH) of specific cysteine residue.

In our recent study we introduced a novel proteomic methodology, noted as "d-SSwitch" to quantitatively profile different modification states (SNO + SS + SH) for targeted cysteine residue (Fig. 1). Adapted from the "d-Switch" approach we developed to measure SNO versus SH, d-SSwitch reduces SNO and SS functionalities to free thiols by selective chemical reactions, then use two isotopologues of the alkylating agent *N*-ethylmaleimide (NEM) to

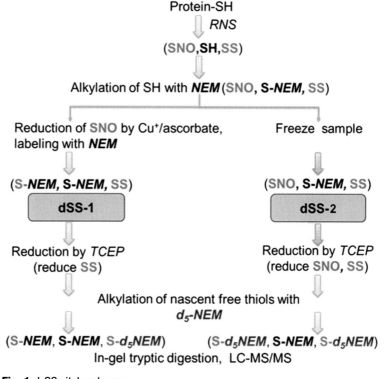

Fig. 1 d-SSwitch scheme

probe one split sample in parallel to simultaneously quantify different chemical states of one protein thiol [30]. Human glutathione S-transferase pi-1 (GSTP1-1) was used in method development and to provide comparison with results from the d-Switch approach. GSTP1-1 has bene proposed to play a regulatory role in cell response to NO and nitrosative stress via *S*-nitrosylation of the most reactive cysteine residue Cys47 [31]; although the weight of evidence is for *S*-oxidation as the response to nitrosative stress, which mediates function [32]. Intra- and intermolecular disulfide formations also contribute to GSTP1-1 mediated signal transduction in cell proliferation and cell death [33]. Using d-SSwith combined with shotgun proteomic analysis we demonstrated that under nitrosative stress *S*-oxidation to disulfide was the dominant modification universal to all NO donors. We also observed that *S*-nitrosylation and disulfide formation result from nitrosothiol-induced nitrosative stress in a concentration-dependent manner and independent of oxygen. These observations made using recombinant protein were extended to living neuronal cells: nitrosative stress induced by nitrosothiol, caused *S*-nitrosylation and *S*-oxidation of intracellular GSTP1 at Cys47 quantitatively comparable to that observed in recombinant protein. Further analysis of the cellular cysteome using d-SSwitch revealed that *S*-oxidation is the major cysteome modification under nitrosative stress.

In this chapter, we describe the detailed experimental procedures of d-SSwitch, the quantitative analysis strategy, and technical notes important in practice. As depicted in Fig. 1, free thiols of treated proteins are blocked with NEM under denaturing condition. The resulting sample is divided into two portions as dSS-1 and dSS-2, followed by selective reduction and labeling with d_5-NEM. In dSS-1 all SH and SNO are labeled with NEM and all SS are labeled with d_5-NEM; whereas in dSS-2 both SNO and SS are labeled with d_5-NEM, only unmodified thiols are coded with NEM. Shotgun proteomic analysis is applied to identify light and heavy NEM-labeled peptides that have identical retention time and ionization efficiency, and peak areas of differentially labeled peptides are acquired from corresponding chromatograms. Simple algebraic derivation using peak area ratios obtained from dSS-1 and dSS-2 provides the quantitative profile of the modified cysteome.

2 Materials

2.1 Nitrosating Agents/NO-Donors

Nitrosocysteine (CysNO) is prepared from cysteine and sodium nitrite in the presence of hydrochloric acid (HCl). Diethylamine NONOate is purchased from Sigma Aldrich. Nitroglycerin (GTN) and GT-094 were synthesized in-house according to published procedures [34, 35]. The HNO donor AcOM-IPA/NO was kindly provided by Dr. Daniela Andrei (Dominican University) [36].

All the NO/HNO donor stock solutions except CysNO solution are prepared in methanol (HPLC grade, Thermo Fisher, Rock Field, IL) freshly before addition to reaction mixture.

1. Cysteine, sodium nitrite ($NaNO_2$), sodium hydroxide (NaOH) (Sigma Aldrich, St. Louis, MO), HCl (Fisher Scientific, Rockford, IL). 210 mM $NaNO_2$ solution is prepared in deionized water and stored at 4 °C; 1 M HCl, 1 M NaOH stock solutions stored at room temperature; 200 mM cysteine in water is freshly prepared from solid before experiment.

2. UV quartz cuvette, 1 mL volume, two sides polished (Sigma Aldrich, St. Louis, MO).

3. Amber-colored glass vial, 1 and 3 mL volume (Fisher Scientific).

2.2 d-SSwitch

1. Ammonium bicarbonate, EDTA, neocuproine, *N*-ethylmaleimide (NEM), cupper (I) chloride (CuCl), and sodium ascorbate (Sigma Aldrich). Tris(2-carboxyethyl)phosphine (TCEP) 500 mM solution at neutral pH (Thermo Fisher, Rockford, IL) (*see* **Note 1**).

2. *N*-ethylmaleimide (ethyl-d5, 98 %) (d_5-NEM, Cambridge Isotope Laboratories, Tewksbury, MA) 50 mM stock solution prepared in anhydrous DMF and stored at −20 °C.

3. Reaction buffer 1: 40 mM ammonium bicarbonate (NH_4HCO_3), 1 mM EDTA, 0.1 mM neocuproine at pH 7.4.

4. Reaction buffer 2: reaction buffer 1 without EDTA and neocuproine.

5. 20 % SDS.

6. Amicon® Ultra centrifugal filter device with 10 kDa mass cutoff (EMD Millipore, Bedford, MT).

7. Amber-colored safe-lock microcentrifuge tubes, 1.5 mL volume (Eppendorf, Hauppauge, NY).

2.3 Cell Culture/ Handling/Cell Lysate Preparation

Human neuroblastoma cell line SH-SY5Y cells were obtained from American Type Culture Collection (ATCC) and maintained in 1:1 mixture of Dulbecco's Eagle's Minimum Essential Medium and Opti-Minimum Essential Medium (DMEM/Opti-MEM) supplemented with 5 % fetal bovine serum (FBS), 1 % penicillin–streptomycin and 5 % CO_2 at 37 °C. The normal growth medium is replaced by Opti-MEM phenol red free, reduced serum medium 1 h prior to the treatment.

1. DMEM, Opti-MEM, Opti-MEM phenol red free with reduced serum, FBS, penicillin-streptomycin (10,000 U/mL) (Life Technologies, Grand Island, NY).

2. Phosphate buffer saline (PBS, 10×, pH 7.4): 10.6 mM potassium phosphate monobasic (KH_2PO_4), 29.7 mM sodium phosphate dibasic (Na_2HPO_4), 1557.1 mM sodium chloride (NaCl). No calcium, magnesium, and phenol red (Life Technologies, Grand Island, NY) tenfold diluted solution was sterilized and used.

3. Lysis buffer: lysis buffer shall be made freshly each time before use. It is comprised of 100 mM Tris–HCl, 50 mM NaCl, 1 mM EDTA, 0.1 mM neocuprine, 1 % NP-40, 20 mM NEM, and protease inhibitors cocktail used following the manufacturer's instruction. The cOmplete Mini protease inhibitor cocktail tablets (EDTA free) is purchased from Roche (Roche Diagnostics, Mannheim, Germany).

4. Amber-colored microcentrifuge tubes, 1.5 mL volume.

2.4 Protein Separation and Visualization

1. NuPAGE® Novex® 4–12 % Bis-Tris protein precast gel, NuPAGE® LDS sample buffer (4×), and NuPAGE® MOPS SDS running buffer (20×) (Life Technologies, Grand Island, NY).

2. Coomassie Brilliant Blue R-250 staining and destaining solutions (Bio-Rad, Herclues, CA).

3. Spectra multicolor broad range protein ladder (Pierce, Thermo Scientific, Rockford, IL).

4. Protein gel chamber system (XCell *Superlock*®, Life Technologies).

2.5 Protein Concentration Measurement

1. BCA protein assay kit (Pierce, Thermo Scientific).

2. Protein standard (1.0 mg/mL bovine serum albumin).

3. Microplate scanning spectrophotometer (Bio-Tek, Winooski, VT).

2.6 Protein In-Gel Digestion

1. In-gel tryptic digestion kit containing lyophilized trypsin protease (Pierce, Thermo Scientific). Upon arrival, prepare the trypsin stock solution by dissolving the enzyme in 20 µL storage solution provided within the kit. The resulting solution is then aliquoted into four 0.5 mL tubes on ice of 5 µL each and stored at −20 °C.

2. Digestion buffer: 25 mM NH_4HCO_3 in milli-Q water.

3. Destaining solution: 25 mM NH_4HCO_3 in a mixture of milli-Q water and acetonitrile (ACN) at 1:1 ratio.

4. Extraction solution: 25 mM NH_4HCO_3 with 1 % formic acid.

2.7 Liquid Chromatography-Mass Spectrometry

1. For recombinant protein digest analysis: Agilent 6310 Ion Trap mass spectrometer with ESI (Agilent Technologies, Santa Clara, CA) coupled to Agilent 1100 series HPLC.

Hypersil BDS C18 column (30 × 2.1 mm, 3 μm, Thermo Scientific).

2. For cell lysate digest analysis: hybrid LTQ-FT linear ion trap mass spectrometer with Nanospray ESI source (Thermo Electron Corp., Bremen, Germany), Nanoflow HPLC (Dionex, Sunnyvale, CA). The trapping cartridge and the nanocolumn used for separation: Zorbax 300 SB-C18 (5 × 0.3 mm, 5 μm) and Zorbax 300 SB-C18 capillary column (150 mm × 75 nm, 3.5 μm) (Agilent Technologies). PicoTip™ emitter (New Objective, Woburn, MA).

3. Mobile phase A: 95/5 (v/v) water/ACN, 0.1 % formic acid; mobile phase B: 95/5 (v/v) ACN/water, 0.1 % formic acid. Optima™ LC/MS grade ACN (Fisher Scientific).

2.8 Data Analysis

1. MassMatrix (http://www.massmatrix.net).

2. Agilent LCMS workstation software (Agilent).

3. Xcalibur Software (Thermo Scientific).

4. GraphPad Prism (GraphPad Software).

2.9 Equipment

1. UV–Vis spectrophotometer (Hewlett Packard,, Palo Alto, CA).

2. Centrifuges (Eppendorf, Fisher Scientific).

3. Water baths.

3 Methods

3.1 CysNO Preparation (See Note 2)

1. CysNO is freshly prepared by acid-catalyzed *S*-nitrosation of the thiol with sodium nitrite. Avoid light during the entire process.

2. Dissolve 29 mg L-cysteine in 1.2 mL water to reach a final concentration of 200 mM. To an amber-colored glass vial, add 1 mL of cysteine solution, 1 mL of 210 mM $NaNO_2$ solution and 0.1 mL of 1 M HCl solution. After vortexing briefly incubate the mixture in water bath at 37 °C for 10 min.

3. After incubation place the entire vial in ice. Adjust the pH with 1 M NaOH solution to 8 and add 1 mM neocuproine.

4. Take 10 μL of the reaction mixture dilute it into 990 μL of water and transfer the resulting solution into a quartz cuvette (zero the background for water before transferring). Measure the UV absorbance of 100-fold diluted CysNO solution at 336 nm.

5. Use the published extinction coefficient of CysNO ($900 M^{-1} \cdot cm^{-1}$), combine with the dilution factor of 100 to calculate CysNO concentration in the original mixture [37].

3.2 Recombinant Protein Sample Preparation

1. Histidine-tagged human GSTP1-1 wild type and C101A mutant were previously expressed in *Escherichia coli* strain TG1 and purified using immobilized Co^{2+} affinity column chromatography. Aliquots of protein stock solutions containing 1 mM EDTA, 1 mM DTT were stored at -80 °C. Detailed procedure has been described [38, 39]. Protein concentrations were determined after fractionation using Pierce BCA assay kit following manufacturer's instruction.

2. Prior to treatment, exchange the protein stock solution with reaction buffer 1 and concentrate the protein stock using Amicon® MWCO 10 kDa ultra centrifugation filter device (*see* **Note 3**).

3.3 Cell Treatment and Lysate Preparation for d-SSwitch

1. For each treatment, culture one dish (10 cm) of SH-SY5Y cells to reach 90 % confluence. Each treatment is in triplicate from three independent runs (*see* **Note 4**).

2. On the day of treatment, remove normal growth medium, gently wash away the remaining media with PBS, add Opti-MEM phenol red free with reduced serum and let the cells settle for 1 h at 37 °C with 5 % CO_2 in air atmosphere.

3. Treat the cells with CysNO at different concentrations (e.g. 10, 100, and 1000 µM) and incubate the cells in the incubator at 37 °C for 20 min.

4. Take pre-made Tris–HCl buffer which contains EDTA, add NEM, neocuproine stock solutions, and NP-40 to reach final concentration of 20 mM, 0.1 mM, and 1 % respectively. Add protease inhibitor cocktail tablet, vortex until it dissolves. Place the resulting lysis buffer in ice until further use.

5. At the end of the cell treatment, immediately remove the medium, gently wash the cells with PBS (avoid direct impact of on cells) and remove PBS completely. To each plate, add 400 µL of lysis buffer, quickly scrape the cells off and collect the suspension into an amber-colored Eppendorf tube.

6. Lyse the cells by sonication. Use a sonic dismembrator (model 500, Fisher Scientific), sonicate the cells at amplitude of 20 % for 2 s, repeat five times. Keep the amber-colored tube in ice for the entire process.

7. Spin the content at 14,000 rpm, 4 °C, for 15 min. Use a pipette to transfer the supernatant (ca. 400 µL) into another amber-colored tube. Take another small fraction (ca. 50 °µL) exchange the lysis buffer with 40 mM $NaHCO_3$ buffer for protein concentration measurement (BCA assay following the manufacturer's instruction).

8. Avoid light through step 3–7. Directly apply the following procedures to lysates without storing samples overnight.

3.4 d-SSwitch Procedure

1. For recombinant protein, incubate the protein (15 μM) with NO or HNO donors at designated concentrations in reaction buffer 1 at 37 °C for 30 min with gentle agitation.

2. After incubation, add NEM stock and 20 % SDS into the same tube and incubate the mixture at 55 °C for 30 min. Vortex the mixture every 6 min for 30 s. Final concentrations of NEM and SDS in the resulting mixture are 20 mM and 5 %.

3. For the cell lysate obtained from NO-donor treated cells, add 5 % SDS and another fraction of NEM stock to reach the final concentration of 25 mM. Incubate the mixture at 55 °C for 30 min with frequent vortexing (same as step 2).

4. Remove excess of NEM from the protein reaction mixture by filtering it through Amicon® MWCO 10 kDa filter. Mix reaction buffer 2 with the content remaining in the filter and filter it through. Repeat for three times to replace reaction buffer 1 with buffer 2.

5. Collect the content left in the filter in a separate tube and divide it equally to two portions. Label them as dSS-1 and dSS-2. Store dSS-2 samples at −20 °C temporally.

6. Make a fresh solution of sodium ascorbate (50 mM) in reaction buffer 2, and a saturated solution of CuCl which gives 100 μM CuCl in water. Centrifuge the suspension, take the clear solution and dilute with reaction buffer 2 to a working solution with appropriate concentration (*see* **Note 5**).

7. To dSS-1 sample, add NEM, sodium ascorbate and CuCl stock solutions with the final concentrations at 5 mM, 5 mM, and 1 μM, respectively. Vortex the mixture briefly and incubate in water bath at 25 °C for 1 h.

8. Remove the treatment in step 7 from dSS-1 sample by filtering through the MWCO 10 kDa filter. Similarly to step 4, wash the remaining content with reaction buffer 2 for three times.

9. Collect the protein content in dSS-1 samples into a clean tube, thaw dSS-2 samples on the ice, incubate both dSS-1 and dSS-2 samples with 50 mM TCEP at 60 °C, for 15 min with mediate agitation.

10. Repeat step 8 to remove excess TCEP from samples. Add 1 mM d_5-NEM in the washing buffer (reaction buffer 2) to assure the presence of d_5-NEM during the removal of TCEP.

11. Collect the remaining dSS-1 and dSS-2 samples from filters into separate tubes and incubate with 5 mM of d_5-NEM in water bath at 25 °C for 1 h.

12. Mix samples resulted from step 11 with 4× SDS-loading buffer, mix thoroughly, and place into a 60 °C water bath incubate for 15 min (*see* **Note 6**).

13. Load samples onto a 1.0 mm, NuPAGE 4–12 % Bis-Tris gel, maximum 45 μL of protein sample per well. Load 3 μL of the protein ladder into the first well. Run the gel at 180 V for approximately 60 min, cut the gel from the precast frame, stain with Coomassie Brilliant Blue in a microwave (heat up the gel with staining solution in a microwave for 5–7 s. Repeat the heating for 2–3 times. Be careful not to heat the gel for too long each time that may break the gel) Destain with destaining solution on a rocker with gentle agitation. Replace the destaining solution with milli-Q water if the gel needs to be stored overnight.

14. For experiments carried out under controlled oxygen level, purge the reaction buffer 1 with N_2 or O_2 for at least 1 h before use. During the incubation seal the reaction vial with proper rubber septa, transfer additional reagents by syringe. Also during the washing steps use the freshly purged buffer. Cap the tube quickly after each addition of reagents.

15. Step 1–9 must be performed with protection from light exposure.

3.5 Protein In-Gel Digestion

1. Excise the desired protein bands with the molecular weight corresponding to GSTP1 (20–26 kDa). Cut the gel into smaller pieces (~1 mm^3).

2. Collect the gel pieces into a 1.5 mL Eppendorf tube.

3. Add 200 μL of the destaining solution described in Section 2.6, incubate in a 37 °C water bath for 30 min. During the incubation vortex the tube 4–5 times. Centrifuge the tube briefly after each vortexing to make sure all the gel pieces are at the bottom in solution. Repeat this step one more time if necessary (*see* **Note 7**).

4. Remove destaining solution from the tube with a pipette. Add 100 μL of ACN to dehydrate the gel pieces for 25 min at RT.

5. Remove ACN from the tube with a pipette. Keep the cap open to air-dry the gel pieces at RT (*see* **Note 8**).

6. While the gel pieces are drying, pull out one aliquot of trypsin stock solution, thaw it on ice. Add 45 μL of milli-Q water to make a trypsin working solution. Place the working solution on ice before further use.

7. Add 60 μL of digestion buffer to the gel pieces. Add 2 μL of trypsin working solution into each sample (~200 ng trypsin/sample) and vortex 2 s for three times. Add another 10 μL of digestion buffer then vortex 2 s × 3 times. Centrifuge samples

briefly make sure all gel pieces are settled at the bottom and covered in solution. Incubate the sample at 30 °C for overnight digestion.

8. One the next day, add 20 μL of extraction solution into the sample, vortex thoroughly, and centrifuge at 8000 rpm for 3 min. Use a pipette with gel loading tip to transfer 40 μL of supernatant to a glass vial for LC-MS/MS analysis.

3.6 d-SSwitch Sample Analysis by LC-Ion Trap MS

1. For samples prepared from recombinant hGSTP1-1 C101A mutant, Agilent 6310 Ion Trap mass spectrometer with electrospray ionization source was used for analysis. The instrument is operated in positive mode with capillary voltage of −3.5 kV and dry temperature of 350 °C. Nebulizer gas and capillary gas flow are at default settings. MS/MS spectra are acquired for two most abundant ions in each scan within a mass range of m/z 100–2000; fragmentation of target precursor ions is introduced by collision-induced dissociation (CID) with ultrapure argon as collision gas.

2. Protein digests in each sample are eluted at 300 μL/min and resolved on a Hypersil C18 column using 10 % MeOH (v/v) with 0.1 % FA as mobile phase A, ACN plus 0.1 % FA as mobile phase B. The gradient program is as follows:

Time (min)	Solvent A (%)	Solvent B (%)
4	90	10
14	80	20
23	62	38
27	10	90
29	90	10
33	90	10

3. In the total ion chromatograms (TIC) of untreated GSTP1-1 tryptic digests, identify the Cys47 containing peptide (residue 46–55: ASCLYGQLPK) by its m/z (doublet-charged) and fragmentation patterns (y ions and b ions) resulted from CID.

4. In the TIC acquired for dSS-1 and dSS-2 samples, extract the ion chromatograms of NEM-modified (m/z 603.2) and d_5-NEM-modified (m/z 605.6) Cys47-containing peptides—each sample contains peptides with both types of modifications. The identity of the modified peptides can be confirmed by b ions and y ions generated after fragmentation, shown in MS/MS spectra. Integrate the peak areas of the extracted ion chromatograms obtained from each sample (Fig. 2). In dSS-1, peak area of NEM labeled peptides represents unmodified plus

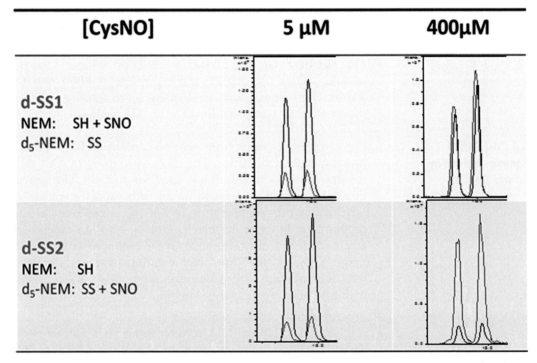

Fig. 2 Extracted ion chromatogram (EIC) for d-SSwitch. Human GSTP1 (C101A mutant) treated with CysNO at 5 and 400 μM subjected to d-SSwitch procedure, followed by in-gel digestion and LC-MS/MS analysis. EIC of Cys47 containing peptides (residue 46–55) labeled by NEM (*brown*) and d_5-NEM (*red*) within each sample are shown

SNO-forming thiols (noted as a); peak area of d_5-NEM labeled peptides represents SS-forming thiols (b). In dSS-2, NEM labeling only represents unmodified thiols (c); peak area of d_5-NEM-modified peptides attributes to both SNO and SS forming thiols (d). The calculation scheme is shown in following equations:

$$\%SS = b/(a + b) \tag{1}$$

$$\%SNO = (a - c)/(a + b) \, \text{or} \, (a - c)/(c + d) \tag{2}$$

$$\%SH = c/(c + d) \tag{3}$$

3.7 d-SSwitch Sample Analysis by NanoLC-HRMS

1. For protein digests prepared from cell lysate, LTQ-FT-ICR (high resolution, HR) MS instrument was used for analysis. The system sensitivity and accuracy are calibrated on a weekly basis using LTQ FT calibration solution (Thermo Scientific). Also run system checks using glu-1-fibrinopeptide B and substance P standards 2–3 times before analyzing the real sample.

2. The separation of peptides is achieved using Dionex nano-HPLC on a reversed phase capillary column. After being introduced onto a trapping cartridge with 100 % mobile phase A (5 % ACN and 0.1 % FA in H_2O) at flow rate of 50 μL/min,

peptides are eluted out and carried through the capillary column at 250 nL/min with increasing % mobile phase B (5 % H_2O and 0.1 % FA in ACN) through a 120-min gradient program:

Time (min)	Solvent A (%)	Solvent B (%)
0	95	5
6	95	5
96	60	40
102	60	40
103	20	80
107	20	80
108	95	5
120	95	5

3. Resolved peptides are introduced into the nano electrospray ionization source via SilicaTip™ nanospray emitter mounted at the end of the column. A spray voltage of 2.13 kV is applied on the emitter and the capillary temperature is set at 200 °C. Data acquisition is completed in positive ion mode using a data-dependent analysis (DDA) method. Survey scan is performed using FTMS at a resolution of 50,000; following MS/MS scans are performed in a mass range of 400–1800 m/z for ten most abundant ions. Fragmentation is induced by CID with an ion isolation width of 3 m/z. The minimum ion threshold is set to 5000 counts.

4. Convert raw data files to mzXML files and submit to MassMatrix (http://www.massmatrix.net) search engine to search against IPI human v3.65 database with NEM and d_5-NEM as variable modifications. Following parameters are applied for peptide/protein identification: enzyme: trypsin; missed cleavage: 2; precursor ion tolerance: ±2.0 Da; product ion tolerance: ±0.8 Da; mass type: monoisotopic; minimum peptide length: 6. Protein ID is accepted with 40 % sequence coverage (*see* **Note 9**).

5. Find NEM- and d_5-NEM-modified peptides in the search results. Look into the raw data file, locate the scans acquired for identified peptides bearing either NEM or d_5-NEM modification, confirm the identity of the peptide by comparing the MS/MS spectra obtained from database search and from data acquisition. Integrate the monoisotopic peak area of the modified peptides in each sample, and use the same strategy used for recombinant protein to quantify the relative amount of SNO and SS formed at a specific cysteine residue (Fig. 3).

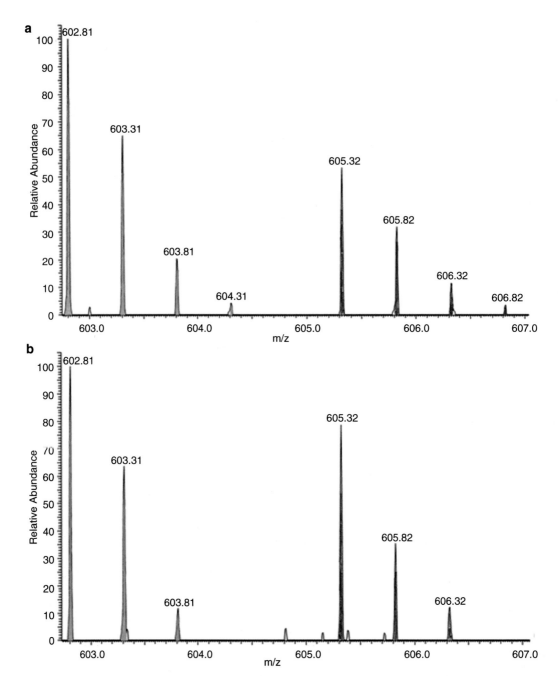

Fig. 3 Assessment of nitrosated, oxidized, and unreacted Cys47 in wt GSTP1 in SH-SY5Y cells. (**a, b**) Shows representative MS spectra for peptide fragments containing Cys47 labeled with NEM (*green peaks*) and d$_5$-NEM (*red* peaks) from the cell lysate samples processed through dSS-1 and dSS-2, respectively. The *green peak* cluster for dSS-1 (**a**) represents the unreacted and nitrosated Cys47 and its *red peak* cluster represents the oxidized Cys47. The *green peak* cluster for dSS-2 (**b**) represents the unreacted Cys47 while its *red* cluster represents the oxidized Cys47 and the nitrosated Cys47

4   Notes

1. We recommend making a highly concentrated stock solution of NEM (e.g. 400 mM) in anhydrous DMF, using that to prepare other NEM solutions each time before use. Keep the NEM-DMF stock solution refriderated and make a new stock every month. Do not weight neocuproine solid into aqueous buffer directly—neocupoine has poor water solubility. Prepare an EtOH stock solution of neocuproine and dilute it into aqueous buffer to reach the desired concentration. Use series dilution if necessary.

2. It has been demonstrated that the dominant product of cysteine nitrosation by nitrous acid remains to be CysNO within 1 h after it's been generated [40]. The CysNO used in our experiments is always prepared freshly and used immediately: the overall stand-on-bench (ice) time is no more than 10 min. Ion chelator neocuproine is also added to prevent CysNO from metal-catalyzed decomposition.

3. Amicon Ultra-0.5 mL 10 K cutoff centrifugal filters are used throughout the entire d-SSwitch procedure for concentrating protein samples and buffer exchange. For buffer exchange, transfer the sample into the filter unit, fill the rest filter volume with exchanging buffer, spin the entire device in a 40° fixed angle centrifuge at $14,000 \times g$ for 5 min. Discard the solvent filtered through and repeat this process for additional two times. After the final wash, reverse the filter and put it into another tube, spin at $1000 \times g$ for 10 min to collect the content.

4. SH-SY5Y cells easily detach from the plate to become floating in medium. Applying large amount of solvent directly onto cells can cause significant loss of cells. Extra caution needs to be taken when changing medium and washing cells with PBS. Cells after 20 passages should not be used for experiment.

5. Always prepare and use fresh ascorbate and CuCl solutions from solid.

6. We suggest running gel electrophoresis on the same day to avoid irreversible protein precipitation during freeze-thaw cycles.

7. We found that if the entire gel is destained well before dissection, only one round of destaining of small gel pieces will be sufficient during the digestion process to wash off the remaining Coomassie blue stain. During the incubation of gel pieces in destaining solution, frequent vortexing will help with fast and efficient destaining.

8. All proteins have been denatured and alkylated during d-SSwitch procedure. Therefore no further reduction and alkylation are necessary during in-gel digestion.

9. MassMatrix is a free proteomic database search engine for tandem mass spectrometric data. It provides improvements in sensitivity over Mascot and SEQUEST with comparably low false positive.

References

1. Miseta A, Csutora P (2000) Relationship between the occurance of cysteine in proteins and the complexity of organisms. Mol Biol Evol 17(8):1232–1239

2. O'Brian CA, Chu F (2005) Post-translational disulfide modifications in cell signaling – role of inter-protein, intra-protein, S-glutathionyl, and S-cysteaminyl disulfide modifications in signal transmission. Free Radic Res 39 (5):471–480. doi:10.1080/10715760500073931

3. Paulsen CE, Carroll KS (2010) Orchestrating redox signaling networks through regulatory cysteine switches. ACS Chem Biol 5 (1):47–62. doi:10.1021/cb900258z

4. Bindoli A, Rigobello MP (2013) Principles in redox signaling: from chemistry to functional significance. Antioxid Redox Signal 18 (13):1557–1593. doi:10.1089/ars.2012.4655

5. Gould N, Doulias PT, Tenopoulou M, Raju K, Ischiropoulos H (2013) Regulation of protein function and signaling by reversible cysteine S-nitrosylation. J Biol Chem 288 (37):26473–26479. doi:10.1074/jbc.R113.460261

6. Butler AR, Flitney FW, Williams DL (1995) NO, nitrosonium ions, nitroxide ions, nitrosothiols and iron-nitrosyls in biology: a chemist's perspective. Trends Pharmacol Sci 16 (1):18–22

7. Tejero J, Basu S, Helms C, Hogg N, King SB, Kim-Shapiro DB, Gladwin MT (2012) Low NO concentration dependence of reductive nitrosylation reaction of hemoglobin. J Biol Chem 287(22):18262–18274. doi:10.1074/jbc.M111.298927

8. Williams DLH (1985) S-Nitrosation and the reactions of S-nitroso compounds. Chem Soc Rev 14(2):171–196

9. Hess DT, Matsumoto A, Kim SO, Marshall HE, Stamler JS (2005) Protein S-nitrosylation: purview and parameters. Nat Rev Mol Cell Biol 6(2):150–166. doi:10.1038/nrm1569

10. Calabrese V, Cornelius C, Rizzarelli E, Owen JB, Dinkova-Kostova AT, Butterfield DA (2009) Nitric oxide in cell survival: a janus molecule. Antioxid Redox Signal 11 (11):2717–2739. doi:10.1089/ARS.2009.2721

11. Wang Y, Chen C, Loake GJ, Chu C (2010) Nitric oxide: promoter or suppressor of programmed cell death? Protein Cell 1 (2):133–142. doi:10.1007/s13238-010-0018-x

12. Burke CL, Stern DF (1998) Activation of Neu (ErbB-2) mediated by disulfide bond-induced dimerization reveals a receptor tyrosine kinase dimer interface. Mol Cell Biol 18 (9):5371–5379

13. Borloo J, Geldhof P, Peelaers I, Van Meulder F, Ameloot P, Callewaert N, Vercruysse J, Claerebout E, Strelkov SV, Weeks SD (2013) Structure of Ostertagia ostertagi ASP-1: insights into disulfide-mediated cyclization and dimerization. Acta Crystallogr D Biol Crystallogr 69 (Pt 4):493–503. doi:10.1107/S0907444912050019

14. Mieyal JJ, Chock PB (2012) Posttranslational modification of cysteine in redox signaling and oxidative stress: focus on s-glutathionylation. Antioxid Redox Signal 16(6):471–475. doi:10.1089/ars.2011.4454

15. Stamler JS, Hausladen A (1998) Oxidative modifications in nitrosative stress. Nat Struct Biol 5(4):247–249

16. Wang YT, Piyankarage SC, Williams DL, Thatcher GR (2014) Proteomic profiling of nitrosative stress: protein S-oxidation accompanies S-nitrosylation. ACS Chem Biol 9 (3):821–830. doi:10.1021/cb400547u

17. Lu XM, Tompkins RG, Fischman AJ (2013) Nitric oxide activates intradomain disulfide bond formation in the kinase loop of Akt1/PKBalpha after burn injury. Int J Mol Med 31 (3):740–750. doi:10.3892/ijmm.2013.1241

18. Lancaster J (2004) Nitroxidation: the dominant process of reactive nitrogen species

chemistry under biological conditions. Free Radic Biol Med 37(S1):S98

19. Townsend DM, Manevich Y, He L, Xiong Y, Bowers RR Jr, Hutchens S, Tew KD (2009) Nitrosative stress-induced s-glutathionylation of protein disulfide isomerase leads to activation of the unfolded protein response. Cancer Res 69(19):7626–7634, doi:0008-5472. CAN-09-0493 [pii] 10.1158/0008-5472. CAN-09-0493

20. Keszler A, Zhang Y, Hogg N (2009) Reaction between nitric oxide, glutathione, and oxygen in the presence and absence of protein: how are S-nitrosothiols formed? Free Radic Biol Med 48(1):55–64, doi:S0891-5849(09)00652-2 [pii] 10.1016/j.freeradbiomed.2009.10.026

21. Nakamura T, Tu S, Akhtar Mohd W, Sunico Carmen R, Okamoto S-i, Lipton Stuart A (2013) Aberrant protein S-nitrosylation in neurodegenerative diseases. Neuron 78 (4):596–614. doi:10.1016/j.neuron.2013. 05.005

22. Zhao QF, Yu JT, Tan L (2014) S-Nitrosylation in Alzheimer's disease. Mol Neurobiol. doi:10. 1007/s12035-014-8672-2

23. Mossuto MF (2013) Disulfide bonding in neurodegenerative misfolding diseases. Int J Cell Biol 2013:318319. doi:10.1155/2013/ 318319

24. Sabens Liedhegner EA, Gao XH, Mieyal JJ (2012) Mechanisms of altered redox regulation in neurodegenerative diseases – focus on S-glutathionylation. Antioxid Redox Signal 16 (6):543–566. doi:10.1089/ars.2011.4119

25. Couvertier SM, Zhou Y, Weerapana E (2014) Chemical-proteomic strategies to investigate cysteine posttranslational modifications. Biochim Biophys Acta 1844(12):2315–2330. doi:10.1016/j.bbapap.2014.09.024

26. Jaffrey SR, Snyder SH (2001) The biotin switch method for the detection of S-nitrosylated proteins. Sci STKE 2001(86):PL1

27. Qu Z, Meng F, Bomgarden RD, Viner RI, Li J, Rogers JC, Cheng J, Greenlief CM, Cui J, Lubahn DB, Sun GY, Gu Z (2014) Proteomic quantification and site-mapping of S-nitrosylated proteins using isobaric iodoTMT reagents. J Proteome Res 13(7):3200–3211. doi:10.1021/pr401179v

28. Fares A, Nespoulous C, Rossignol M, Peltier JB (2014) Simultaneous identification and quantification of nitrosylation sites by combination of biotin switch and ICAT labeling. Methods Mol Biol 1072:609–620. doi:10. 1007/978-1-62703-631-3_41

29. Leichert LI, Gehrke F, Gudiseva HV, Blackwell T, Ilbert M, Walker AK, Strahler JR, Andrews PC, Jakob U (2008) Quantifying changes in the thiol redox proteome upon oxidative stress in vivo. Proc Natl Acad Sci U S A 105(24):8197–8202. doi:10.1073/pnas. 0707723105

30. Sinha V, Wijewickrama GT, Chandrasena RE, Xu H, Edirisinghe PD, Schiefer IT, Thatcher GR (2010) Proteomic and mass spectroscopic quantitation of protein S-nitrosation differentiates NO-donors. ACS Chem Biol 5 (7):667–680. doi:10.1021/cb100054m

31. Cesareo E, Parker LJ, Pedersen JZ, Nuccetelli M, Mazzetti AP, Pastore A, Federici G, Caccuri AM, Ricci G, Adams JJ, Parker MW, Lo Bello M (2005) Nitrosylation of human glutathione transferase P1-1 with dinitrosyl diglutathionyl iron complex in vitro and in vivo. J Biol Chem 280(51):42172–42180. doi:10.1074/jbc. M507916200

32. Townsend DM, Manevich Y, He L, Hutchens S, Pazoles CJ, Tew KD (2009) Novel role for glutathione S-transferase pi Regulator of protein S-glutathionylation following oxidative and nitrosative stress. J Biol Chem 284 (1):436–445, doi:M805586200 [pii] 10. 1074/jbc.M805586200

33. Laborde E (2010) Glutathione transferases as mediators of signaling pathways involved in cell proliferation and cell death. Cell Death Differ 17(9):1373–1380. doi:10.1038/cdd. 2010.80

34. Pathi SS, Jutooru I, Chadalapaka G, Sreevalsan S, Anand S, Thatcher GR, Safe S (2011) GT-094, a NO-NSAID, inhibits colon cancer cell growth by activation of a reactive oxygen species-microRNA-27a: ZBTB10-specificity protein pathway. Mol Cancer Res 9(2):195–202. doi:10.1158/1541-7786.MCR-10-0363

35. Meah Y, Brown BJ, Chakraborty S, Massey V (2001) Old yellow enzyme: reduction of nitrate esters, glycerin trinitrate, and propylene 1,2-dinitrate. Proc Natl Acad Sci U S A 98 (15):8560–8565. doi:10.1073/pnas. 151249098

36. Andrei D, Salmon DJ, Donzelli S, Wahab A, Klose JR, Citro ML, Saavedra JE, Wink DA, Miranda KM, Keefer LK (2010) Dual mechanisms of HNO generation by a nitroxyl prodrug of the diazeniumdiolate (NONOate) class. J Am Chem Soc 132(46):16526–16532. doi:10.1021/ja106552p

37. Cook JA, Kim SY, Teague D, Krishna MC, Pacelli R, Mitchell JB, Vodovotz Y, Nims RW, Christodoulou D, Miles AM, Grisham MB, Wink DA (1996) Convenient colorimetric and fluorometric assays for S-nitrosothiols. Anal Biochem 238(2):150–158. doi:10. 1006/abio.1996.0268

38. Chang M, Shin YG, van Breemen RB, Blond SY, Bolton JL (2001) Structural and functional consequences of inactivation of human glutathione S-transferase P1-1 mediated by the catechol metabolite of equine estrogens, 4-hydroxyequilenin. Biochemistry 40 (15):4811–4820

39. Chang M, Bolton JL, Blond SY (1999) Expression and purification of hexahistidine-tagged human glutathione S-transferase P1-1 in Escherichia coli. Protein Expr Purif 17 (3):443–448

40. Arnelle DR, Stamler JS (1995) NO+, NO, and NO- donation by S-nitrosothiols: implications for regulation of physiological functions by S-nitrosylation and acceleration of disulfide formation. Arch Biochem Biophys 318 (2):279–285. doi:10.1006/abbi.1995.1231

Neuromethods (2016) 114: 73–96
DOI 10.1007/7657_2016_102
© Springer Science+Business Media New York 2016
Published online: 14 May 2016

Label-Free LC-MS/MS Comparative Analysis of Protein S-Nitrosome in Synaptosomes from Wild-Type and *APP* Transgenic Mice

Monika Zaręba-Kozioł, Maciej Lalowski,
and Aleksandra Wysłouch-Cieszyńska

Abstract

Posttranslational S-nitrosylation (SNO) of cysteine thiols in proteins is one of the important mechanisms of nitric oxide-based signaling in vivo. A role for protein S-nitrosylation has been proposed in different tissues both under physiological and pathophysiological conditions. Protein SNOs play an especially important role in the nervous system. Increased protein SNO has been observed as an effect of nitrosative/oxidative stress in many neurodegenerative disorders including Alzheimer disease, Parkinson disease, Huntington disease, and amyotrophic lateral sclerosis and in aging processes as a consequence of the activity of environmental factors. S-nitrosylation of a protein in a certain (patho-) physiological state is the result of a complicated crosstalk of variety of factors. Precise mechanisms of in vivo formation of S-nitrosoproteins are still under investigation. There are no easy and reliable methods to theoretically predict sites of posttranslational S-nitrosylation, such as developed for other types of posttranslational modifications. Furthermore, because of the lability of SNO bond, experimental methods for SNO detection are demanding. A breakthrough in the identification of SNO proteins was the elaboration the biotin switch technique (BST) and SNO site identification method (SNOSID). These techniques were used by us recently, to identify SNO targets among synaptosomal proteins in Alzheimer's disease mouse model. Despite its utility in identifying SNO-Cys modification in proteins, the BST and SNOSID methods are constrained by several limitations. Each step of these procedures represents a potential source of methodological errors. This chapter presents a detailed protocol for identification of SNO sites in synaptosomal proteins from mouse brains using BST or SNOSID affinity methods and mass spectrometry.

Keywords: Protein S-nitrosylation, Alzheimer disease, Label-free analysis, Differential proteomics, LC-MS/MS, Mouse models, Aβ peptide, Biotin switch method, SNOSID, Synaptosomes

1 Introduction

S-nitrosylated proteins (protein SNOs) constitute one of the experimentally detected end products of the nitric oxide (NO) reactivity in living organisms [1]. NO was discovered as a signaling molecule in the central nervous system in 1988, soon after the initial finding of its crucial role as the endothelial derived relaxing factor, EDRF [2]. It was later explored that brain contains one of the highest activities of NO-forming enzyme (NO synthase, NOS) in all examined tissues [3]. Nitric oxide is a freely diffusible radical molecule, rapidly reacting with various endogenous substrates forming, i.e., iron and copper adducts in prosthetic groups of proteins, peroxynitrite $(ONOO^-)$ in the reaction with reactive oxygen species (ROS) [4, 5]. It forms S-nitrosothiols, both with endogenous low-molecular-weight thiols like cysteine and glutathione or with the thiol groups in proteins [5, 6]. Gow et al. have demonstrated that the presence of protein SNOs is strictly related to the production of NO by appropriate NO synthases in various cell types [7, 8]. However, precise mechanisms of in vivo formation of protein SNOs are still under investigation. NO itself is chemically not able to directly modify the side chain thiols of cysteine residues at physiologically significant rates. Thus, protein S-nitrosylation is an indirect product of secondary reactions of proteins with different bioactive nitric oxide species, which are formed after the "activation" of NO by its primary targets [7, 8]. A detailed discussion on the chemical biology of S-nitrosothiols has been recently presented by Broniowska and Hogg [9]. An enzymatic regulation of protein SNO formation was suggested to be similar to other posttranslational modifications of proteins, i.e. phosphorylation. Proteins such as GAPDH or SIRT1 can act as "transnitrosylases" for other protein targets [10]. A protein denitrosylase activity has been attributed to thioredoxin [11, 12]. On the other hand, protein S-NO formation is unique among PTMs since it relies on many nonenzymatic reactions with low-molecular weight compounds and is sensitive to the overall redox status of the cells [13]. Thus, SNO of a protein in a certain (patho-) physiological state is a resultant of a complicated cross talk of multiple factors [14]. As a consequence, there are no easy and reliable methods to theoretically predict sites of post translational S-nitrosylation, such as those developed for other types of PTMs. Furthermore, experimental methods of SNO detection are quite demanding. Endogenous protein SNOs are usually present at low concentrations. The S-NO bond is labile and prone to exchange, i.e., with endogenous free thiols when the SNO target protein is devoid of the context of its cellular milieu. The stability of protein SNO in vivo depends also on the proper folding of the modified protein, which is often challenged during the analytical detection procedures of SNOs.

A breakthrough in the identification of SNO proteins was the elaboration by Jaffrey et al. of the biotin switch technique (BST) [15]. BST relies on initial blocking of any free cysteine thiols present in a protein, followed by the selective reduction of protein SNOs by ascorbate to generate free cysteine thiols in the presence of other thiol derivatives. The free thiols selectively released from the SNOs may then be labelled using a wide variety of available thiol reactive compounds, which allow either for the visualization or enrichment of solely those proteins that were originally

S-nitrosylated in vivo. The free thiol modifier originally used by Jaffrey et al. was the biotin-HPDP, which reacts with -SH groups to form a mixed disulfide biotin derivative. Western blot analysis with a biotin recognizing antibody was used to visualize the modified proteins on SDS-PAGE. Alternatively, the commonly used avidin-based affinity chromatography method was utilized to enrich the biotin labelled proteins, and individual proteins were detected using appropriate protein target-specific antibodies. A further application of BST is the release of enriched proteins from avidin chromatography resin using thiol reducing reagents and identification of either individual SNO targets, or whole cellular "S-nitrosomes" by liquid chromatography mass spectrometry (LC-MS-MS/MS). During the release of proteins from avidin affinity resin, the protein/biotin mixed disulfide is reduced and the characteristic biotin label is removed. Thus, BST allows for identification of the protein SNO targets, but does not provide information on the precise site of S-nitrosylation. A partial solution to this problem was the development of the SNOSID (SNO site identification) technique [16]. Similarly to BST, SNOSID involves the selective exchange of CysSNO residues to CysS (S-biotin derivative), but additionally it involves trypsin digestion of all proteins prior to avidin affinity chromatography. This leads to enrichment, only of biotinylated tryptic peptides that were originally nitrosylated. After disulfide bond reduction these peptides are released from the affinity column and sequenced by tandem LC-MS/MS. In this manner, SNOSID allows pinpointing not only the S-nitrosylated proteins, but also precise sites of SNO modification.

Despite their utility in identifying SNO-Cys modification in proteins, the BST and SNOSID methods are constrained by several limitations. Each step of these techniques is a potential source of methodological errors, which are discussed in this chapter. For example, some protein free thiols can be resistant to complete blocking, resulting in SNO-independent biotinylation. The specificity and yield of SNO reduction by ascorbate have been questioned by experiments, which proposed that under some reaction conditions the ascorbate is capable of reducing mixed disulfide bridges [17, 18]. This has been later challenged by observations that thiol-dependent reduction of dehydroascorbate to ascorbate, a scenario supported by extensive in vitro and in vivo experimentation, is thermodynamically favored [19]. Presence in solution of free redox active metal ions, such as iron or copper may compromise the BST specificity by initiating trans-nitrosylation reactions and inducing production of hydroxyl radicals. The sensitivity of SNO bond to indirect sunlight has also been demonstrated [19]. Our own experience has shown that the best quality of used reagents is important for maximizing the yield of each reaction in this multistep procedure. Though criticized, BST and SNOSID remain mainstay assays for detecting SNO proteins in complex biological systems and these methods were a basis for deciphering the role of protein S-nitrosylation in the pathology of various diseases, including cancer, heart condition, and neurodegenerative disorders [20–22].

Protein SNOs play an important role in the nervous system. Increased protein SNO has been observed as an effect of nitrosative/oxidative stress in many neurodegenerative disorders including AD, PD, HD and ALS, and in aging processes as a consequence of activity of environmental factors [23–26]. N-methyl-D-aspartate receptor (NMDAR) and caspase enzyme activities can be lowered by S-nitrosylation in neurons, thereby facilitating neuroprotection [27]. This finding led to development of nitro-memantine, a nitric oxide donor and selective NMDAR interacting drug. It selectively S-nitrosylates the NMDA receptor and prevents its hyperactivation, observed, i.e., in Alzheimer's disease [28]. On the contrary, S-nitrosylation of protein-disulfide isomerase, dynamin-related protein 1, glyceraldehyde dehydrogenase, cyclo-oxygenase-2, N-ethylmaleimide sensitive protein, Parkin, Gospel, cyclin-dependent kinase-5, mitochondrial complex I, stargazin, and serine racemase, has been related to severe neuropathological alterations in the brain due to induction of protein misfolding/ aggregation, mitochondrial dysfunction, bioenergetic compromise, synaptic injury, and subsequent neuronal loss [29–40].

Most of the work on endogenous S-nitrosylation in the nervous system has focused on a single-protein analysis, but the number of targets, which may contribute to neurodegeneration via disruption of different signaling pathways, is quickly increasing. It is expected that many more S-nitrosylated proteins will be found to play a role in neurodegenerative diseases.

Because of the low sensitivity of analytical methods to detect SNOs, the initially undertaken, global proteomic studies managed to identify only cellular targets of SNO induced by nitric oxide donor treatment, and those may not always be the physiologically important ones. The potential for identification of various SNO target proteins has increased significantly with the improvement of mass spectrometry based protein identification techniques coupled with bioinformatics. Currently, the sensitivity of SNO-proteome measurements allow to identify endogenously formed protein SNOs.

Our recent contribution to the subject is the profiling of endogenous SNO of brain synaptosomal proteins from wild type and transgenic mice overexpressing mutated human Amyloid Precursor Protein (hAPP) [41, 42]. We utilized the original BST and SNOSID procedures, but we have optimized the reaction conditions for our biological model and complemented the experimental scheme with additional control reactions. All reagents utilized in BST, in parallel to their use for treating biological samples, were also utilized to react with recombinant S100BSNO protein, which is S-nitrosylated at a single Cys85 residue, while the second Cys64 thiol is free. In the first BST step Cys64 is thiomethylated using MMTS and the Cys85 SNO is still preserved. This is followed by biotinylation of Cys85 after reduction of SNO with ascorbate and treatment of the released thiol with biotin-HPDP. Formation of appropriate S100B protein derivatives was monitored using reversed-phase HPLC and whole-protein mass measurement, and served as an estimate of the overall yield of the SNO to S-S-biotin exchange reaction in our experiments.

The main goal of our recently published work was to get insight into the SNO of protein targets important for Alzheimer's disease progression and related to an overexpression of Aβ peptide [43]. One of the key features of patients with neurodegenerative disorders including AD is impaired signaling at the neuronal synapse, often correlated with alterations in PTM of synaptic proteins [44–49]. At the same time, all cell types in the nervous system are inseparable i.e. astrocytes and microglia play a crucial role both in the physiology and pathophysiology of neurons [50]. This is also true for the NO dependent activity in brain tissue [51]. For example, glial cells play an important role in protecting neurons from nitrosative stress [52]. Neuronal injury is observed at much higher NO concentrations for neurons in the presence of glial cells, than for neurons cultured alone [53]. It is expected that different cell types behave very differently and similar proteins are most probably regulated, i.e., by PTMs in a cell-specific manner. Thus, bulk proteomic studies of PTMs in brain tissue lysates cannot provide the detailed information of subcellular processes.

To elucidate in vivo SNO regulation of proteins involved in synaptic functions, we employed freshly isolated brain synaptosomes derived from a mouse model of AD [43]. Synaptosomes are a well-recognized model in the studies on synaptic signaling pathways. They contain complete presynaptic terminals, with postsynaptic membranes and densities, as well as other components necessary to store, release, and retain neurotransmitters. Furthermore, synaptosomes contain viable mitochondria, enabling production of ATP and active energy metabolism [54, 55].

In this work we present the technical details of preparation of synaptosomes from mouse brains, with a special emphasis on the use of BST or SNOSID affinity methods to yield enriched fractions of protein or peptide SNOs, respectively.

Furthermore, we describe our solutions to be used in mass spectrometry serving for identification of S-NO targets characteristic in a given biological state. It has to be strongly stressed that the aim of our work was not to obtain precise quantitative information on protein SNOs, which would require a different experimental approach. Rather, we tried to prove that this process is highly confined to specific key molecules and/or pathways, and is modified upon AD symptoms progression in mice. We searched for "differential" targets that were modified by SNO only in one of the biological states examined (hAPP mice) and not in the other (FVB wild-type mice).

Liquid chromatography-tandem mass spectrometry (LC-MS-MS/MS) has been used to analyze synaptosomal peptides enriched by SNOSID from either FVB or hAPP mice. In a typical LC-MS-MS/MS run repetitive recording of MS spectra is interleaved with the selection and analysis of peptide peaks for MS/MS fragmentation. Peptide MS/MS spectra are acquired under automated instrument control based on intensities of parental peptide ions. The spectra are then matched to database sequences, and protein identifications are deduced from the list of identified peptides. A drawback of these analyses is that minor differences in experimental conditions may change the liquid chromatography retention times of peptides, or alter which peptides are selected for MS/MS fragmentation; small differences in fragmentation may cause some spectra to be misidentified by database search software. Such variations decrease the ability of LC MS-MS/MS to accurately analyze the proteome in a more complex mixture [56, 57]. To overcome these difficulties, for a single biological sample a standard LC-MS-MS/MS run aimed at peptide sequence identification was supplemented with a profile type LC-MS experiment. It provided an information about the LC retention time (LC rt), mass-to-charge ratio (m/z), and most precise peak intensity data for all ionizable peptides present in the sample. In-house developed software was used to generate 2D heat-maps with the two axes representing the m/z and LC retention times of peptide ions. Additionally, peptide signals on the 2D heat-maps obtained for different biological samples were correlated and labelled with appropriate peptide sequences (as described in the procedures below). Sequence labels used for signal assignments in a 2D heat-map for a single biological replicate were derived from a merged list of all peptide sequences acquired in any LC-MS-MS/MS experiment measured for either FVB or hAPP synaptosomes. This significantly increased the number of proper sequence assignments and thus provided a more reliable proteomic representation of the analyzed systems (on average more than 75 % of 2D heat-map signals had assigned sequences). An SNO site was defined as differential if a signal was present in all 2D heat-maps for three biological replicates of one state and absent in all three biological replicates of the other.

Data obtained by our procedures suggest a role for S-nitrosylation in the regulation of at least 138 synaptic proteins. Among this group, we identified 38 proteins solely S-nitrosylated in hAPP and not in the wild type FVB mice. The LC-MS/MS identifications were verified for ten of the synaptosomal proteins using BST coupled with Western blot analysis with highly specific protein recognizing antibodies.

2   Materials

HEPES (Sigma), EDTA (Sigma), MMTS (Fluka), Complete inhibitors cocktail (Roche), Ficoll (Sigma), Neocuproine (Sigma), DMF (Roth), SDS (Sigma), biotin-HPDP (Thermo Scientific), Sodium Ascorbate (Sigma), DTT (Roth), NaCl (Sigma), Neutravidin (Thermo Scientific), Sequencing grade modified trypsin (Promega), isopropyl 1-thio-β-D-galactopyranoside (Sigma) TFA (Roth), Acetonitrile (Waters),GSH (Sigma), Bradford reagent (Sigma) ECL chemiluminescence system (Amersham), PVDF (Milipore) IPTG (Roth), Nano Aquity Liquid Chromatography system (Waters, Milford, MA), LTQ-FTICR mass spectrometer (Thermo Scientific), pre-column Symmetry C18, 180 μm × 20 mm, and 5 μm (Waters), BEH130 column C18, 75 μm × 250 mm, 1.7 μm (Waters), MascotDistiller software (version 2.3, MatrixScience, Boston, MA).

2.1   Mouse Model

Female wild-type FVB and transgenic mice overexpressing human APP with London mutation (V717I) were obtained thanks to the courtesy of Professor F. Van Leuven (Experimental Genetics Group, Center for Human Genetics, Flemish Institute for Biotechnology, Katholieke Universiteit Leuven). Generation of these AD mouse models, as well as their phenotype has been described elsewhere [41, 58, 59].

2.2   Isolation of Synaptosomes from Mouse Brain Tissue

Recipe 2.2.1: Buffer A containing 5 mM Hepes pH 7.4, 0.32 M sucrose, 0.2 mM EDTA, 20 mM MMTS and Complete inhibitors cocktail.

Recipe 2.2.2: 3 %, 6 %, 12 % Ficoll dissolved in buffer A (**Recipe 2.2.1**).

2.3   Synthesis of S-Nitrosylated Recombinant S100B Protein

Recipe 2.3.1: The synthetic gene encoding human S100B was cloned into pAED4 plasmid and expressed in *Escherichia coli* utilizing the T7, AmpR expression system. Bacterial cells (HMS174 (DE3)) were grown in LB medium at 37 °C. Expression was induced by the addition of 0.4 mM isopropyl 1-thio-β-D-galactopyranoside at $A_{600} = 0.8$ and the bacterial culture was grown for an additional 2 h. The overexpressed protein was purified as described previously [60]. Reduced, lyophilized recombinant S100B protein was dissolved in 50 mM Tris–HCl pH = 8.0 and 100 mM $CaCl_2$ to a final concentration of 20 μM. Freshly prepared 100 mM GSNO solution was added to a final concentration of10 mM and the reaction was kept for 30 min in the dark. The reaction mixture was then diluted 5 times with water, and 100 mM EDTA pH = 8.0 solution was added to complex all calcium ions. Proteins were purified on an analytical HPLC column in a gradient of 52–65 %

acetonitrile/0.1 % TFA. Fractions of purified S100B SNO were lyophilized and used as positive control in BST and SNOSID methods.

Recipe 2.3.2: After every step of SNOSID method and 200 μl of the reaction mixtures were injected onto an analytical chromatography column using a gradient of 50–64 % acetonitrile in 14 min at 1 ml/min rate, which well separated the reduced and modified forms of the protein. The quantity of various forms of the protein was elucidated by measuring peak areas. The identity of individual protein peaks separated during HPLC was confirmed by ESI-MS.

Recipe 2.3.3: MS spectra were measured on a Q-TOF mass spectrometer. Desalted protein samples (HPLC peaks) were diluted 100 times in 0.1 % formic acid and 50 % acetonitrile/water solution before a syringe injection into the spectrometer source. Raw spectra were deconvoluted to obtain the protein masses using Maxent 1 program (Waters).

2.4 Enrichment of S-Nitrosylated Proteins Using Biotin Switch or SNOSID Procedure

Recipe 2.4.1: HEN Buffer 250 mM Hepes pH 7.7 1 mM EDTA, 0.1 mM Neocuproine.

Recipe 2.4.2: MMTS Stock Prepare a 2-M solution in DMF.

Recipe 2.4.3: HENS Buffer Adjust HEN Buffer (**Recipe 2.4.1**) 1 to 5 % SDS by addition of a 1:5 volume of 25 % (w/v) SDS solution.

Recipe 2.4.4: Blocking Buffer Nine volumes HENS Buffer (**Recipe 2.4.3**) adjust to 20 mM MMTS with MMTS Stock (**Recipe 2.4.2**).

Recipe 2.4.5: Ascorbate Solution Prepare a 50 mM solution of sodium ascorbate in deionized water (protect from light).

Recipe 2.4.6: Biotin-HPDP Prepare biotin-HPDP as a 4-mM suspension in DMF freshly before use.

Recipe 2.4.7: Neutralization Buffer 20 mM Hepes, pH 7.7, 100 mM NaCl 1 mM EDTA.

Recipe 2.4.8: Wash Buffer Prepare neutralization buffer (**Recipe 2.4.7**) with 600 mM NaCl.

Recipe 2.4.9: Elution Buffer 25 mM NH_4CO_3 50 mM DTT pH 8.0 freshly before use.

Recipe 2.4.10: Prepare 500 mM iodoacetamide in deionized MQ water.

3   Methods

3.1   Harvesting of Adult Mouse Brains

FVB wild-type and transgenic APP mice were brought up under exactly the same conditions at the Animal House in Mossakowski Medical Research Centre, Polish Academy of Sciences. For a single proteomic experiment four mice of the same age (14 months +/− 10 days) were decapitated using sharp scissors without anesthesia. Brains were immediately removed from the skulls (in our hands the harvesting of four brains takes less than 5 min), and washed once with HEN buffer (**Recipe 2.4.1**) to remove any traces of blood. Every brain was separately placed in a Dounce homogenizer.

3.2   Isolation of Synaptosomes from Mouse Brain Tissue

The scheme for synaptosome isolation is presented in Fig. 1. Every brain was separately homogenized in 6 ml of buffer A (**Recipe 2.2.1**) containing 5 mM Hepes pH 7.4, 0.32 M sucrose, 0.2 mM EDTA, a protease inhibitor mixture and freshly prepared 20 mM MMTS (**Recipe 2.4.2**). Excess MMTS immediately blocks all released endogenous free thiols. All four homogenates were initially gently centrifuged ($2500 \times g$ for 5 min) to remove nuclear pellet and the cleared homogenates were then centrifuged at $12,000 \times g$ for 5 min. Obtained pellets were resuspended in 500 μl of buffer A, placed onto a discontinuous Ficoll gradient (4 %, 6 % and 13 %) and centrifuged at $70,000 \times g$ for 45 min. Clear synaptosomal fractions formed between the layers of 6 % and 13 % Ficoll. Synaptosomes were gently collected, resuspended in 500 μl of buffer A and centrifuged at $20,000 \times g$ for 20 min. Purity of synaptosomes separated in the pellet was confirmed using Western blot analysis to show the absence of a typical nuclear protein laminin A and presence of presynaptic protein synaptophysin. Synaptosomes from all 4 mice were mixed in a 15 ml Falcon tube and used immediately, without freezing in the subsequent steps of BST or SNOSID procedures.

3.3   Use of S100B-SNO Protein as a Positive Control of Enrichment Procedures

Our experience has shown that the quality of different batches of reagents used in BST, such as MMTS, ascorbate, or Biotin-HPDP, is not equal. To control the effectiveness of every reaction in the multistep BST procedure we used a Cys85 S-nitroso form of recombinant S100B protein as a positive control (**Recipe 2.3.1**, final concentration 0.8 mg/ml). Solutions of recombinant S100B SNO protein were treated with exactly the same reagents as synaptosomal proteins. The reaction products were followed by reversed-phase HPLC using Vydac C18 analytical column and UV detection at 220 nm. Identity of appropriate S100B derivatives was confirmed by measuring the mass of whole proteins and comparing the obtained results with theoretical masses of S100B derivatives (**Recipe 2.3.3**). Relative concentration of substrates and products was established by direct measurement of areas of HPLC peaks.

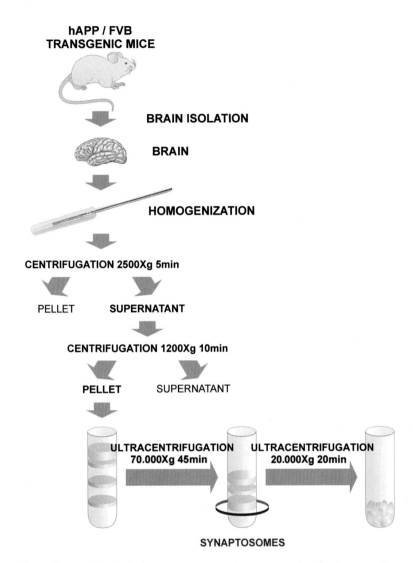

Fig. 1 Protocol for isolating synaptosomes from mouse brain. The procedure requires multistep ultracentrifugation to collect synaptosomal protein fractions

3.4 Enrichment of S-Nitrosylated Proteins Using Biotin Switch or SNOSID Procedure

BST and SNOSID are multireaction procedures. Thus, to maximize the overall yield of the method it was critical to optimize the effectiveness of every individual step. We found it crucial to experimentally adjust the reaction times and concentrations of reagents used specifically for proteins in our biological model. Different reagent concentration, times of experiments, and different concentrations of sample were used to obtain highest specificity and derivatization yields. The most important steps in our procedure are presented in Fig. 2. Our protocol requires approximately two consecutive days for sample processing, including the overnight incubation time for protein trypsinolysis. Additional 1–2 days are

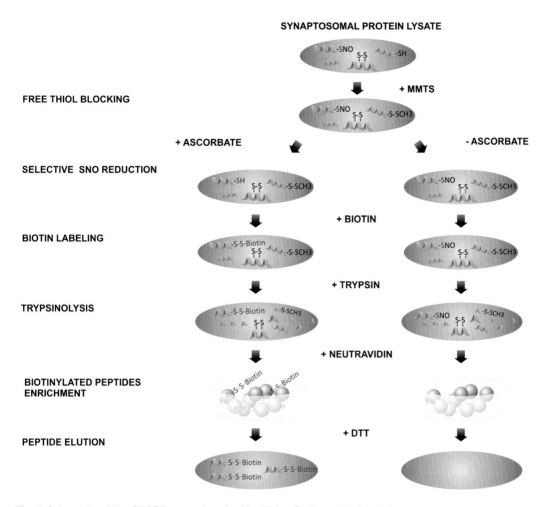

Fig. 2 Schematic of the SNOSID procedure for identifying S-nitrosylated proteins

needed to perform LC-MS-MS/MS experiments and initial data analysis and interpretation. The time of further analysis depends on the complexity and repeatability of the obtained results.

Workflow of S-Nitrosylation Analysis

1. Blocking of free thiol groups in synaptosomal protein fraction.

2. Selective exchange of the protein S-nitroso groups to protein S-S-biotin mixed disulfide derivatives.

3. Trypsinolysis of the whole synaptosomal protein fraction (only for SNOSID procedure).

4. Avidin affinity chromatography of biotinylated proteins or peptides obtained in BST or SNOSID methods, respectively.

5. Alkylation of enriched tryptic peptides for LC-MS-MS/MS analysis (only for SNOSID procedure).

6. Collection of mass spectrometry data.

7. Analysis of mass spectrometry data.

8. Validation of MS-based results using protein-specific Western blot analysis.

9. Bioinformatic analysis of experimental data.

3.4.1 Blocking of Free Thiol Groups in Synaptosomal Protein Fraction

Mouse synaptosomal fractions were dissolved in HEN buffer containing 250 mM Hepes pH 7.7, 1 mM EDTA and 0.1 mM neocuproine (**Recipe 2.4.1**). Total protein concentration in the mixture was measured using Bradford assay [61]. For a single proteomic experiment 6 mg of synaptosomal protein sample was dissolved in the final volume of 5 ml of HEN buffer. The protein mixture was immediately treated with two volumes of thiol blocking solution (**Recipe 2.4.4**) to avoid the rearrangement of the endogenously modified thiol groups with all present unmodified thiols. Free thiols react with excess of MMTS to form S-S-methyl mixed disulfides. The reaction is performed for 20 min at 50 °C in the dark. The presence of metal chelators EDTA and neocuproine (effective copper chelator) prevents side reactions with metal ions. Combination of heat and 5 % SDS is needed to better denature proteins and provide optimal access of MMTS even to the thiols deeply buried in the protein structure. Effective blocking of free thiols before their release from S-nitroso cysteines is a prerequisite to avoid false positive results of experiments. To remove excess reagents, seven volumes of −20 °C cold acetone were added to the sample. After 20 min at −20 °C precipitated proteins were collected by centrifugation at $6000 \times g$ for 5 min. After removal of supernatant the protein pellet was gently washed three times with 20 ml of cold acetone (−20 °C).

3.4.2 Selective Exchange of the S-Nitroso Group to an S-S-Biotin Mixed Disulfide Derivative

Both in BST and SNOSID the selective exchange of the S-nitroso group to an S-S-biotin mixed disulfide derivative is accomplished by removal of the nitroso groups from thiols in a transnitrosation reaction with sodium ascorbate and simultaneous derivatization of the released thiols using biotin-HPDP. The specificity of BST is based on the fact that ascorbate will convert SNOs to free thiols without reducing other cysteine-based oxidative modifications such as mixed disulfides or S-oxides. In our experimental workflow as much as half of the analyzed sample was treated with biotin-HPDP without ascorbate denitrosation (Fig. 2) and further treated and analyzed exactly the same as the ascorbate treated counterpart.

More precisely, proteins precipitated after thiol blocking (Sect. 3.4.1) were resuspended in 5 ml HENS buffer (**Recipe 2.4.3**). The obtained protein solution was divided into two equal parts. One part was treated simultaneously with biotin-HPDP (**Recipe 2.4.6**, final concentration 400 μM) and freshly prepared sodium ascorbate (**Recipe 2.4.5**, final concentration 10 mM). The second part (negative control) was treated only with Biotin-HPDP

(**Recipe 2.4.6**, final concentration 400 µM) devoid of ascorbate. The positive control protein—S100B SNO was treated in the same way. All samples were incubated in the dark for 1.5 h at room temperature. It is crucial to protect samples from any sources of sunlight during this step, because indirect sunlight may induce ascorbate-dependent biotinylation. After reaction completion proteins were precipitated with seven volumes of cold acetone for 20 min at −20 °C, and collected by centrifugation at 6000 × g for 5 min. The clear supernatant was removed and protein pellet gently washed with acetone (3 × 20 ml). Proteins were resuspended in 5 ml of HEN buffer (**Recipe 2.4.1**) and protein concentration was measured using Bradford method.

3.4.3 Trypsinolysis of Proteins (Only for SNOSID Procedure)

SNOSID procedure was used by us for all LC-MS/MS-based differential proteomic analyses of SNO sites. The two synaptosomal protein solutions treated with HPDP-biotin +/− ascorbate, as described in Sect. 3.4.2, were separately digested under identical conditions using sequencing grade modified trypsin for 16 h at 37 °C. The final protease:protein ratio was 1:150. Digestion reaction was terminated by addition of one tablet of Complete—protease inhibitor mixture dissolved in 1 ml of neutralization buffer and incubation for 30 min at room temperature.

3.4.4 Avidin Affinity Chromatography of Biotinylated Proteins (BST) or Peptides (SNOSID)

For a single enrichment experiment, 400 µl of neutravidin agarose 50 % slurry was placed in a salinized Eppendorf tube. The slurry was centrifuged for 2 min at 2000 × g. After removal of the supernatant it was gently vortexed with 1 ml of fresh neutralization buffer (**Recipe 2.4.7**) and centrifuged. This procedure was repeated three times**.** After prewashing 400 µl of fresh neutralization buffer was added to the resin to form slurry which can be easily transferred using a disposable 1 ml pipette tip. 200 µl of the slurry was put into each sample containing whole biotin labeled proteins after step in Sect. 3.4.2 or biotin-labeled tryptic peptides after step in Sect. 3.4.3. It is critical that bead volumes in each sample are precisely the same. The beads were gently agitated for 1 h at room temperature. Afterwards, beads were washed five times with 1 ml of wash buffer containing high NaCl (**Recipe 2.4.8**) to remove non-specifically bound proteins or peptides. Each time the affinity resin was gently vortexed, and centrifuged for 2 min at 2000 × g. The supernatant was discarded and the pellet with proteins (BST) or peptides (SNOSID) bound to neutravidin resin was incubated with 150 µl of elution buffer containing the disulfide reducing reagent DTT (**Recipe 2.4.9**) for 20 min at room temperature with continuous rotation. Samples were then centrifuged at 1000 × g for 2 min and the supernatant was collected to a silanized Eppendorf tube. DTT reduces the mixed disulfide bonds formed by peptide cysteine thiols and the biotin label. Thus, the supernatant contains enriched proteins or peptides of interest with fully reduced cysteine thiol groups that are very sensitive to oxidation.

3.4.5 Derivatization of Tryptic Peptides for LC-MS/MS Analysis

Peptides eluted from neutravidin were immediately alkylated by 45-min incubation with iodoacetamide (**Recipe 2.4.10**), final concentration 200 mM) at 50 °C in the dark. After reaction completion, solutions were concentrated to a volume of circa 45 µl using SpeedVac and diluted with 1 % TFA/water (v/v) solution to the final volume of 50 µl.

3.4.6 Collection of Mass Spectrometry Data

Nano Aquity Liquid Chromatography system (Waters, Milford, MA) coupled to LTQ-FTICR mass spectrometer (Thermo Scientific) was used by us in all experiments.

50 µl of alkylated tryptic peptides solution obtained in procedure in Sect. 3.4.5 (initially starting from four mouse brains of either FVB or hAPP mice) was placed in a silanized glass vial and loaded into a cooled (10 °C) autosampler tray. The UPLC system was coupled directly to the ion source of the LTQ-FITCR mass spectrometer.

For a single biological replicate two portions of the same sample were used for different type of MS experiments. Firstly, 20 µl of sample was automatically transferred to a pre-column (Symmetry C18, 180 µm × 20 mm, and 5 µm Waters) using 0.1 % formic acid in water as a mobile phase. Afterwards, the peptide mixture was separated on a reversed-phase BEH130 column (C18, 75 µm × 250 mm, 1.7 µm, Waters), using a gradient of acetonitrile (5–30 % acetonitrile/0.1 % formic acid over 70 min) with a flow rate of 0.3 µl/min) and analyzed by a profile type LC-MS experiment, which provides data on LC retention time, mass to charge ratio and intensity of all ionizable peptides in solution. Peak intensities for peptide ions measured in such experiments were the basis of label-free quantitative information. The second 20 µl sample portion taken from the same vial was separated with exactly the same column and LC gradient as in the profile experiment, but analyzed using LC-MS-MS/MS to obtain fragmentation data, necessary for peptide sequence identification. In LC-MS-MS/MS runs, we utilized data dependent acquisition (DDA) mode, selecting five most intense signals in each MS spectrum for fragmentation. Dynamic exclusion was activated, with m/z tolerance of 0.05–1.55 and duration of 15 s. Up to five fragmentation events were allowed for every parent ion. Every set of the two MS runs described above was separated by at least one blank full gradient run to reduce the carry-over of peptides from previous samples.

The whole proteomic experiment aimed at identification of SNO protein targets in either FVB or hAPP mice consisted of three biological replicates (all together 12 mouse brains were used for every strain).

3.4.7 Analysis of Mass Spectrometry Data

Initial Analysis of an MS Data for a Single Biological Replicate

MS data obtained for a single biological replicate were only briefly analyzed after each run to assess the correctness of performed enrichment experiments. LC-MS-MS/MS data file was processed with MascotDistiller software (version 2.3, MatrixScience, Boston, MA) generating a Mascot input .mgf format file. For Mascot searches, the complete assembly of the mouse proteome was used, derived from UniProtKB/Swiss-Prot reviewed database version 2013_10 (45889 sequences). Mascot search parameters were set as follows: taxonomy *Mus musculus*, fixed modification—cysteine carbamidomethylation, variable modification—methionine oxidation, parent ion mass tolerance—40 ppm, fragment ion mass tolerance—0.8 Da, number of missed cleavages—1, enzyme specificity—semi-trypsin. To estimate false-positive discovery rate (FDR) values the decoy search option was enabled. In a properly performed experiment there should be no cysteine peptides identified in samples treated with biotin-HPDP without ascorbate reduction (negative control runs) while a large majority of peptide sequences identified after SNOSID enrichment with the ascorbate reduction step should contain at least one cysteine. Usually, several hydrophobic peptides without cysteine residues in their sequences are found on the lists. These are probably peptides that nonspecifically bind to the avidin affinity resin and are not taken under consideration in our analysis.

Generation of Selected Peptide List (SPL) of Identified Sequences for All LC-MS-MS/MS Experiments

Thorough inspection of proteomic data was performed only after obtaining MS data for three biological replicates, for both FVB and hAPP mice. LC-MS-MS/MS data files acquired in all six qualitative runs were preprocessed with MascotDistiller (version 2.3, MatrixScience, Boston, MA) and merged together. The resulting .mgf file was searched against *Mus musculus* reviewed database with Mascot as described above. The output of this program was a .dat format file containing a list of peptide sequences, together with such parameters as the peptide's LC retention time (LC Rt), mass/charge ratio (m/z) of the parental ion, sequence and ion charge information, and a Mascot score, which is an estimate of peptide identification quality. A *Mus musculus* protein identifier was assigned to each identified sequence.

The Mascot-generated peptide list was further filtered using *in-house* Mscan software to select only peptides with Mascot scores above 30. FDR analysis by Mscan demonstrated that for such peptides the FDR values did not exceed 0.29 %. The obtained shortlist of selected peptides (SPL) was further used to tag peptide peaks in 2D heat-maps generated on the basis of the MS profile data (described below).

Generation of 2D
Heat-Maps Representing
LC-MS Data

LC-MS data obtained directly from the mass spectrometer for each biological replicate were converted to a general .ucsf format using an MsConvert data conversion tool. This file format is recognized by Msparky (http://proteom.ibb.waw.pl/msparky—an in-house modification of a commonly used graphical NMR assignment and integration program—Sparky NMR (http://www.cgl.ucsf.edu/home/sparky). Msparky displays LC-MS data as 2D peptide heat-maps (with peptide LC Rt and m/z as vertical and horizontal axes, respectively). A typical fragment of a 2D heat-map obtained for our experiments is presented in Fig. 3. It is important to note that each peptide on the map is represented as a group of peaks corresponding to its isotopic envelope (a feature characteristic for high-resolution MS data, which originates from the presence of different numbers of isotopes, mainly 13C instead of 12C in fractions of analyzed molecules).

Overlay of Qualitative Data
(SPL) on Quantitative
Profile Datasets (2D Heat-
Maps)

Another feature of Msparky combines the information on identified peptide sequences summarized in the SPL (Sect. 3.4.7.3) with the 2D heat-maps generated for each biological replicate. It matches the sequence information with intensity data for peptide signals of the same m/z and LC Rt. For each group of isotopic envelopes, only the mass spectral peak representing the monoisotopic peptide mass was used for labeling. Automatic labeling of peptide signals provided by Msparky was always followed by manual data inspection (i.e., the program allows for efficient correction of the unavoidable differences in LC retention times between different MS runs). Acceptance criteria for manual data inspection included: m/z value deviation, 20 ppm; LC retention time deviation, 10 min, and envelope root mean squared error (a deviation between the expected isotopic envelope of the peak heights and their experimental values)—0.7. The spacing between isotopic envelope peaks in a signal had to match the charge state of the related peptide ion described in SPL.

In our hands approximately 75 % of isotopic envelopes observed on 2D maps are labelled by peptide sequences. Figure 3 shows a peptide signature on the SPL list (peptide sequence GLYG-PEQLPDCLK (+2)) and the appropriate fragment of a 2D heat-map representing the isotopic envelope peaks of the same peptide. The SPL with obtained quantitative values was then reduced so that cysteine was represented by single peptide entry with one quantitative value. Two groups of peptides lists are the final effect of our work, one for the three biological replicates of the hAPP mice and one for the FVB mice. An SNO site was defined as differential between the two biological states using a very strong criterion, only if a signal was present in all 2D heat-maps for three biological replicates of one state and absent in all three biological replicates of the other. Venny analysis was used to generate the list of differential SNO sites (http://bioinfogp.cnb.csic.es/tools/venny/index.html) [62].

a

b

Fig. 3 2D heat-map of tryptic peptides from synaptosomal fractions of mice brains after SNOSID enrichment. (a) In two panels of different magnification the heat map peaks detected in a typical LC-MS of digested fraction of S-nitrosylated synaptosomal proteins is presented. Their retention times are shown in vertical axis, m/z values in horizontal axis, and the peak amplitudes are color coded in a hypsometric rendering with their values increasing from red to blue. The map represents smoothed raw data with no further processing. Each peptide is represented by a characteristic group of peaks, an isotopic envelope, which originates from the presence of different isotopes, mainly 13C instead of 12C in a fraction of the molecules. The upper panel shows a full range dataset with m/z values between 300 Th and 1500 Th and retention times between 0 and 70 min. *Lower panel*—a magnified section 500–700 in the m/z and 30–40 min in the RT domain. The magnified panel shows that even peptides with close m/z and retention time values generate well resolves isotopic envelopes. Monoisotopic peaks of isotopic envelopes assigned to peptides from Selected Peptides List are tagged by identifier from Swiss-Prot database, peptide sequence and charge. Crosses indicate two peaks of the isotopic envelope included into analysis. (b) The part of a table with peptide sequence of peptides which are verified at the SD heat map. The table contains peptide sequence, Swiss-Prot identifier of a protein and chromatographic data values: retention time, delta retention time, m/z value and delta m/z, and high of the peak. *Red color* indicates peptides which were rejected in the verification process

The obtained lists constitute the final effect of S-nitrosome investigation using our procedure. They describe synaptosomal protein SNO sites specific for both hAPP and FVB mice.

The list of differential SNO sites provided us with unique information about changes of protein SNO induced by overexpression of Aβ peptide—a characteristic feature of AD progression.

3.4.8 Validation of LC MS/MS Results Using Protein Specific Western Blots

Western blot analysis was used to validate the results of MS based identifications of selected, differential SNO-proteins. We analyzed different fractions obtained during BST enrichment of S-nitrosylated synaptosomal proteins from FVB and hAPP mice (*see* section 3.4.4). Protein fractions were separated using 12 % SDS-PAGE and transferred to PVDF membrane (0.22 μm). For each protein we performed three biological replicates. The PVDF membranes were first blocked with casein-based buffer (Sigma) and incubated with specific primary antibodies followed by secondary HRP-conjugated antibodies. Protein bands were detected using the ECL chemiluminescence system (Amersham Biosciences). The fractions containing enriched S-nitrosylated proteins harvested from FVB and hAPP synaptosomes were quantified densitometrically with GelQuant software. Heteroschedastic two-tailed *t*-test was used to statistically assess the changes in endogenous protein S-nitrosylation. Ten differentially and two non-differentially S-nitrosylated proteins from different functional classes were chosen for MS data validation.

Figure 4 shows the results of Western blot analysis with specific antibodies. (This figure was originally published by Zaręba-Kozioł et al. *in Global analysis of S-nitrosylation sites in the wild-type (APP) transgenic mouse brain clues for synaptic pathology.* Mol Cell Proteomics, 2014. **13**(9): p. 2288–305.) [43]. The total expression of studied proteins was unchanged in the hAPP and FVB brains. Positive signals were observed only in fractions derived from hAPP brain synaptosomes after BST procedure, but not in the FVB brain, confirming the MS-based identification of differentially SNO-proteins. In internal control in which S-nitrosylation does not change, positive signals were observed in both fractions derived from hAPP and FVB brains.

3.4.9 Bioinformatic Analysis of Experimental Data

Functional analysis of the data allows portraying the interconnectivity between proteins that are post-translationally modified in the context of disease protein. Therefore, SNO-datasets were functionally analyzed using ClueGO (http://www.ici.upmc.fr/cluego/, *see* **Note 13**). To maximize the functional assignments of mouse differentially S-nitrosylated proteins we first searched for their human orthologs utilizing NCBI homologene function (http://www.ncbi.nlm.nih.gov/sites/homologene/). Subsequently, in order to focus on functional enrichments within the synapse milieu, we built up as

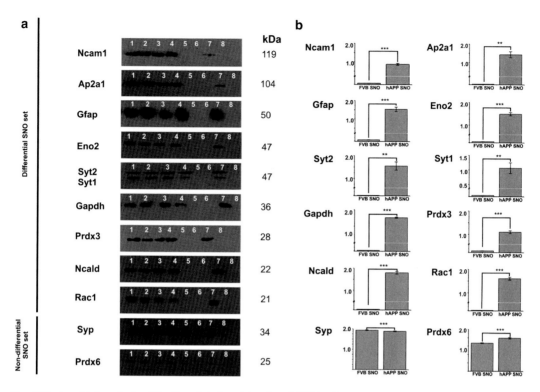

Fig. 4 Western blot analysis of S-nitrosylated proteins from hAPP brain synaptosomes. (**a**) Synaptosomal SNO-proteins enriched using BST were detected with specific antibodies. Differential SNO set: Ncam1—Neural cell adhesion molecule, Ap2a1—AP-2 complex subunit alpha-1, Gfap—Glial fibrillary acidic protein, Eno2—Gamma enolase, Syt1, Syt2—Synaptotagmin-1 and 2, Gapdh—glyceraldehyde-3-phosphate dehydrogenase, Ncald—neurocalcin-delta, Prxd3—peroxiredoxin 3, Rac1—Ras-related C3 botulinum toxin substrate 1 precursor. Non-differential SNO set: Syp—synaptophysin, Prxd6—peroxiredoxin 6. Lane 1—total FVB mouse brain lysate, lane 2—total hAPP brain lysate, lane 3—soluble fraction (FVB), lane 4—soluble fraction (hAPP); lane 5—proteins enriched on neutravidin resin using BST (FVB), lane 6—neutravidin resin after elution (FVB), lane 7—proteins enriched on neutravidin resin using BST (hAPP), lane 8—neutravidin resin after elution (hAPP). kDa—molecular weight in kilo-Daltons. (**b**) Densitometric quantitation of lanes 5 (Neutr_FVB) and 7 (Neutr_hAPP); $n = 3$ experiments, P values from t-test. **$P \leq 0.01$; ***$P \leq 0.001$

a "synaptic reference set" (includes more than 5600 unique human genes) comprising of human orthologs of MS-measured mouse synaptosomal proteins (Malinowska et al. submitted) combined with non-redundant genes derived from expert-curated databases of synaptic proteins (SynsysNet; http://bioinformatics.charite.de/synsys/ and SynaptomeDB; http://psychiatry.igm.jhmi.edu/SynaptomeDB/ [63, 64]. P values for term enrichment in ClueGO were determined using right-sided hypergeometric test. The relationships between the terms based on the similarity of their associated genes were assessed according to kappa score values (≥ 0.3). To connect mouse differentially S-nitrosylated proteins with human APP we similarly established the connectivity between their human orthologs and APP using *GeneMania*

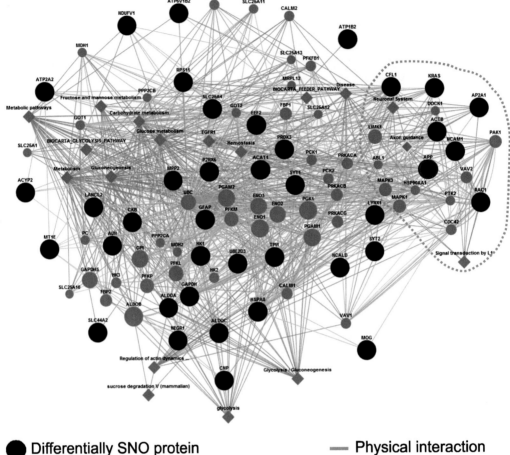

Fig. 5 Functional network linking differentially S-nitrosylated proteins to APP. 89 nodes in the network were connected via 892 links from various interaction, pathway and genetic data. Eighteen different attributes were assigned to the functional network. SNO—S-nitrosylation, Pathway Commons—a depository of common biological pathways (http:/www.pathwaycommons.org/). Functional module, *Axonal guidance* derived from Pathway Commons and Gene Ontology biological process annotation links seven SNO-proteins and ten bridging nodes to APP

(*see* **Note 14**) [65]. To achieve a maximum connectivity we utilized physical, genetic, predicted and pathway interaction datasets stored in the database, and visualized the network using Cytoscape plug-in (Fig. 5). In subsequent step, we filtered the generated network with Gene Ontology biological process ontology criteria to ensure the stringent and highest connectivity of network nodes within ontology classes. In order to portray the dynamics of S-nitrosylation in the AD synaptosomes we included the number of measured SNO-sites as one of the parameters in network depiction (*see* **Note 15**).

Moreover, the functional connectivity between the nodes can be established by FunCoup (FunCoup database v.3.0) (http://FunCoup.sbc.su.se/), mining the data from other exhaustive interaction databases, i.e., Hippie (http://cbdm.mdc-berlin.de/tools/hippie/information.php) or utilizing a commercial database, Ingenuity Pathways (IPA).

4 Notes

1. The SNO bond is labile under standard conditions used for electrospray protein ionization. It is important to adjust the instrument parameters, i.e., cone voltage or temperature to avoid the induced release of NO from SNO.

2. Synaptosomes should be used immediately, without freezing in the further steps of BST or SNOSID procedures.

3. To prevent artifact S-nitrosylation decomposition or unspecific SNO reduction by endogenous free thiols released during synaptosomes isolation excess of MMTS should be added to the homogenization buffer.

4. Ascorbate solution, biotin–HPDP, and elution buffer should be prepared freshly before use. Additionally, ascorbate solution should be protected from light.

5. The quality of different batches of reagents used in BST, such as MMTS, ascorbate, or biotin-HPDP, is not equal. The effectiveness of every reaction in this multistep procedure should be tested using a recombinant, S-nitrosylated protein as a positive control. We used an Cys85 S-nitroso form of recombinant S100B protein as a positive control.

6. The BST is technically challenging and labor intensive; each step contains potential source error; a rigorous experimentalist should include proper negative and positive controls to add confidence to their results.

7. Effective blocking of free thiols by MMTS is required to minimize "background" and to maximize assay sensitivity.

8. The reaction between SNOs and ascorbate must be performed in the darkness because exposure of samples to indirect sunlight resulted in artifactual ascorbate-dependent biotinylation. The negative control without ascorbate should be performed to verify the ascorbate dependency of the biotin modification.

9. During avidin affinity chromatography of biotinylated proteins (BST) or pepides (SNOSID) it is critical that bead volumes in each sample should be precisely the same.

10. We prefer to collect proteins or peptides eluted from neutravidin agarose in silanized tubes to prevent from sticking to the tubes.

11. Peptides eluted from neutravidin resin in the SNOSID procedure contain reduced cysteine residues. The free cysteine thiols should be immediately alkylated using iodoacetamide to avoid unwanted air oxidation of the peptides (i.e., uncontrolled formation of mixed peptide disulfides).

12. In our analysis the SPL with obtained quantitative values was reduced so that each cysteine was represented by single peptide entry with one quantitative value.

13. ClueGO is a Cytoscape plug-in capable of implementing Gene Ontology/KEGG/Reactome pathway hierarchies for clustering of term distributions. For our calculations we utilized the 1.4 version of the software. Please note that latest version 2.1.4 of the program is compatible is with Cytoscape 3.0.

14. The *GeneMania* database currently indexes 2104 association networks containing 535,774,338 interactions mapped to 161,629 genes from eight organisms.

15. In cases when the number of SNO-sites in a peptide after neutravidin affinity could not be unambiguously assigned, for simplicity, we chose the highest possible number of SNO-sites, according to a number of available Cys in a given sequence.

Acknowledgments

This study was supported by a grant from Foundation for Polish Science TEAM program (TEAM/2011-7/1), CEPT (POIG.02.02.00-14-024/08-00) and Ministry of Science and Education (2543/B/P01/2007/33).

References

1. Hess DT et al (2005) Protein S-nitrosylation: purview and parameters. Nat Rev Mol Cell Biol 6:150–166

2. Garthwaite J, Charles SL, Chesswilliams R (1988) Endothelium-derived relaxing factor release on activation of nmda receptors suggests role as intercellular messenger in the brain. Nature 336:385–388

3. Bredt DS et al (1991) Nitric-oxide synthase protein and messenger-rna are discretely localized in neuronal populations of the mammalian cns together with nadph diaphorase. Neuron 7:615–624

4. Drapier JC, Bouton C (1996) Modulation by nitric oxide of metalloprotein regulatory activities. Bioessays 18:549–556

5. Chiueh CC, Rauhala P (1999) The redox pathway of S-nitrosoglutathione, glutathione and nitric oxide in cell to neuron communications. Free Radic Res 31:641–650

6. Pryor WA, Squadrito GL (1995) The chemistry of peroxynitrite: a product from the reaction of nitric-oxide with superoxide. Am J of Physiology 268:L699–L722

7. Gow AJ et al (2002) Basal and stimulated protein S-nitrosylation in multiple cell types and tissues. J Biol Chem 277:9637–9640

8. Gow A et al (2007) S-nitrosothiol measurements in biological systems. J Chromatog B 851:140–151

9. Broniowska KA, Hogg N (2012) The chemical biology of S-nitrosothiols. Antioxid Redox Signal 17:969–980

10. Stamler JS, Hess DT (2010) Nascent nitrosylases. Nat Cell Biol 12:1024–1026

11. Benhar M et al (2010) Identification of S-nitrosylated targets of thioredoxin using a quantitative proteomic approach. Biochemistry 49:6963–6969

12. Wu C et al (2013) Functional proteomics approaches for the identification of transnitrosylase and denitrosylase targets. Methods 62:151–160

13. Smith BC, Marletta MA (2012) Mechanisms of S-nitrosothiol formation and selectivity in nitric oxide signaling. Curr Opin Chem Biol 16(5–6):498–506

14. Martínez-Ruiz A et al (2013) Specificity in S-nitrosylation: a short-range mechanism for NO signaling? Antioxid Redox Signal 19:1220–1235

15. Jaffrey SR, Snyder SH (2001) The biotin switch method for the detection of S-nitrosylated proteins. Sci STKE 2001(86):pl1

16. Hao G et al (2006) SNOSID, a proteomic method for identification of cysteine S-nitrosylation sites in complex protein mixtures. Proc Natl Acad Sci U S A 03:1012–1017

17. Wang X et al (2008) Copper dependence of the biotin switch assay: modified assay for measuring cellular and blood nitrosated proteins. Free Radic Biol Med 44:1362–1372

18. Giustarini D et al (2008) Is ascorbate able to reduce disulfide bridges? A cautionary note. Nitric Oxide 19:252–258

19. Forrester MT et al (2009) Detection of protein S-nitrosylation with the biotin-switch technique. Free Radic Biol Med 46:119–126

20. Murray CI et al (2011) Site-mapping of in vitro S-nitrosation in cardiac mitochondria: implications for cardioprotection. Mol Cell Proteomics 10(3):M110.004721

21. Derakhshan B, Wille PC, Gross SS (2007) Unbiased identification of cysteine S-nitrosylation sites on proteins. Nat Protoc 2:1685–1691

22. Nakamura T et al (2013) Aberrant protein s-nitrosylation in neurodegenerative diseases. Neuron 78:596–614

23. Nakamura T, Lipton SA (2011) Redox modulation by S-nitrosylation contributes to protein misfolding, mitochondrial dynamics, and neuronal synaptic damage in neurodegenerative diseases. Cell Death Differ 18:1478–1486

24. Nakamura T, Lipton SA (2011) S-nitrosylation of critical protein thiols mediates protein misfolding and mitochondrial dysfunction in neurodegenerative diseases. Antioxid Redox Signal 14:1479–1492

25. Schonhoff CM et al (2006) S-nitrosothiol depletion in amyotrophic lateral sclerosis. Proc Natl Acad Sci U S A 103:2404–2409

26. Haun F et al (2013) S-nitrosylation of dynamin-related protein 1 mediates mutant huntingtin-induced mitochondrial fragmentation and neuronal injury in Huntington's disease. Antioxid Redox Signal 19:1173–1184

27. Choi YB et al (2000) Molecular basis of NMDA receptor-coupled ion channel modulation by S-nitrosylation. Nat Neurosci 3:15–21

28. Nakamura T, Lipton SA (2010) Preventing Ca (2+)-mediated nitrosative stress in neurodegenerative diseases: possible pharmacological strategies. Cell Calcium 47:190–197

29. Uehara T et al (2006) S-nitrosylated protein-disulphide isomerase links protein misfolding to neurodegeneration. Nature 441:513–517

30. Nakamura T et al (2010) S-nitrosylation of Drp1 links excessive mitochondrial fission to neuronal injury in neurodegeneration. Mitochondrion 10:573–578

31. Kornberg MD et al (2010) GAPDH mediates nitrosylation of nuclear proteins. Nat Cell Biol 12:1094–1100

32. Tian J et al (2008) S-nitrosylation/activation of COX-2 mediates NMDA neurotoxicity. Proc Natl Acad Sci U S A 105:10537–10540

33. Huang Y et al (2005) S-nitrosylation of N-ethylmaleimide sensitive factor mediates surface expression of AMPA receptors. Neuron 46:533–540

34. Sunico CR et al (2013) S-Nitrosylation of parkin as a novel regulator of p53-mediated neuronal cell death in sporadic Parkinson's disease. Mol Neurodegener 8:29

35. Chung KKK et al (2004) S-nitrosylation of Parkin regulates ubiquitination and compromises Parkin's protective function. Science 304:1328–1331

36. Sen N et al (2009) GOSPEL: a neuroprotective protein that binds to GAPDH upon S-nitrosylation. Neuron 63:81–91

37. Ozawa K et al (2013) S-nitrosylation regulates mitochondrial quality control via activation of parkin. Sci Rep 3:2202

38. Clementi E et al (1998) Persistent inhibition of cell respiration by nitric oxide: crucial role of S-nitrosylation of mitochondrial complex I and

protective action of glutathione. Proc Natl Acad Sci U S A 95:7631–7636

39. Selvakumar B, Huganir RL, Snyder SH (2009) S-nitrosylation of stargazin regulates surface expression of AMPA-glutamate neurotransmitter receptors. Proc Natl Acad Sci U S A 106:16440–16445

40. Mustafa AK et al (2007) Nitric oxide S-nitrosylates serine racemase, mediating feedback inhibition of D-serine formation. Proc Natl Acad Sci U S A 104:2950–2955

41. Moechars D et al (1999) Early phenotypic changes in transgenic mice that overexpress different mutants of amyloid precursor protein in brain. J Biol Chem 274:6483–6492

42. Moechars D, Lorent K, Van Leuven F (1999) Premature death in transgenic mice that overexpress a mutant amyloid precursor protein is preceded by severe neurodegeneration and apoptosis. Neuroscience 91:819–830

43. Zaręba-Kozioł M et al (2014) Global analysis of S-nitrosylation sites in the wild type (APP) transgenic mouse brain-clues for synaptic pathology. Mol Cell Proteomics 13:2288–2305

44. Wasling P et al (2009) Synaptic retrogenesis and amyloid-beta in Alzheimer's disease. J Alzheimers Dis 16:1–14

45. Shankar GM, Walsh DM (2009) Alzheimer's disease: synaptic dysfunction and Abeta. Mol Neurodegener 4:48

46. Butterfield DA, Sultana R (2007) Redox proteomics identification of oxidatively modified brain proteins in Alzheimer's disease and mild cognitive impairment: Insights into the progression of this dementing disorder. J Alzheimers Dis 12:61–72

47. Di Domenico F et al (2009) Glutathionylation of the pro-apoptotic protein p53 in Alzheimer's disease brain: implications for AD pathogenesis. Neurochem Res 34:727–733

48. Di Domenico F et al (2011) Quantitative proteomics analysis of phosphorylated proteins in the hippocampus of Alzheimer's disease subjects. J Proteomics 74:1091–1103

49. Sultana R, Butterfield DA (2009) Oxidatively modified, mitochondria-relevant brain proteins in subjects with Alzheimer disease and mild cognitive impairment. J Bioenerg Biomembr 41:441–446

50. Cerbai F et al (2012) The neuron-astrocyte-microglia triad in normal brain ageing and in a model of neuroinflammation in the rat hippocampus. PLoS One 7:e45250

51. Santos RM et al (2012) Nitric oxide inactivation mechanisms in the brain: role in bioenergetics and neurodegeneration. Int J Cell Biol 2012:391914

52. Grima G, Grima G, Benz B, Do KQ (2001) Glial-derived arginine, the nitric oxide precursor, protects neurons from NMDA-induced excitotoxicity. Eur J Neurosci 14:1762–1770

53. Bal-Price A, Brown GC (2001) Inflammatory neurodegeneration mediated by nitric oxide from activated glia-inhibiting neuronal respiration, causing glutamate release and excitotoxicity. J Neurosci 21:6480–6491

54. Bai F, Witzmann FA (2007) Synaptosome proteomics. Subcell Biochem 43:77–98

55. Witzmann FA et al (2005) A proteomic survey of rat cerebral cortical synaptosomes. Proteomics 5:2177–2201

56. Nilsson T et al (2010) Mass spectrometry in high-throughput proteomics: ready for the big time. Nat Methods 7:681–685

57. Tabb DL et al (2010) Repeatability and reproducibility in proteomic identifications by liquid chromatography-tandem mass spectrometry. J Proteome Res 9:761–776

58. Dewachter I et al (2000) Modeling Alzheimer's disease in transgenic mice: effect of age and of presenilin1 on amyloid biochemistry and pathology in APP/London mice. Exp Gerontol 35:831–841

59. Dewachter I et al (2002) Neuronal deficiency of presenilin 1 inhibits amyloid plaque formation and corrects hippocampal long-term potentiation but not a cognitive defect of amyloid precursor protein [V717I] transgenic mice. J Neurosci 22:3445–3453

60. Zhukova L et al (2004) Redox modifications of the C-terminal cysteine residue cause structural changes in S100A1 and S100B proteins. Biochimica Et Biophysica Acta-Molecular Cell Research 1742:191–201

61. Bradford MM (2007) A rapid and sensitive method for the quantitation of microgram quantities of protein utilizing the principle of protein-dye binding. Anal Biochem 72:248–254

62. Oliveros JC. 2007. VENNY. An interactive tool for comparing lists with Venn Diagrams. http://bioinfogp.cnb.csic.es/tools/venny/index.html

63. Pirooznia M et al (2012) SynaptomeDB: an ontology-based knowledgebase for synaptic genes. Bioinformatics 28:897–899

64. von Eichborn J et al (2013) SynSysNet: integration of experimental data on synaptic protein-protein interactions with drug-target relations. Nucleic Acids Res 4:D834–D840

65. Warde-Farley D et al (2010) The GeneMANIA prediction server: biological network integration for gene prioritization and predicting gene function. Nucleic Acids Res 38:W214–W220

Neuromethods (2016) 114: 97–110
DOI 10.1007/7657_2015_90
© Springer Science+Business Media New York 2015
Published online: 08 November 2015

Proteomics Identification of Redox-Sensitive Nuclear Protein Targets of Human Thioredoxin 1 from SHSY-5Y Neuroblastoma Cell Line

Mohit Raja Jain, Changgong Wu, Qing Li, Chuanlong Cui, Huacheng Dai, and Hong Li

Abstract

Oxidative and nitrosative modifications of cysteines are very important posttranslational modifications (PTMs) that can regulate the proper folding and functions of proteins. Given the frequent correlation of oxidative and nitrosative stress with neurodegenerative diseases, it is important to accurately identify redox-sensitive cysteines and their oxidative PTMs. Oxidative PTM levels of specific cysteines are determined by the relative contributions of oxidative and reductive pathways. The actors within these pathways include both small redox molecules and oxidoreductases, such as NADPH oxidases and thioredoxins. In this chapter, we will demonstrate the use of mass spectrometry techniques to identify redox-sensitive nuclear proteins that can be selectively reduced by thioredoxin 1 (Trx1) in neuroblastoma cells. This comprehensive workflow can be applied to the identification of redox-sensitive signaling proteins involved in modulating neurodegenerative diseases.

Keywords: Thioredoxin 1 (Trx1), Posttranslational modification (PTM), Mass spectrometry (MS)

1 Introduction

Oxidative and nitrosative modifications of protein amino acid residues, especially cysteines, are known to play important roles both in modulating redox signaling pathways and in marking damaged proteins for degradation [1, 2]. When cellular redox environments are in flux, selective oxidative PTMs can be cytoprotective, conditioning cells for survival at higher levels of oxidative stress. However, prolonged and severe oxidative insults can non-specifically add oxidative PTMs onto many sensitive cysteines, rendering proteins inactive and marking them for proteolytic degradation [3–5].

To counter such damaging effects of oxidative stress, cells can use numerous antioxidant enzymes to either reverse oxidative PTMs or alleviate oxidative damages to proteins [6]. Thioredoxins (Trxs) are antioxidant proteins that can facilitate the reduction of other proteins by a cysteine thiol-disulfide exchange mechanism [7]. Human Trx1 is a 12-kDa oxidoreductase enzyme containing a dithiol-disulfide active site at Cys^{32} and Cys^{35} within a CXXC motif which is present in many Trx1-like proteins [8]. These two cysteines are keys to the ability of Trx1 to reduce specific target proteins. Wild type Trx1 contains a conserved $C^{32}XXC^{35}$ motif, whereby the Cys^{32} thiol initiates the reduction of a target disulfide bond by forming an intermolecular disulfide with one of the oxidized target cysteines. The transient Trx1-target protein mixed disulfide is then rapidly reduced by donation of a proton from Cys^{35} to the remaining cysteines within the target disulfide; thus the target protein is concurrently released. The resulting Trx1 intramolecular Cys^{32}-Cys^{35} disulfide can itself regenerate into a catalytically active sulfhydryls by thioredoxin reductases [7].

Not all oxidized proteins are reduced by Trx1; identification of Trx1 redox target proteins may aid the identification of novel stress-response signaling molecules and pathways in neurodegenerative diseases. In this chapter we will describe a mass spectrometry (MS)-based approach to identify nuclear Trx1 targets in human neuroblastoma cells, using an affinity capture strategy with a $Trx1^{C35S}$ mutant (Fig. 1). By introducing a point mutation (Cys^{35} to Ser^{35}) in a nucleus-targeted Trx1, the rapid dissociation of Trx1 from its reduction targets is ablated, trapping the target protein in a mixed disulfide with Trx1. After enrichment of putative Trx1 target through affinity capture with an anti-Trx1 tag antibody, subsequent identification of nuclear Trx1 targets is achieved using tandem MS. In this chapter we provide the protocol to overexpress both the wild type Trx1 ($Trx1^{WT}$) and mutant Trx ($Trx1^{C35S}$) in SH-SY5Y neuroblastoma cells, as well as the details regarding the extraction of nuclear proteins, affinity capture of putative Trx1 targets, and identification of Trx1 targets using MS.

2   Materials

2.1   Cell Culture and Transfection

1. Human neuroblastoma SHSY-5Y cells.

2. Culture medium: 1:1 mixture of Dulbecco's Modified Eagle Medium (DMEM) and F12 medium supplemented with 0.1 mM nonessential amino acids, 1 % penicillin/streptomycin, and 10 % fetal bovine.

3. pcDNA3-EGFP plasmid encoding wild type Trx1 ($nTrx1^{WT}$).

4. pcDNA3-EGFP plasmid encoding mutant Trx1 ($nTrx1^{C35S}$).

5. Lipofectamine 2000.

6. Opti-MEM® I reduced serum medium.

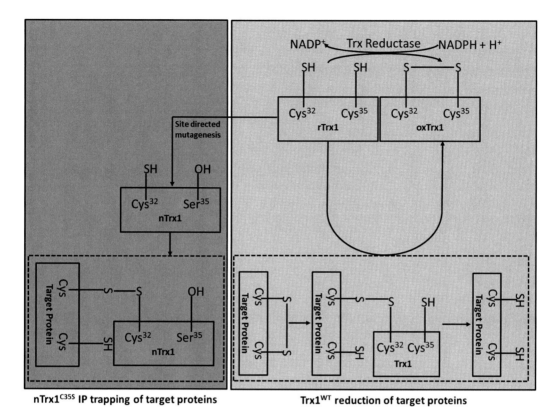

Fig. 1 A workflow describing how the nTrx1C35S mutant traps target proteins of thioredoxin 1. Normally, Trx1[WT] will reduce the oxidized proteins via a redox switch shown in *right panel*; however, nTrx1[C35S] binds to target proteins via mixed disulfide bonds as Trx1[WT] but cannot consummate the reduction and release of the target proteins (*left panel*). Therefore, the disulfide-linked proteins can be isolated through immunoprecipitation for proteomic identification

      7. Appropriate tissue culture plates and supplies.

2.2 Nuclear Protein Extraction

1. Phosphate buffer saline (PBS): 1.5 mM Potassium phosphate monobasic (KH_2PO_4), 155 mM Sodium chloride (NaCl), 2.7 mM Sodium phosphate dibasic (Na_2HPO_4-$7H_2O$).

2. Hypotonic lysis buffer: 10 mM HEPES (pH 7.9), 10 mM KCl, 1.5 mM $MgCl_2$, 0.5 mM DTT, protease inhibitors and phosphatase inhibitors.

3. NP-40 Detergent.

4. Hypertonic buffer: 20 mM HEPES (pH 7.9), 1.5 mM $MgCl_2$, 420 mM NaCl, 25 % v/v glycerol, 0.5 mM DTT, 0.2 mM EDTA, and protease inhibitors and phosphatase inhibitors.

5. Ultrasonic homogenizer with 5/32″ Micro-Tip (Omni International, Kennesaw, GA).

6. Ice bath.

7. Microcentrifuge 5415 R (Eppendorf, Hauppauge, NY).

8. Appropriate lab ware supplies, e.g., Eppendorf tubes, 15 ml centrifuge tubes, pipettes, centrifuge etc.

2.3 Immuno-precipitation

1. Anti-GFP antibody (A11122, Life Technologies, Grand Island, NY).

2. Protein-A agarose beads (15918-014, Life Technologies).

3. Wash buffer: 10 mM Tris (pH 8.0x), 25 mM NaCl, 0.5 % Triton X-100, 25 mM EDTA.

4. Elution buffer: 100 mM Tris (pH 6.8), 4 % SDS, 0.2 % bromophenol blue, 20 % glycerol with 5 % β-mercaptoethanol.

5. GeneMate Rotator.

6. Cold room or refrigerated cabinet set at 4 °C.

7. Heat block set at 100 °C.

8. Appropriate lab ware supplies, e.g., Eppendorf tubes, 15 ml centrifuge tubes, pipettes, centrifuge etc.

2.4 In-Gel Digestion

1. Mini-PROTEAN® Tetra Cell electrophoresis chamber.

2. 12 % Mini-PROTEAN® TGX Gel, 10 well, 50 μl/well.

3. Running buffer: 25 mM Tris, 192 mM glycine, 0.1 % SDS, pH 8.3.

4. Precision Plus Protein™ All Blue Standards.

5. PowerPac™ basic power supply.

6. Fixing solution: 50 % methanol, 40 % H_2O, 10 % acetic acid.

7. GelCode™ Blue Safe Protein Stain.

8. GeneCatcher disposable gel excision tips.

9. Surgical scalpel blades No. 11.

10. 100 mM ammonium bicarbonate, pH approx. 8.0.

11. Acetonitrile (ACN, HPLC-MS grade).

12. DTT solution: 10 mM DTT (prepared in 100 mM ammonium bicarbonate).

13. Iodoacetamide solution: 55 mM iodoacetamide (prepared in 100 mM ammonium bicarbonate).

14. Extraction buffer (60 % acetonitrile, 2 % formic acid).

15. Water (HPLC-MS grade).

16. Trypsin (mass spectrometry grade).

17. Centrifuge speed vacuum.

18. Incubator shaker set at 37 °C.

19. Appropriate lab ware supplies, e.g., Eppendorf tubes, 15 ml centrifuge tubes, pipettes, centrifuge etc.

2.5 Peptide Desalting	1. PepClean C$_{18}$ spin columns (Thermo Scientific Pierce, Rockford, IL).
	2. Activation solution: 50 % ACN.
	3. Equilibration solution: 5 % ACN, 0.5 % trifluoroacetic acid (TFA) in water.
	4. Microcentrifuge.
	5. Elution solution: 75 % ACN in water.

2.6 Reversed-Phase High Performance Liquid Chromatography (RPHPLC)

1. Solvent A: 2 % ACN, 0.1 % Formic Acid, 97.9 % Water.
2. Solvent B: 85 % ACN, 0.1 % Formic Acid, 14.9 % Water.
3. UltiMate® 3000 RSLCnano HPLC system.
4. Acclaim PepMap100 trap column (C$_{18}$, 5 μm, 100 Å, 300 μm i.d. × 5 mm length).
5. Acclaim PepMap100 capillary nano column (C$_{18}$, 3 μm, 100 Å, 300 μm i.d. × 150 mm length).

2.7 Mass Spectrometry

1. Thermo Fisher Scientific Q Exactive Mass Spectrometer.
2. Thermo Fisher Scientific Nanospray Flex™ Ion Source.

2.8 Data Analysis Software

1. Proteome Discoverer (Version 1.4) (Thermo Fisher Scientific, Waltham, MA, http://www.thermoscientific.com/).
2. Mascot (Matrix Science Inc., Boston, MA, http://www.matrixscience.com).
3. Scaffold (Proteome Software Inc., Portland, OR, http://www.proteomesoftware.com).
4. Excel (Microsoft Corporation, Redmond, WA, http://office.microsoft.com).

3 Methods

3.1 Cell Culture and Transfection

1. For each transfection, using 10 cm culture dish, plate 2×10^6 SHSY-5Y cells in 15 ml of culture medium and incubate for ~24 h at 37 °C in 5 % CO$_2$ so that cells will be 50–80 % confluent at the time of transfection.
2. For each transfection, in a sterile Eppendorf® tube, gently mix 2.5 μg of nTrx1^{WT} plasmid DNA into 300 μl of Opti-MEM® I Reduced Serum Medium.
3. Just before use, gently mix Lipofectamine® 2000. In a sterile Eppendorf® tube, gently mix 7.5 μl Lipofectamine® 2000 into 300 μl of Opti-MEM® I Reduced Serum Medium. Incubate at room temperature for 5 min.

4. Combine the diluted nTrx1^{WT} plasmid DNA with diluted Lipofectamine®. Total volume is ~600 µl. Mix gently and incubate for 30 min at room temperature.

5. Meanwhile remove the culture medium from cells and add 15 ml of fresh culture medium.

6. Add the ~600 µl of plasmid DNA–Lipofectamine® complexes to cells and medium. Mix gently by rocking the plate back and forth.

7. Repeat **steps 2–6** using nTrx1^{C35S} plasmid DNA.

8. Incubate cells at 37 °C in a CO_2 incubator for 6 h.

9. Change the cell culture medium after 6 h and continue the incubation for a total of 48 h.

3.2 Nuclear Protein Extraction

1. Aspirate the culture medium from each plate.

2. Wash cells on the plate with ice-cold PBS twice to remove any residual media (*see* **Note 1**).

3. For each transfection, carefully harvest the cells in 1 ml of hypotonic lysis buffer with a corning cell scraper in sterile Eppendorf tube.

4. Incubated on ice for 15 min.

5. Add 5 µl of NP-40 detergent.

6. Gently mix the contents of Eppendorf tube by inversion.

7. Incubate on ice for another 15 min.

8. Centrifuge the resulted cellular extract at 800 × *g* for 10 min at 4 °C.

9. Transfer the cytoplasmic components (supernatant) to fresh Eppendorf tube and store at −70 °C.

10. Gently resuspend the resulted nuclei-enriched pellet in 1 ml of hypotonic lysis buffer supplemented with 5 µl of NP-40 detergent.

11. Centrifuge at 800 × *g* for 10 min at 4 °C and remove the supernatant.

12. Repeat **steps 10** and **11** one time.

13. Resuspend nuclear pellet in 50 µl hypertonic buffer for 30 min on ice with vigorous vortexing at each 10 min interval.

14. Lyse the nuclei by a 10 sec sonication pulse followed by 30 s incubation in the ice bath. Repeat three times. Incubate the cell lysate on the ice bath for 10 min.

15. Centrifuge the resulting solution at 16,000 × *g* at 4 °C for 15 min.

16. Transfer the supernatant to fresh Eppendorf tube and proceed to immunoprecipitation.

3.3 Immuno-
precipitation

1. Perform immunoprecipitation on each sample separately.

2. Place 1.0 mg of nuclear protein extract in a fresh Eppendorf tube (*see* **Note 2**).

3. Add anti-GFP antibody to a final concentration of 0.2 μg/ml and incubate the tubes at 4 °C overnight with gentle rotation (*see* **Note 3**).

4. Meanwhile, place 100 μl of protein-A agarose beads in fresh Eppendorf tube (*see* **Note 3**).

5. Add 1 ml of PBS to tube and gently mix by inverting the tube. Centrifuge at 3000 × *g* for 2 min at 4 °C and discard the supernatant.

6. Repeat **step 4** two more times.

7. Resuspend the protein-A agarose beads in 250 μl of hypertonic buffer.

8. Transfer the protein-A agarose beads to the tube from **step 2** and continue the incubation for 1 h with gentle rotation.

9. Centrifuge at 3000 × *g* for 2 min at 4 °C and remove the supernatant to fresh Eppendorf tube.

10. Resuspend the protein-A agarose beads from **step 8** in 1 ml of wash buffer.

11. Centrifuge at 3000 × *g* for 2 min at 4 °C and discard the supernatant.

12. Repeat **steps 9** and **10** two more times.

13. Elute the proteins from beads by adding 50 μl of elution buffer and incubating on a heat block at 100 °C for 5 min.

14. Cool down the tube at room temperature for 5 min.

15. Centrifuge 16,100 × *g* for 5 min and transfer the supernatant to a fresh Eppendorf tube.

3.4 In-Gel Digestion

1. Remove the Mini-PROTEAN® TGX Gel from the pouch (*see* **Note 4**).

2. Prepare the Mini-PROTEAN® Tetra Cell electrophoresis chamber as per manufacturer's instructions.

3. Load the sample from **step 15** in Section 3.3 in a well on gel.

4. Repeat **step 3** for all other sample categories using a new well for each sample. Leave an empty well between each sample.

5. Run gel at constant voltage of 200 V.

6. Continue electrophoresis till the dye front is about 4 cm from the top of gel cassette (*see* **Note 5**).

7. Remove the gel from cassette and rinse in distilled water.

8. Transfer the gel to fixing solution in a clean container and incubate at room temperature with gentle shaking for 60 min.

9. Transfer the gel to a clean container containing water and incubate at room temperature for 10 min.

10. Discard the solution and add fresh water and incubate for 10 min.

11. Repeat **step 10** for one more time.

12. Remove the solution completely.

13. Add a sufficient volume of GelCode™ blue safe protein stain to completely cover the gel. Incubate on an orbital shaker overnight at room temperature (*see* **Note 6**).

14. Decant the staining solution and replace with 200 ml of ultrapure water. Incubate on an orbital shaker for 15 min at room temperature.

15. Replace the water and continue for 1 h at room temperature.

16. Carefully transfer the gel onto a clean glass plate on light box and excise whole stained protein lane with a clean scalpel.

17. Process all the stained protein lanes from the gel in parallel.

18. Cut excised each gel lane into 1 mm^3 pieces (*see* **Note 7**).

19. Transfer the gel pieces into a 15 ml centrifuge tube and spin down to bring the gel pieces to bottom of tube.

20. Add 5 ml of 100 mM ammonium bicarbonate/acetonitrile (7:3, vol/vol) solution and incubate with gentle rotation for 30 min (*see* **Note 8**).

21. Decant the solution and add 5 ml of fresh 100 mM ammonium bicarbonate/acetonitrile (7:3, vol/vol) solution and incubate with gentle rotation for another 30 min.

22. Repeat **step 21** till the gel pieces are completely clear.

23. Remove the solution completely and add 5 ml of acetonitrile and incubate at room temperature for 15 min with occasional vortexing. Gel pieces will become opaque and shrink.

24. Using a pipette tip, remove acetonitrile completely.

25. Add 5 ml of the DTT solution to completely cover gel pieces. Incubate for 30 min at 56 °C in an incubator (*see* **Note 9**).

26. Cool down the tubes to room temperature and repeat **steps 23** and **24**.

27. Add 5 ml of the Iodoacetamide solution to completely cover gel pieces. Cover the tube with aluminum foil to block the light. Incubate for 30 min at room temperature (*see* **Note 10**).

28. Repeat **steps 23** and **24**.

29. Prepare a trypsin solution of 20 ng/μl in 10 mM ammonium bicarbonate containing 10 % (vol/vol) acetonitrile (*see* **Note 11**).

30. Add trypsin solution to gel pieces from **step 24** in such a way that gel pieces are completely submerged.

31. Incubate the tubes on ice for 30 min.

32. Check if all of trypsin solution is absorbed. Add more trypsin solution, if needed. Make sure that gel pieces are completely covered with trypsin buffer (*see* **Note 12**).

33. Incubate the tubes with gel pieces at 37 °C overnight.

34. Add 2 ml of extraction buffer to each tube and incubate at 37 °C for 15 min (*see* **Note 13**).

35. Using a pipette with fine gel loader tip, transfer the supernatant to a fresh Eppendorf tube.

36. Dry the peptide solution completely in a centrifuge speed vacuum (*see* **Note 14**).

3.5 Peptide Desalting

1. Resuspend the peptides in each tube in 200 μl of the equilibration solution.

2. Sonicate in a water bath for 15 s, vortex, and then spin briefly.

3. Activate the C_{18} resin by adding 200 μl of activation solution to the PepClean C_{18} spin columns.

4. Centrifuge at $1500 \times g$ for 1 min.

5. Repeat **steps 3** and **4** once more.

6. Equilibrate the C_{18} resin by adding 200 μl of the equilibration solution.

7. Centrifuge at $1500 \times g$ for 1 min.

8. Repeat **steps 6** and **7** once more.

9. For each sample, transfer the peptides from **step 2** completely onto an equilibrated C_{18} spin column. Centrifuge at $1000 \times g$ for 1 min.

10. Collect the flow through and load the flow through again onto the spin column. Centrifuge at $1000 \times g$ for 1 min. Repeat **step 9** once more.

11. Add 200 μl of the equilibration solution to the column. Centrifuge at $1500 \times g$ for 1 min.

12. Repeat **step 11** twice.

13. Add 30 μl of the elution solution onto the column to elute the bound peptides.

14. Collect the eluted peptides by centrifugation at $1500 \times g$ for 1 min in a fresh Eppendorf tube.

15. Repeat **steps 13** and **14** twice and collect all the eluted peptides from each sample.

16. Dry the desalted peptides in centrifuge speed vacuum.

3.6 Reversed-Phase High Performance Liquid Chromatography and Mass Spectrometry (RPHPLC-MS)

1. Reconstitute the peptides from each sample in 15 μl of RPHPLC solvent A. Vortex at high speed for 30 s and then spin briefly.

2. Sonicate the samples in a water bath for 15 s and vortex at high speed for 15 s and then spin briefly.

3. Sonicate the samples one more time in a water bath for 15 s and vortex at high speed for 15 s.

4. Centrifuge the tubes at $16,100 \times g$ for 5 min.

5. Transfer each sample solution into the bottom of an autosampler vial and place all the vials in cooled autosampler tray.

6. Equilibrate the RPHPLC C_{18} column for 30 min with 2 % solvent B at flow rate of 0.300 μl/min.

7. Set up the RPHPLC gradient method as follows:

Time (min)	Solvent A (%)	Solvent B (%)
0	98	2
6	98	2
7	95	5
157	70	30
172	50	50
182	5	95
197	5	95
198	98	2

8. Set up Q Exactive acquisition method with following parameters:

Full MS	
Microscans	1
Resolution	70,000
AGC target	$1e^6$
Maximum IT	100 ms
Number of scan ranges	1
Scan range	300–1700 m/z
Spectrum data type	Profile
dd-MS²/dd-SIM	
Microscans	1
Resolution	17,500
AGC target	$1e^5$
Maximum IT	50 ms
Loop count	1
MSX count	1

(continued)

TopN	20
Isolation window	2.0 m/z
Isolation offset	0.0 m/z
Fixed first mass	—
NCE/stepped NCE	27
Spectrum data type	Profile
dd Settings	
Underfill ratio	1.0 %
Intensity threshold	$2.0e^4$
Apex trigger	–
Charge exclusion	Unassigned, 6–8, >8
Peptide match	Preferred
Exclude isotopes	on
Dynamic exclusion	60.0 s
If idle	Do not pick others

9. Tune and calibrate the Q Exactive mass spectrometer (*see* **Note 15**).

10. Load 6 μl of the reconstituted peptides from **step 5** onto a C_{18} trapping column at a flow rate of 5 μl/min (*see* **Note 16**).

11. Peptides bound on trap column are washed to remove the salt and other impurities and subsequently resolved in a high resolution C_{18} PepMap column at a flow rate of 0.3 μl/min according to gradient from **step 7**.

12. Peptides eluted are introduced directly to Q Exactive mass spectrometer through the Thermo Scientific Nanospray Flex Ion Source and analyzed according to parameters from **step 8**.

13. Analyze peptides from each sample in same way.

3.7 Protein Identification and Data Analysis

Peptide identification is performed using Mascot search engine through Proteome Discoverer (v. 1.4) and Scaffold (v 4.4.1) softwares against the Human protein sequences from UniRef100 protein database.

1. Separately analyze each of the raw files generated after **step 11** from Section 3.6 using the Thermo Proteome Discoverer (V 1.4) platform with Mascot (2.4.1) as a search engine against all of the human protein sequences in the UniRef100 protein database.

2. Use following Mascot search parameters: trypsin, two missed cleavages, precursor mass tolerance: 10 ppm, fragment mass tolerance: 0.1 Da, dynamic modifications: methionine oxidation and carbamidomethylation of cysteines.

3. Engage the decoy search option for Mascot through the percolator.

4. After search is completed, separately import the results (.msf files) for each *.raw file from Proteome Discoverer onto Scaffold (version Scaffold 4.4.1; Proteome Software, Portland, OR).

5. In Scaffold, filter the proteins to achieve a false discovery rate (FDR) of less than 1.0 % at both protein and peptide level. Scaffold uses Peptide Prophet algorithm (22), with a Scaffold delta-mass correction and Protein Prophet algorithm (23) to calculate FDR. Proteins that contained similar peptides and could not be differentiated based on the MS/MS analysis alone are grouped to satisfy the principles of parsimony.

6. Using export function of Scaffold, export the list of identified proteins and accession no. to excel.

7. Upload the accession no. of identified proteins to Ingenuity Pathway Analysis (IPA) software (http://www.ingenuity.com) to define their putative subcellular localization and function.

8. Export the putative subcellular localization and function of identified proteins from IPA software to excel.

9. Repeat **steps 6–8** for all the samples.

10. Consider the proteins which have nuclear localization.

11. Compare the list of nuclear proteins from wild type Trx1 (nTrx1WT) and mutant Trx1 (nTrx1C35S).

12. Nuclear proteins which are only identified in mutant Trx1 (nTrx1C35S) are redox-sensitive nuclear protein targets of Human Thioredoxin 1 (*see* **Note 17**).

4 Notes

1. Remove the culture medium completely from the plate as any remaining medium (serum, salt etc.) would interfere with subsequent mass spectrometry analysis.

2. Amount of protein used for each analysis may be more of less depending upon the availability.

3. Adjust the final concentration of anti-GFP antibody and protein-A agarose based upon manufacturer's recommendation.

4. Use clean gel apparatus and staining tray to process the gel to avoid the contamination with common contaminants like keratins, BSA etc.

5. Do not run the gel till the end as it would create more bigger gel lane to process in subsequent analysis and contribute to common contaminations.

6. Only use the protein stain which is mass spectrometry compatible and should not modify the protein irreversibly.

7. Ideally this step should be done under the laminar flow hood to minimize the air borne contamination, e.g., keratins. Do not cut the pieces too small as it may clog the tips.

8. Volume of 100 mM ammonium bicarbonate/acetonitrile (7:3, vol/vol) solution should be at least three times the volume of gel pieces to help the destaining of gel pieces.

9. DTT solution should be prepared fresh just before the step.

10. Iodoacetamide solution should be prepared fresh just before the step.

11. Trypsin solution should be prepared fresh just before the step.

12. Make sure the pH of solution is ~8.0, otherwise trypsin would not work optimally.

13. Volume of extraction buffer should be at least two times the volume of gel pieces.

14. Do not overdry the peptides in this step, otherwise it may be difficult to solubilize.

15. It is important to tune and calibrate the mass spectrometer to get the good data.

16. Volume of sample loaded in this step may be varied depending upon the concentration of peptides in the sample.

17. It is important to have biological repeats for each sample to get most confident identification of putative nuclear target of Trx1. It is recommend to have at least three biological repeats of both wild type and mutant sample.

Acknowledgements

The chapter described is supported in part by a grant P30NS046593 from the National Institute of Neurological Disorders and Stroke. The content is solely the responsibility of the authors and does not necessarily represent the official views of the National Institute of Neurological Disorders and Stroke or the National Institutes of Health. The authors report no conflict of interest.

References

1. Aiken CT, Kaake RM, Wang X, Huang L (2011) Oxidative stress-mediated regulation of proteasome complexes. Mol Cell Proteomics 10: R110.006924

2. Iyer AKV, Rojanasakul Y, Azad N (2014) Nitrosothiol signaling and protein nitrosation in cell death. Nitric Oxide 42C:9–18

3. Höhn TJA, Grune T (2014) The proteasome and the degradation of oxidized proteins: Part III. Redox regulation of the proteasomal system. Redox Biol 2:388–394

4. Jung T, Grune T (2013) The proteasome and the degradation of oxidized proteins: Part I. Structure of proteasomes. Redox Biol 1:178–182

5. Jung T, Höhn A, Grune T (2013) The proteasome and the degradation of oxidized proteins: Part II. Protein oxidation and proteasomal degradation. Redox Biol 2C:99–104

6. Ye Z-W, Zhang J, Townsend DM, Tew KD (2015) Oxidative stress, redox regulation and diseases of cellular differentiation. Biochim Biophys Acta 1850(8):1607–1621

7. Collet J-F, Messens J (2010) Structure, function, and mechanism of thioredoxin proteins. Antioxid Redox Signal 13:1205–1216

8. Holmgren A (1968) Thioredoxin. 6. The amino acid sequence of the protein from escherichia coli B. Eur J Biochem 6:475–484

Neuromethods (2016) 114: 111–125
DOI 10.1007/7657_2015_95
© Springer Science+Business Media New York 2015
Published online: 18 November 2015

Identification and Quantification of K63-Ubiquitinated Proteins in Neuronal Cells by High-Resolution Mass Spectrometry

Gustavo Monteiro Silva, Wei Wei, Sandhya Manohar, and Christine Vogel

Abstract

Protein ubiquitination is a widespread modification serving many roles in neuronal development and function. Moreover, the accumulation of ubiquitinated proteins is a prominent feature of neurodegeneration and oxidative stress related diseases. The emerging diversity of ubiquitin signals beyond protein degradation—based on distinct types of polyubiquitin chains—necessitates tools that specifically and quantitatively investigate its different functions. Polyubiquitin chains linked by lysine 63 (K63) relate to neurodegenerative diseases, but most of their targets and functions have not yet been elucidated. K63-linked ubiquitin has been implicated in DNA repair signaling, endocytosis, and inclusion body clearance. In addition, we recently identified an important role of K63 ubiquitin in regulating translation in response to oxidative stress in yeast. The change in K63 ubiquitination in response to hydrogen peroxide is conserved in mouse hippocampal HT22 cells, highlighting the importance of this modification in higher eukaryotes. In this chapter, we discuss cutting-edge methodologies available to investigate protein ubiquitination in a proteome-wide and quantitative manner, and we present a method to simultaneously isolate and identify the specific targets of K63 ubiquitin. This method relies on the use of a selective K63 ubiquitin isolation tool with subsequent analysis of protein content by high-resolution mass spectrometry. The proposed workflow can be combined with additional methods for ubiquitin analysis and applied to several research models. This approach can also provide the scientific basis for the development of new tools to isolate and identify targets of other ubiquitin linkages.

Keywords: K63 ubiquitin, Mass spectrometry, Neurons, Oxidative stress

1 Introduction

In eukaryotes, ubiquitination is a prevalent modification that affects a diverse array of proteins and has a variety of functional consequences [1]. In this chapter, we will describe a method to investigate ubiquitinated proteins from neuronal cells using selective enrichment and high-resolution mass spectrometry. The accumulation of ubiquitinated proteins is a hallmark of many neurodegenerative diseases—including Alzheimer's (AD), Parkinson's (PD), and Huntington's (HD)—and is also associated with the presence of abnormal protein aggregates and identified in neurofibrillary tangles and plaques [2]. The Ubiquitin Proteasome System (UPS) is the most important proteolytic machinery in the cell. It is responsible for the intracellular degradation of damaged, unfolded, and unneeded proteins, thus preventing the accumulation of protein aggregates. Due to its widespread role in proteostasis, impairment of the UPS is arguably one of the primary causes of aggregate formation. These aggregates are toxic for the cells, and can in turn, inhibit proteasome activity, generating a feedforward loop [3–5].

Ubiquitination occurs via the formation of an isopeptide bond between a lysine residue of a target protein and the C-terminal glycine residue of ubiquitin. A cascade of three enzyme classes catalyzes ubiquitin transfer to substrates: E1 ubiquitin activating enzymes, E2 ubiquitin conjugating enzymes, and E3 ubiquitin ligases. The human genome is thought to encode a dozen E1s, ~40 E2s, and >600 putative E3 genes—which confer target specificity [6, 7]. Several mutations in these genes correlate with neurodegenerative diseases and changes in neuronal development. For example, in familial cases of PD, mutations in *PARK2* (which encodes parkin, an E3 ligase) are responsible for the pathology of approximately half of all patients [8]. Moreover, mutations in ubiquitin itself have been found in AD patients [9], and disruptions in the E3 *UBE3A* and E2 *UBE2A* ubiquitin enzymes can cause Angelman's Syndrome [10] and mental retardation [11], respectively.

Although impairment of ubiquitin homeostasis and proteasome activity impacts neuronal viability and development [2, 12, 13], the role of accumulated ubiquitin conjugates during neurodegeneration remains elusive. The challenge of understanding the function of ubiquitin results from the emerging complexity of ubiquitin signals and the dearth of technologies available to investigate them. Upon ligation of ubiquitin to its substrate, additional ubiquitin moieties can be added at one of ubiquitin's seven lysine residues (K6, K11, K27, K29, K33, K48, or K63) or at the N-terminus, forming a poly-ubiquitin chain [1]. The lysine at which polyubiquitin branch formation occurs imparts varied cellular signals, ranging from target degradation to signaling in DNA repair [1]. Therefore, investigating the topography of polyubiquitin chains, the specificity of targets, and the variability of ubiquitin signals is key to understanding the diverse functionality of ubiquitin in neurons and neurodegeneration. While K48-linked polyubiquitin serves as the primary and canonical signal for substrate degradation by the 26S proteasome [14], the biological function of many ubiquitin linkages (K6, K27, K29, and K33) remains largely unexplored [15]. K11-linked polyubiquitin also signals for protein degradation, and its most studied role is in cell cycle control mediated by the E3 APC/C (anaphase-promoting complex) [16, 17]. In contrast, K63-linked polyubiquitin can serve a multitude of nonproteasomal functions in endocytosis, DNA repair, the inflammatory response, and others [18–22]. Our group recently discovered a new mechanism of translation regulation mediated by K63 ubiquitin during the response to oxidative stress [21]. Under oxidative stress, K63 ubiquitin modifies ribosomal proteins, impacting both polysome levels and protein synthesis [21]. The K63 ubiquitin redox pathway was initially characterized in baker's yeast, but we showed that K63 ubiquitin levels are also altered in hippocampal cells [21]. Anecdotal evidence has shown the accumulation of K63 ubiquitin in the hippocampus of AD patients [23] and has suggested a role for this linkage in Lewy's body biogenesis, as well as in the formation of inclusion bodies [24, 25]. Particularly, Olzmann and colleagues were able to demonstrate that parkin-mediated K63 ubiquitination directs misfolded proteins to aggresomes leading to clearance by autophagy [8]. These findings show that K63-linked ubiquitination may have many functional and relevant roles in neurodegenerative diseases, though the targets and molecular mechanisms have not yet been elucidated.

Comprehensive examination of the ubiquitinated proteome requires the ability to identify the protein substrates, the modification sites within the amino acid sequence, and the polyubiquitin chain linkage type. Establishing this information using liquid chromatography-mass spectrometry (LC-MS)-based techniques is challenging since current methods can either detect the targets of ubiquitination or quantify the linkage type, but not both at the same time. This impasse occurs because tryptic digestion of ubiquitinated proteins prior to MS analysis also cleaves the polyubiquitin chain, separating the linkage information from the protein target. One established MS-based approach, called ubiquitin remnant profiling, identifies ubiquitinated substrates irrespective of their linkage type, yielding site-specific information on the lysine that was modified (Fig. 1a). This method relies on the fact that the C-terminus of ubiquitin (Arg-Gly-Gly) contains an arginine residue susceptible to tryptic digestion, resulting in a diglycyl remnant (diGG) at the original site of ubiquitin conjugation [26]. Hence, identification of the diGG remnant is a strong indication of an ubiquitination event at a particular lysine residue. Unfortunately, any information on the type of ubiquitin chain is lost during tryptic digestion. Applying this method to human cell lines, different groups successfully quantified ~20,000 diGG sites distributed across ~5,000 proteins [27, 28].

A second approach, called ubiquitin linkage profiling, quantifies the abundances of specific polyubiquitin linkage types by targeted mass spectrometry, but does not monitor which proteins are modified (Fig. 1b) [29–31]. Targeted mass spectrometry analysis by Selected Reaction Monitoring (SRM) uses the fact that tryptic digestion of polyubiquitin chains produces peptides that are characteristic of the linkage type, and these can be quantified in the MS analysis. Using this method, Dammer and colleagues quantified the

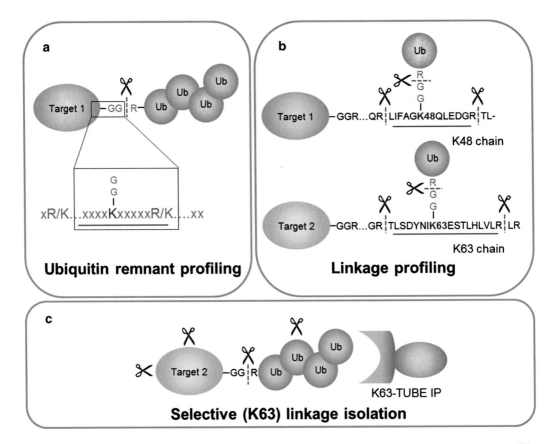

Fig. 1 Mass spectrometry-based methods to investigate protein ubiquitination. (**a**) Ubiquitin remnant profiling identifies precise sites of protein ubiquitination. (**b**) Linkage profiling quantifies the amount of distinct ubiquitin chains using signature linkage peptides (underlined in red). (**c**) Selective linkage isolation (described here) provides a method to identify targets of specific ubiquitin linkage type

ubiquitin types in HEK293 cells. They found K48-, K63-, and K29-linked polyubiquitin chains present at 52 %, 38 %, and 8 %, respectively, and minor contributions by K11-linked chains (2 %), and K33-, K27-, K6-linked chains (with <0.5 %) [32]. Moreover, the authors showed that K48, K63, and K11 increase in the frontal cortex of AD patients, reinforcing the role of distinct ubiquitin chains in neurodegeneration [32].

Both ubiquitin remnant profiling and ubiquitin linkage profiling provide insight into two different aspects of ubiquitination, i.e. the targets the modification or its type, but not at the same time. Here, we describe a method that both identifies and quantifies the specific targets of K63-linked polyubiquitin in neuronal cells (Fig. 1c). The method relies on the selective isolation of K63 ubiquitin targets coupled to the identification and quantification of these targets by mass spectrometry. To isolate the K63 ubiquitin targets we use the K63-TUBE system, which is a polypeptide composed of several ubiquitin binding domains with high affinity for K63 ubiquitin chains. K63-TUBE is conjugated to a FLAG epitope and the modified proteins are purified using an anti-FLAG M2 agarose resin (Sigma) prior to the mass spectrometry analysis. Importantly, this method can be extended to provide site-specific information, to study other linkage types, or to model the dynamics of the modification event in many experimental systems.

2 Materials

2.1 Cell Culture

1. Mouse hippocampal cell HT22—Salk Institute http://salkinnovations.testtechnologypublisher.com/technology/7060

2. Dulbecco's Modified Eagle's Medium—high glucose (Sigma, Cat. D5796)

3. Penicillin—Streptomycin (Life Technologies, Cat. 15140)

4. Fetal Bovine Serum (FBS), Advantage (Atlanta Biologicals, Cat. S11050)

5. Culture medium—SILAC™ Protein ID & Quantitation Media Kit, with Lysine and D-MEM-Flex (Life Technologies, Cat. MS10030)

6. Trypsin-EDTA (0.05 %), phenol red (Life Technologies)

7. Dulbecco's Phosphate-Buffered Saline (PBS), no calcium, no magnesium (Life Technologies)

8. CytoOne tissue culture plates—(USA Scientific Cat. CC7682-3394)

9. Bright-Line Hemacytometer (Sigma-Aldrich cat. Z359629)

10. 0.4 % Trypan Blue Solution (Life Technologies)

11. Disposable cell lifter (Fisherbrand Cat. 08-100-240)

2.2 Protein Extraction

1. Lysis buffer: 50 mM Phosphate buffer pH 7.4, 150 mM NaCl, 25 mM chloroacetamide (CAM), protease inhibitor (cOmplete Mini EDTA-free, Roche), 5 mM EDTA, 50 nM K63-TUBE Flag peptide (LifeSensors).

2. FLAG® K63-TUBE 1 (LifeSensors—Cat. UM604)

3. Misonix Ultrasonic Liquid Processor S-4000

4. Refrigerated Centrifuge (Eppendorf 5424R)

5. Bradford reagent (Biorad)

6. Genesys 20 Spectrophotometer (Thermo Scientific)

7. General labware

2.3 K63 Ubiquitin Pull Down

1. Dynabeads® Protein G for immunoprecipitation (Life Technologies)

2. Monoclonal Anti-FLAG® M2 antibody (Sigma)

3. MagneSphere® Magnetic Stand (Promega)

4. Immunoprecipitation (IP) Wash buffer—PBS containing 0.05 % Tween-20

5. IP Elution buffer: 8 M Urea buffer, 50 mM Tris–HCl pH 8.0

6. Cold room or refrigerated chamber

7. GyroMini Nutating Mixer (Labnet International)

8. Digital Dry Block Heater (VWR international)

2.4 Protein Digestion

1. Trypsin from porcine pancreas (Sigma, Cat. T6567)

2. Digestion buffer (8 M Urea, 50 mM Tris–HCl pH 8.0)

3. Dithiothreitol (DTT, Sigma)

4. Chloroacetamide (CAM, Sigma)

5. Acetonitrile (ACN, LC-MS grade, Sigma)

6. Formic acid (FA, LC-MS grade, Sigma)

7. Vacufuge plus Speed Vac concentrator (Eppendorf)

8. Hypersep C18 SpinTip (Thermo Scientific)

9. C18 Activation/Elution buffer—60 % ACN, 0.1 % FA.

10. C18 Wash Buffer solution—0.1 % FA

11. C18 Equilibration Buffer solution—5 % ACN, 0.1 % FA

12. Polypropylene vial and pre-slit screw top (Agilent Cat.5190-2242, Cat. 51832077)

2.5 Liquid Chromatography and Mass Spectrometry

1. LTQ Orbitrap Velos Mass Spectrometer (Thermo Scientific)

2. 2D NanoPlus Liquid Chromatography System (Eksigent)

3. Buffer Solvent A: 0.1 % FA HPLC quality water

4. Buffer Solvent B: ACN, 0.1 % FA Formic Acid (Sigma, Cat. 34668)

5. NanoViper Acclaim PepMap100 Trap column (C18, particle size 3 µm, pore 100 Å, 75 µm i.d. × 20 mm length—Thermo Scientific, P/N 164535)

6. NanoViper Acclaim PepMap100 Capillary column (C18, particle size 2 µm, pore 100 Å, 75 µm i.d. × 250 mm length—Thermo Scientific, P/N 164536)

7. Butterfly Portfolio heater 20 × 4 cm (Phoenix S&T, Cat. PST-BHP-20)

2.6 Data Analysis

1. MaxQuant quantitative proteomics software package and Perseus data processing tool (www.maxquant.org)

3   Methods

3.1   Cell Culture

1. Prepare SILAC heavy and light media according to manufacturer's protocol and cell line requirements. For every liter of medium, supplement with 10 % FBS, antibiotics, 2–4 mM glutamine, 4.5 g/L glucose, and phenol red. The SILAC kit from *Life Technologies* uses L-lysine for metabolic incorporation. Add and label one of the bottles with light L-lysine (Lys0) and the other with heavy (U-^{13}C$_6$) L-lysine (Lys6) isotopes.

2. Culture one 100 × 20 mm plate of HT22 mouse to confluence in DMEM medium supplemented with 10 % FBS and antibiotics.

3. Release cells from plate by adding 1 mL of trypsin and incubate at 37 °C for 2 min. Resuspend cells in 10 mL of PBS, divide into two separate centrifuge tubes and spin for 3 min at 3000 × *g*. Remove PBS and wash the cells again to remove remaining medium.

4. Plate equal amounts of cells in SILAC heavy and light medium and culture them for at least eight generations to achieve high incorporation of the SILAC isotopes (*see* **Note 1**).

5. For the final passage, detach the cells using trypsin, resuspend in PBS and spin according to Section 3.1, **step 3**. Remove the supernatant and resuspend the cells in the respective medium. Count living cells using a hemacytometer according to manufacturer's instruction by taking a 25 μL aliquot and mix 9:1 with Trypan Blue. Seed a total of 400,000 cells per plate and let them attach and grow for 24–48 h (*see* **Note 2**).

6. Treat one set of plates (light or heavy) and use the other as the control set to be treated only with the vehicle solvent. A total of 1–2 mg of total protein will be required for the isolation of K63 ubiquitin targets (*see* **Note 3**).

7. Quickly wash the cells with PBS and add 1 mL of trypsin for 2 min to each plate to detach the cells. Resuspend cells in PBS, combine both cells suspensions, and centrifuge for 3 min at 3000 × *g* at 4 °C to obtain the cell pellet. Remove PBS and store at −20 °C until ready to use (*see* **Note 4**).

3.2   Immuno-precipitation and MS Preparation of K63 Ubiquitin Targets

1. From this point on, be mindful of keratin and other common contaminants during sample processing for mass spectrometry. Change gloves frequently, use clean and detergent-free labware and tubes. Preferentially use filter tips throughout the entire protocol. Avoid touching hair and skin, and avoid woolen clothes.

2. Resuspend cells in 750 μL of lysis buffer containing 50 nM K63-FLAG TUBE (*see* **Note 5**).

3. Sonicate samples twice for 15 s at amplitude 5 using the microtip. Rest the sample on ice for 1 min between sonication. Clear the cellular lysate from debris and precipitated material by centrifuging for 30 min at $15,000 \times g$ at 4 °C. Cell lysis protocol may vary depending on cell line.

4. Transfer supernatant to a fresh tube and determine protein concentration using the Bradford reagent.

5. Incubate the lysate for one extra hour at 4 °C under agitation to allow further binding of K63-TUBE to the K63 ubiquitin chains. If NP-40 buffer was used for extraction, dilute NP-40 to <0.1 %. When diluting, adjust K63-TUBE, chloroacetamide, and protease inhibitor concentration accordingly (*see* **Note 6**).

6. Conjugate the anti-FLAG antibody to the protein-G beads while the sample is incubating at 4 °C. Gently agitate the Protein G Dynabeads until the solution is homogeneous and transfer 50 μL of beads to a fresh tube. Do not vortex the beads. Place the tube in the magnetic rack and remove the storage buffer.

7. Wash the beads with 400 μL of PBS containing 0.05 % Tween-20 and place tube in the magnetic rack for 1 min. Remove the supernatant and then remove the tube from the magnetic rack.

8. For 1 mg of protein extract, saturate the 50 μL beads (30 mg/mL) with 10.5 μg of anti-FLAG antibody (7 μg of AB/mg of beads) dissolved in 300 μL of IP wash buffer. Rotate for 15 min at room temperature (RT), place sample in the magnetic rack and discard the supernatant. Add more 400 μL of wash buffer and let beads sit at 4 °C until ready to use.

9. Remove the wash buffer and incubate the beads with the protein extract for 2 h at 4 °C under agitation. Place the tubes in the magnetic rack and save the unbound fraction for troubleshooting.

10. Wash the beads with IP wash buffer for 5 min at RT. Transfer beads to a fresh tube and wash four times with plain PBS buffer without Tween. Detergent in the sample might imbue into the column and impact peptide binding and resolving power.

11. Elute the bound proteins with 50 μL 8 M Urea buffered in 50 mM Tris–HCl pH 8.0 for 30 min at 37 °C. Repeat elution, combine both eluates, and proceed with the preparation for digestion (*see* **Note 7**).

12. Reduce disulfide bonds by incubating the samples with freshly made 5 mM DTT for 45 min at 37 °C.

13. Alkylate the thiol groups by incubating the samples with 15 mM CAM for 30 min in the dark at RT. Agitate sample

every 10 min. Quench the remaining CAM with 10 mM DTT incubation for 15 min at 37 °C.

14. Dilute urea to <1 M with 50 mM Tris–HCl pH 8.0. Mix well and add 3 μg of trypsin. Vortex the samples and incubate overnight at 37 °C under agitation.

15. Stop the digestion reaction by adding formic acid to 1 % final concentration. Protocol can be safely paused at this point and samples stored at −80 °C.

16. Concentrate samples by speedvac to 10 to 20 μL at RT. This step should take around 3.5 h to complete. Do not increase the temperature, because it can promote peptide degradation. As an alternative, lyophilization can be performed to reduce sample volume.

3.3 Sample Clean Up Using C18 Microchromatography Tips

1. For sample clean up, all centrifugation steps should be performed at $2500 \times g$ for 1 min at RT. Buffers are prepared as per the manufacurer's instructions.

2. Resuspend samples in 150 μL of C18 Equilibration Buffer.

3. Use 2-mL centrifuge tubes to collect flow-through and add the centrifuge adaptor provided in the kit to each tube. Activate the C18 clean up tip with 150 μL of C18 Activation Buffer and centrifuge. Remove the flow-through and repeat this step.

4. Wash C18 tip twice by centrifugation with 150 μL of C18 Wash Buffer.

5. Change C18 tip to a new tube, load the sample and centrifuge. Collect the flow-through and load it again. Save the flow-through for troubleshooting.

6. Wash C18 tip twice with 150 μL of C18 Wash Buffer.

7. Move C18 tip to a new tube once again and elute with 150 μL of C18 Elution Buffer. Collect the flow-through in a clean tube, and elute it again (*see* **Note 8**).

8. Combine both eluates and dry the samples by speedvac to eliminate ACN.

9. Resuspend sample thoroughly in 10 μL of equilibration buffer, vortex well and centrifuge.

10. Transfer sample to an LC vial with pre-slit cap and visually inspect for bubbles at the bottom of the vial. Place the vials into a 4 °C cooled autosampler or freeze it at −80 °C until ready to use.

3.4 LC-MS/MS

1. Pre-heat the C18 analytical column at 55 °C and acclimatize the mass spectrometer in FT mode before loading the samples. Run a blank or an alternative method to equilibrate the entire system (*see* **Note 9**).

2. Load 5 μL of the reconstituted sample onto the C18 trap column at a flow rate of 2 μL/min. After washing the samples in the trap column, perform peptide separation in the C18 analytical column.

3. Run the following gradient for 150 min at 250 nL/min:

Time (min)	%B
0	2
2	7
110	28
126	40
130	60
136	98
146	98
150	2

4. Eluted peptides are in-line injected and analyzed in a LTQ Orbitrap Velos equipped with an Ion Max Nanosource.

5. Design a double-play mass spectrometry method with the following settings:

Scan settings	Scan event 1	Scan event 2
Mass Analyzer	FTMS	Ion Trap
Mass Range	Normal	Normal
Resolution	60,000	–
Scan type	Full	Full
Polarity	Positive	Positive
Data type	Profile	Centroid
Scan Ranges (m/z)	400–1500	

AGC settings	FT	IT
Full MS	1E6	3E4
Max ion time (ms)	500	10
Microscans	1	1

Isolation and activation settings	
Isolation width	2 m/z
Activation type	CID
Collision energy (%)	35
Activation time (ms)	10

6. Dynamic Exclusion is set to one repeat count, 45 s duration, and 90 s exclusion duration.

7. Activate 'predict ion injection time'. Enable 'charge state screening', 'monoisotopic precursor selection', and reject 'unassigned' and 'singly-charged ions'.

8. Minimum signal threshold should be set to 2,000 counts and the top 20 most intense ions should be selected for data-dependent analysis.

3.5 Protein Identification and Quantification

1. Analyze the raw files using MaxQuant suite for quantitative proteomics.

2. Before running the search engine, add and configure the fasta file on the sequence tab in the Andromeda database. We recommend Uniprot fasta files.

3. Select the raw files and create an 'experimental design' file. If technical replicates were run, combine them by providing the same name in the Experiment column inside the "experimental design template" file. Do not combine biological replicates.

4. Load your experimental design template into MaxQuant and select the parameters below for SILAC-labeled samples (*see* **Note 10**). If other isotopic/isobaric labels have been used, provide the proper configuration for the heavy and light label (*see* **Note 11**).

MaxQuant parameters	
Variable modifications	Oxidation (M), Acetyl (N-term), GG (K)
Fixed modification	Carbamidomethylation (C)
Multiplicity	2
Heavy labels	Lys6
Max labeled amino acids	3
Enzyme	Trypsin/P
Missed cleavages	2

(continued)

(continued)

MaxQuant parameters	
Max modification per peptide	5
First search ppm	20
Main search ppm	6
Decoy	Reverse
Peptide, protein, and site FDR	0.1
Minimum peptide length	7
Minimum unique + razor peptide	1
Re-quantify	yes

5. MaxQuant generates several tab-delimited output files. Protein identification and quantification results are located in the protein_groups.txt file. Peptide information is contained in the peptides.txt and evidence.txt files. The tables.pdf and parameters.txt files describe the output files' content and the parameters used for the search, respectively.

6. For quantification purposes use Heavy/Light ratios provided as expression values (*see* **Note 12**).

7. Remove reverse and contaminant hits from your file. Set the minimum ratio threshold for the expression value changes, and select the K63 targets that are varying significantly in response to the treatment (*see* **Note 13**).

4   Notes

1. Cells with Heavy and Light amino acids can be frozen and preserved in liquid nitrogen for later use to accelerate the label incorporation step.

2. The number of plates depends on the cell line used. For HT22, we use five plates for each SILAC condition to explore the role of oxidative stress in K63 ubiquitin dynamics. Leaving cells growing for more than 2 days will increase the basal level of K63 ubiquitin conjugates. Keep this in mind when selecting the time for treatment and the controls to be used. The amount of plates, cells, and protein should be determined empirically according to the cell line used and treatment performed.

3. A replicate with swapped SILAC labels should be performed to identify experimental biases.

4. Alternatively, remove medium, quickly rinse the plate with PBS, and add lysis buffer containing 0.5 % NP-40. Scrape the cells off

the plates using disposable cell lifters. Collect the cell suspension and proceed with sonication.

5. The concentration of K63-TUBE beads should also be determined empirically. Increased concentration of K63-TUBE might improve protein yield at the cost of linkage selectivity. Selectivity and efficiency of the pull-down should be tested by western blotting using the anti-K63 ubiquitin antibody (Millipore, Cat. 05-1308) and the anti-K48 ubiquitin (Millipore, Cat. 05-1307). The K63-TUBE should prevent deubiquitination [33], but keep chloroacetamide and EDTA in the buffer to further inhibit cysteine-and metallo-DUBs.

6. Save a 5 % aliquot of the cell lysate (50–100 μg of protein) and test for isotope incorporation. Follow the protocol from **step 12** of Section 3.2 and analyze 1 μg of sample by LC-MS/MS. To estimate incorporation rates, calculate the median of Heavy (H) and Light (L) raw intensities and apply the following formula: Incorporation = $[H/L]/(1 + [H/L])$.

7. Use freshly-made urea solution, and do not incubate samples in temperatures higher than 37 °C. At high temperatures, proteins can be carbamylated by urea, modifying the masses of peptides. Alternative elution methods include: (1) 100 % triflourethanol, (2) 0.2 M Glycine pH 2.5, and (3) 0.15 % Trifluoroacetic acid. Avoid elution with 3× FLAG peptide to prevent exogenous peptide contamination.

8. High concentrations of ACN are not recommended as they can release lipids and hydrophobic compounds from the C18 column back into the solution.

9. Always keep the MS instrument tuned and calibrated according to the manufacturer's guidelines. In addition, test the spray stability to achieve maximum performance.

10. Tryptic digest of a ubiquitinated protein generates a diGG-lysine signature peptide [34] that has to be included in mass spectrometry data search as variable modification.

11. For further information on MaxQuant protein quantification, consult Cox et al. [35].

12. Do not use the ratios of the raw intensities as the protein total ratio is calculated and normalized for the combination of peptides selected. Also, do not use normalized Ratios for immunoprecipitation samples as the normalization assumes that the majority of the proteins in the sample is not changing in abundance.

13. False discovery rates (FDR) can be estimated by comparing the distribution of the experimental expression rates to the rates from naked agarose beads experiments or anti-FLAG precipitation in the absence of K63-TUBE peptide.

Acknowledgements

This work was supported in part by the US National Science Foundation EAGER grant MCB-1355462 (CV), by the US National Institutes of Health K99 award ES025835 (GMS), and the Zegar Family Foundation Fund for Genomics Research at New York University (CV).

We thank R. Ratan (Burke Medical Research Institute) for providing the HT22 cells.

References

1. Komander D (2009) The emerging complexity of protein ubiquitination. Biochem Soc Trans 37:937–953. doi:10.1042/BST0370937

2. Tai HC, Schuman EM (2008) Ubiquitin, the proteasome and protein degradation in neuronal function and dysfunction. Nat Rev Neurosci 9:826–838. doi:10.1038/nrn2499

3. Dennissen FJ, Kholod N, van Leeuwen FW (2012) The ubiquitin proteasome system in neurodegenerative diseases: culprit, accomplice or victim? Prog Neurobiol 96:190–207. doi:10.1016/j.pneurobio.2012.01.003

4. Ciechanover A, Brundin P (2003) The ubiquitin proteasome system in neurodegenerative diseases: sometimes the chicken, sometimes the egg. Neuron 40:427–446

5. Bence NF, Sampat RM, Kopito RR (2001) Impairment of the ubiquitin-proteasome system by protein aggregation. Science 292:1552–1555. doi:10.1126/science.292.5521.1552

6. Li W et al (2008) Genome-wide and functional annotation of human E3 ubiquitin ligases identifies MULAN, a mitochondrial E3 that regulates the organelle's dynamics and signaling. PLoS One 3:e1487. doi:10.1371/journal.pone.0001487

7. van Wijk SJ, Timmers HT (2010) The family of ubiquitin-conjugating enzymes (E2s): deciding between life and death of proteins. FASEB J 24:981–993. doi:10.1096/fj.09-136259

8. Olzmann JA et al (2007) Parkin-mediated K63-linked polyubiquitination targets misfolded DJ-1 to aggresomes via binding to HDAC6. J Cell Biol 178:1025–1038. doi:10.1083/jcb.200611128

9. Tan Z et al (2007) Mutant ubiquitin found in Alzheimer's disease causes neuritic beading of mitochondria in association with neuronal degeneration. Cell Death Differ 14:1721–1732. doi:10.1038/sj.cdd.4402180

10. Kishino T, Lalande M, Wagstaff J (1997) UBE3A/E6-AP mutations cause Angelman syndrome. Nat Genet 15:70–73. doi:10.1038/ng0197-70

11. Nascimento RM, Otto PA, de Brouwer AP, Vianna-Morgante AM (2006) UBE2A, which encodes a ubiquitin-conjugating enzyme, is mutated in a novel X-linked mental retardation syndrome. Am J Hum Genet 79:549–555. doi:10.1086/507047

12. Yi JJ, Ehlers MD (2007) Emerging roles for ubiquitin and protein degradation in neuronal function. Pharmacol Rev 59:14–39. doi:10.1124/pr.59.1.4, 59/1/14 [pii]

13. Hallengren J, Chen PC, Wilson SM (2013) Neuronal ubiquitin homeostasis. Cell Biochem Biophys 67:67–73. doi:10.1007/s12013-013-9634-4

14. Hershko A, Ciechanover A (1998) The ubiquitin system. Annu Rev Biochem 67:425–479. doi:10.1146/annurev.biochem.67.1.425

15. Kulathu Y, Komander D (2012) Atypical ubiquitylation - the unexplored world of polyubiquitin beyond Lys48 and Lys63 linkages. Nat Rev Mol Cell Biol 13:508–523. doi:10.1038/nrm3394

16. Matsumoto ML et al (2010) K11-linked polyubiquitination in cell cycle control revealed by a K11 linkage-specific antibody. Mol Cell 39:477–484. doi:10.1016/j.molcel.2010.07.001

17. Brown NG et al (2014) Mechanism of polyubiquitination by human anaphase-promoting complex: RING repurposing for ubiquitin chain assembly. Mol Cell 56:246–260. doi:10.1016/j.molcel.2014.09.009

18. Hamilton AM, Zito K (2013) Breaking it down: the ubiquitin proteasome system in neuronal morphogenesis. Neural Plast 2013:196848. doi:10.1155/2013/196848

19. Spence J, Sadis S, Haas AL, Finley D (1995) A ubiquitin mutant with specific defects in DNA repair and multiubiquitination. Mol Cell Biol 15:1265–1273

20. Peng J et al (2003) A proteomics approach to understanding protein ubiquitination. Nat Biotechnol 21:921–926. doi:10.1038/nbt849

21. Silva GM, Finley D, Vogel C (2015) K63 poly-ubiquitination is a new modulator of the oxidative stress response. Nat Struct Mol Biol 22:116–123. doi:10.1038/Nsmb.2955

22. Deng L et al (2000) Activation of the IκB kinase complex by TRAF6 requires a dimeric ubiquitin-conjugating enzyme complex and a unique polyubiquitin chain. Cell 103:351–361

23. Paine S et al (2009) Immunoreactivity to Lys63-linked polyubiquitin is a feature of neurodegeneration. Neurosci Lett 460:205–208. doi:10.1016/j.neulet.2009.05.074

24. Liu C et al (2007) Assembly of lysine 63-linked ubiquitin conjugates by phosphorylated alpha-synuclein implies Lewy body biogenesis. J Biol Chem 282:14558–14566. doi:10.1074/jbc.M700422200

25. Tan JM et al (2008) Lysine 63-linked ubiquitination promotes the formation and autophagic clearance of protein inclusions associated with neurodegenerative diseases. Hum Mol Genet 17:431–439. doi:10.1093/hmg/ddm320

26. Xu G, Paige JS, Jaffrey SR (2010) Global analysis of lysine ubiquitination by ubiquitin remnant immunoaffinity profiling. Nat Biotechnol 28:868–873. doi:10.1038/nbt.1654

27. Kim W et al (2011) Systematic and quantitative assessment of the ubiquitin-modified proteome. Mol Cell 44:325–340. doi:10.1016/j.molcel.2011.08.025

28. Udeshi ND et al (2013) Refined preparation and use of anti-diglycine remnant (K-epsilon-GG) antibody enables routine quantification of 10,000s of ubiquitination sites in single proteomics experiments. Mol Cell Proteomics 12:825–831. doi:10.1074/mcp.O112.027094

29. Ziv I et al (2011) A perturbed ubiquitin landscape distinguishes between ubiquitin in trafficking and in proteolysis. Mol Cell Proteomics 10:M111 009753. doi:10.1074/mcp.M111.009753

30. Kirkpatrick DS et al (2006) Quantitative analysis of in vitro ubiquitinated cyclin B1 reveals complex chain topology. Nat Cell Biol 8:700–710. doi:10.1038/ncb1436

31. Xu P et al (2009) Quantitative proteomics reveals the function of unconventional ubiquitin chains in proteasomal degradation. Cell 137:133–145. doi:10.1016/j.cell.2009.01.041

32. Dammer EB et al (2011) Polyubiquitin linkage profiles in three models of proteolytic stress suggest the etiology of Alzheimer disease. J Biol Chem 286:10457–10465. doi:10.1074/jbc.M110.149633

33. Hjerpe R et al (2009) Efficient protection and isolation of ubiquitylated proteins using tandem ubiquitin-binding entities. EMBO Rep 10:1250–1258. doi:10.1038/embor.2009.192, embor2009192 [pii]

34. Udeshi ND, Mertins P, Svinkina T, Carr SA (2013) Large-scale identification of ubiquitination sites by mass spectrometry. Nat Protoc 8:1950–1960. doi:10.1038/nprot.2013.120

35. Cox J et al (2009) A practical guide to the MaxQuant computational platform for SILAC-based quantitative proteomics. Nat Protoc 4:698–705. doi:10.1038/nprot.2009.36

Neuromethods (2016) 114: 127–141
DOI 10.1007/7657_2015_89
© Springer Science+Business Media New York 2015
Published online: 08 November 2015

Phosphorylation Site Profiling of NG108 Cells Using Quadrupole-Orbitrap Mass Spectrometry

Fang-Ke Huang, Guoan Zhang, and Thomas A. Neubert

Abstract

Reversible protein phosphorylation regulates a wide variety of physiological processes and pathogenesis in the nervous system. Highly regulated protein phosphorylation events can be both necessary and sufficient to mediate responses of excitable cells to extracellular signals. Abnormal patterns of phosphorylation contribute to neuronal pathologies such as neurodegenerative diseases. The characterization of a large number of phosphorylated proteins in the nervous system should elucidate molecular mechanisms underlying normal physiology and pathogenesis. Global phosphoproteomic analysis based on mass spectrometry (MS) is a powerful tool to identify phosphorylated proteins and locate phosphorylation sites. Here we present a protocol for phosphorylation site profiling of hybrid mouse neuroblastoma/rat glioma NG108 cells. We provide technical details on sample preparation, fractionation, phosphopeptide enrichment, liquid chromatography, mass spectrometry, database searching, and functional annotation of phosphorylated proteins.

Keywords: Mass spectrometry (MS), Phosphoproteomics, NG108, Functional annotation, Pathway analysis, Neurotrophin signaling pathway, Phosphorylation, Posttranslational modification (PTM), Titanium dioxide (TiO_2), Hydrophilic interaction chromatography (HILIC)

1 Introduction

Advances in genetics and genomics drive discoveries in neuroscience and yield insights into neuronal diversity and function, as well as disease. While there are fewer than 30,000 genes in the human genome, the proteome is much larger and more dynamic [1]. This is because: (1) most mammalian genes show alternative splicing of transcripts, leading to different isoforms; (2) proteins are often posttranslationally modified by proteolysis, phosphorylation, glycosylation, ubiquitination, etc. The combination of various isoforms and different posttranslational modifications (PTMs) creates a large pool of diverse protein species from the same gene. Therefore, study of the proteome provides a perfect complement to genetics and genomics in understanding nervous system function.

As many as 300 PTMs are known to occur physiologically on proteins, and provide essential mechanisms by which cells diversify protein function and dynamically coordinate their signaling networks [2]. Reversible protein phosphorylation is one of the most common PTMs for cell regulation and signaling. It is estimated that a third of eukaryotic proteins are phosphorylated, a result of carefully regulated protein kinase and phosphatase activities [2].

Classical biochemical assays to characterize protein phosphorylation include phosphoimaging based on the incorporation of a ^{32}P-radiolabel and phosphorylation-specific staining using Pro-Q Diamond. However, these methods do not provide information regarding the phosphorylation sites and are often tedious and low-throughput [3]. Global phosphoproteomic analysis based on MS is a powerful technique to identify phosphorylated proteins and locate phosphorylation sites in a high-throughput manner.

And while there is a large dynamic range in the types of phosphorylated species present, they are typically present at low abundances, presenting challenges for MS-based phosphorylation analysis [3]. Despite increasing scanning speeds of modern MS instruments, sample complexity reduction and phosphopeptide enrichment are necessary steps for global phosphoproteomic analysis [4]. The most common strategy to reduce sample complexity uses two-dimensional separations at the peptide level, where the second separation is predominantly nanoflow reversed-phase chromatography (RPLC) [5]. Hydrophilic interaction chromatography (HILIC) is a separation technique where retention increases with increasing polarity (hydrophilicity) of peptides and has been shown to provide the highest degree of orthogonality to RPLC among all commonly used peptide separation modes [6]. Titanium dioxide (TiO$_2$) chromatography and immobilized metal affinity chromatography (IMAC) are both efficient ways to enrich phosphopeptides before MS analysis [7, 8].

By combing global phosphoproteomic analysis with functional annotation and pathway analysis, a systems-level understanding of signaling events at the cellular level can be achieved.

In this chapter, we applied a protocol for phosphorylation site profiling of hybrid mouse neuroblastoma/rat glioma NG108 cells. We provide technical details on sample preparation, HILIC fractionation, phosphopeptide enrichment, MS analysis using a hybrid quadrupole-Orbitrap mass spectrometer, and database searching. We also demonstrate how to perform functional annotation and pathway analysis of phosphorylated proteins.

2   Materials

2.1   Protein Extraction

1. NG108 cells.

2. Costar 6-well clear TC-treated multiple-well plates # 3516 (Corning Incorporated, Corning, NY).

3. Phosphate buffered saline (PBS) (http://cshprotocols.cshlp.org/content/2006/1/pdb.rec8247).

4. Cell lysis buffer: 150 mM sodium chloride (NaCl), 20 mM tris (hydroxymethyl)aminomethane-hydrogen chloride buffer (Tris–HCl, pH 8.0), 1 % Triton X-100, 0.2 mM ethylenediaminetetraacetic acid (EDTA), 2 mM tyrosine phosphatase inhibitor sodium orthovanadate (Na_3VO_4), 2 mM serine/threonine phosphatase inhibitor sodium fluoride (NaF), and protease inhibitor cocktail tablets (cOmplete Mini, Roche, Mannheim, Germany). Fresh phosphatase and protease inhibitors should be added just before use.

5. Cell Lifter # 3008 (Corning Incorporated, Corning, NY).

6. 1.5 mL graduated microcentrifuge tubes # 05-408-129 (Fisher Scientific, Waltham, MA).

7. Quick Start™ Bradford 1× Dye Reagent # 500-0205 (Bio-Rad, Hercules, CA).

8. KIMTECH Science Kimwipes # 34155 (Kimberly-Clark Professional, Roswell, GA).

9. Acetone HPLC Grade #A949-1 (Fisher Scientific, Waltham, MA).

2.2 In-Solution Digestion

1. Urea buffer: 9 M Urea, 5 mM NaF, 1 mM Na_3VO_4, 50 mM Tris–HCl 8.0, two tablets of protease inhibitors cocktail (cOmplete Mini, Roche) per 10 mL lysis buffer. Make 0.5 mL aliquots and store at −80 °C.

2. Ultrasonic water bath FS36 (Fisher Scientific, Waltham, MA).

3. Dithiothreitol (DTT) # 161-0611 (Bio-Rad, Hercules, CA).

4. ThermoMixer 21516-170 (Eppendorf, Hauppauge, NY).

5. Iodoacetamide #I1149 (Sigma, St. Louis, MO).

6. Ammonium bicarbonate # 09830 (Sigma, St. Louis, MO).

7. Trypsin Gold, mass spectrometry grade # V5280 (Promega, Madison, WI).

2.3 Desalting

1. Trifluoroacetic acid (TFA) # 28904 (Thermo Fisher Scientific, Waltham, MA).

2. Whatman pH indicator paper # 2613991 (Thermo Fisher Scientific, Waltham, MA).

3. Sep-Pak tC18 1 cc Vac Cartridge (100 mg Sorbent) # WAT03820 (Waters, Milford, MA).

4. HPLC grade Acetonitrile (ACN) # A955-4 (Fisher Chemical, Waltham, MA).

5. HPLC grade Water # W5-4 (Fisher Chemical, Waltham, MA).

6. Savant SpeedVac SC110 Vacuum Concentrator (Thermo Fisher Scientific, Waltham, MA).

7. Washing buffer: 2 % ACN and 0.1 % TFA.

8. Elution buffer: 80 % ACN and 0.1 % TFA.

2.4 HILIC Chromatography

1. Agilent 1100 Series HPLC System (Agilent Technologies, Santa Clara, CA).

2. GILSON FC 203 B Fraction Collector (Gilson, Middleton, WI).

3. Isopropanol # A451-4 (Fisher Chemical, Waltham, MA).

4. Mobile phase A: 0.1 % TFA.

5. Mobile phase B: 0.1 % TFA, 99.9 % ACN.

6. Mobile phase C: 100 % Isopropanol.

7. Mobile phase D: 75 % ACN.

8. BSA peptides (5 μg/μL, in 80 % mobile phase B and 20 % mobile phase A).

9. TSKgel Amide-80 column packed with 5 μm spherical silica particles (Tosoh, Grove City, OH).

2.5 TiO$_2$ Phosphopeptide Enrichment and Cleanup

1. Solid Phase Extraction Disk with octyl group (C8) # 2214 (3 M Empore, St. Paul, MN).

2. Titansphere TiO$_2$ 10 μm # 5020-75010 (GL Science, Japan).

3. 0.5 mL Microcentrifuge tubes # 05-408-120 (Fisher Scientific, Waltham, MA).

4. 10 μL Low-Retention pipet tips # 02-717-134 (Fisher Scientific, Waltham, MA).

5. Ammonium hydroxide solution (NH$_3$·H$_2$O) # 338818 (Sigma, St. Louis, MO).

6. Conditioning buffer A: 0.1 % TFA and 99.9 % ACN.

7. Conditioning buffer B: 3 % TFA and 60 % ACN.

8. Washing buffer A: 3 % TFA and 30 % ACN.

9. Washing buffer B: 0.1 % TFA and 80 % ACN.

10. Elution buffer A: 3 % NH$_3$·H$_2$O.

11. Elution buffer B: 1.5 % NH$_3$·H$_2$O and 50 % ACN.

2.6 LC-MS/MS Analysis

1. EASY-nLC 1000 Liquid Chromatograph #LC120 (Thermo Fisher Scientific, Waltham, MA).

2. Q Exactive™ Hybrid Quadrupole-Orbitrap Mass Spectrometer (Thermo Fisher Scientific, Waltham, MA).

3. PeproSil-Pur C18-AQ 3 μm resin # r15.aq. (Dr. Maisch GmbH, Ammerbuch-Entringen, Germany).

4. PicoFrit™ SELF/P column # PF360-75-10-N-5 (New Objective, Woburn, MA).

5. 11 mm Plastic Crimp/Snap Top Autosampler Vials # C4011-13 (Thermo Fisher Scientific, Waltham, MA).

6. Mobile phase A: 0.1 % TFA.

7. Mobile phase B: 0.1 % TFA, 99.9 % ACN.

2.7 Bioinformatics

1. MaxQuant software (Version 1.2.7.0, Max Planck Institute of Biochemistry).

2. A personal computer (PC) with 48 GB of RAM, Intel Xeon Processor E5645 (12 M Cache, 2.40 GHz, 5.86 GT/s Intel QPI), and 1TB SSD Drive.

3 Methods

3.1 Protein Extraction and Precipitation

1. Wash cell culture plate (6-well format) with ice-cold PBS (2 mL/well) two times.

2. After the second wash, place plates on ice and tilt the surface from horizontal by an angle (*see* **Note 1**).

3. Remove residual PBS and add 50 μL lysis buffer to each well (*see* **Note 2**).

4. Use cell lifter to scrape cells to the side of the well where lysis buffer resides.

5. Collect cell lysates into 1.5 mL microcentrifuge tubes.

6. Estimate the volume of the cell lysate and adjust the concentration of Triton X-100 to 1 % (v/v) by adding 20 % Triton X-100 to the cell lysate (*see* **Note 3**).

7. Incubate the cell lysate on ice for 15 min.

8. Centrifuge at $15,000 \times g$ for 10 min at 4 °C to remove cell debris.

9. Transfer the supernatant fluid to a new 1.5 mL microcentrifuge tube.

10. Measure protein concentration using Bradford protein assay.

11. Aliquot one volume of supernatant fluid (protein concentration >3 mg/mL) into each 1.5 mL microcentrifuge tube (*see* **Note 4**).

12. Add four volumes of cold acetone (−20 °C) to each tube to precipitate proteins.

13. Keep the mixture at −20 °C overnight. You will see protein pellet the next morning.

14. Spin down the pellet at $3000 \times g$ for 2 min (*see* **Note 5**).

15. Reverse tubes and dry on top of a Kimwipe.

16. Protein pellet can be kept at −80 °C.

3.2 In-Solution Digestion

1. Thaw out aliquots of Urea buffer at room temperature (*see* **Note 6**).

2. Add appropriate amount of Urea buffer to protein pellet (120 μL Urea buffer per mg protein).

3. Brief sonication in ultrasonic water bath can help to dissolve the protein pellet.

4. Add 1 M DTT to a final concentration of 5 mM. Incubate at 50 °C for 20 min with mixing at 700 RPM in ThermoMixer (*see* **Note 7**).

5. Add 1 M freshly made iodoacetamide to a final concentration of 14 mM. Incubate at room temperature for 20 min in the dark.

6. Add additional 1 M DTT to increase the DTT concentration an additional 5 mM (total concentration 10 mM) and incubate at room temperature for 20 min.

7. Dilute the solution with 25 mM ammonium bicarbonate to make urea final concentration to 1.6 M.

8. Add 1 M $CaCl_2$ to a final concentration of 1 mM.

9. Add trypsin with an enzyme:substrate ratio of 1:200.

10. Incubate with mixing at 300 RPM at 37 °C overnight in ThermoMixer.

3.3 Desalting

1. Add 10 % TFA to a final concentration of 0.4 %. Make sure the pH of the solution is below 2 using pH indicator paper.

2. Centrifuge at $20,000 \times g$ for 5 min at room temperature and discard the pellet.

3. Add 1 mL of 100 % ACN to activate the Sep-Pak tC18 1 cc Vac Cartridge.

4. Allow the solution to drain by gravity (*see* **Note 8**).

5. Equilibrate the cartridge by passing through 1 mL washing buffer.

6. Add peptides onto the cartridge and allow them to bind to the sorbent until all solution passes through.

7. Wash the cartridge with 2 mL washing buffer.

8. Elute with 1 mL of 80 % ACN/0.1 % TFA.

9. Dry the eluate by vacuum centrifugation in a SpeedVac with the heat off.

10. A white fluffy powder should remain after drying.

3.4 HILIC Chromatography

1. Resuspend dried peptides in 100 μL 20 % mobile phase A and 80 % mobile phase B.

2. Sonicate in an ultrasonic water bath for 5 min to help dissolve the peptides.

3. Centrifuge the sample at $20,000 \times g$ for 5 min and keep the supernatant fluid.

4. Inject 100 μL 20 % mobile phase A and 80 % mobile phase B into the sample loop to equilibrate it (*see* **Note 9**).

Table 1
Gradient used for separation of tryptic peptides by HILIC chromatography

Time (min)	Duration (min)	Buffer B (%)	Flow (mL/min)
0	N/A	80	0.3
5	5	80	0.3
40	35	60	0.3
55	15	0	0.3
65	5	0	0.3

5. Load the entire peptide sample into the sample loop and start the gradient (Table 1) for HILIC fractionation.

6. Collect 18 four-min fractions from min 0 to 64.

7. Dry down each fraction by vacuum centrifugation in the SpeedVac with the heat off.

3.5   TiO$_2$ Phosphopeptide Enrichment and Cleanup

1. Dissolve peptides in 20 µL conditioning buffer B in the tube [9].

2. Sonicate in an ultrasonic water bath for 5 min.

3. Centrifuge at 20,000 × g for 5 min and keep the supernatant fluid.

4. Make C8 StageTips (*See* **Note 10**) [10].

5. Place the StageTips on top of 10 µL tips. Insert the assembly into an empty 0.5 mL microcentrifuge tube.

6. Wet StageTips with 20 µL conditioning buffer A by spinning at 750 × g for 2 min or until all liquid passes through.

7. Equilibrate the StageTips with 20 µL conditioning buffer B.

8. Place the StageTips on top of new 10 µL tips. Insert the assembly into new 0.5 mL microcentrifuge tubes.

9. Load samples into StageTips and centrifuge at 750 × g for 3 min or until all liquid passes through and collect the flow through.

10. Make new StageTips (*see* **Note 10**). Add ~ 1 µL (3 mm length) TiO$_2$ beads in conditioning buffer A on top of the C8 plug.

11. Wash it with 20 µL conditioning buffer A at 320 × g until all liquid passes through. Record the time for all liquid to pass through. A rate of 3–5 µL/min is recommended. Adjust the relative centrifugal force if necessary.

12. Equilibrate StageTips with 20 µl of conditioning buffer B by spinning at 320 × g until all liquid passes through.

13. Place StageTips on another 10 µL tip. Load peptide samples into the StageTips. Centrifuge at 320 × g until all liquid passes through.

14. Save flow-through samples and store at −80 °C (*see* **Note 11**).

15. Wash with 20 µL conditioning buffer B at 320 × g until all liquid passes through.

16. Wash with 20 µL washing buffer A at 320 × g until all liquid passes through.

17. Wash with 20 µL washing buffer B at 320 × g until all liquid passes through.

18. Place StageTips on new 10 µL tips and transfer the tips to new 0.5 mL tubes.

19. Elute with 20 µL elution buffer A at 320 × g until all liquid passes through.

20. Elute with 20 µL elution buffer B at 320 × g until all liquid passes through.

21. Dry down the combined eluent by vacuum centrifugation in a Speedvac with heat off (*see* **Note 12**).

3.6 LC-MS/MS Analysis

1. Reconstitute dried samples in 10 µL 4 % FA and 2 % ACN.

2. Centrifuge samples at 20,000 × g for 5 min.

3. Load the supernatant fluid into autosampler vials (*see* **Note 13**).

4. A nanoflow EASY-nLC 1000 HPLC instrument was coupled directly to a Q Exactive mass spectrometer with a nanoelectrospray ion source [11]. Chromatography columns were packed in-house with PeproSil-Pur C18-AQ 3 µm resin.

5. Enriched phosphopeptides were loaded onto a C_{18}-reversed phase column (15 cm long, 75 µm inner diameter) and separated with a linear gradient listed in Table 2 over 140 min.

Table 2
Gradient used for separation of phosphopeptides by C_{18}-reversed phase column

Time (min)	Duration (min)	Buffer B (%)	Flow (nL/min)
0	N/A	0	250
1	1	3	250
121	120	30	250
131	10	60	250
140	9	90	250

6. The Q Exactive was operated in data-dependent acquisition mode. The method dynamically chose the top ten most abundant precursor ions from the survey scan (300–1650 Th) for HCD fragmentation. Survey scans were acquired at a resolution of 70,000 at m/z 200 and a target value of 3×10^6 ions with a maximum fill time of 20 ms. MS/MS scans were set to a resolution of 17,500 at m/z 200 using 1×10^6 ions as the target value and a maximum fill time of 60 ms. Spectra were acquired with a normalized collision energy of 27 eV and a dynamic exclusion duration of 45 s.

3.7 Bioinformatics

3.7.1 Protein and Phosphorylation Site Identification

1. The data analysis was performed with MaxQuant software (Version 1.2.7.0, Max Planck Institute of Biochemistry) supported by the Andromeda search engine [12, 13]. MS/MS data were used to search a UniProt Mouse/Rat FASTA database. Mass tolerance for searches was set to maximum 7 ppm for peptide masses and 20 ppm for HCD fragment ion masses. Searches were performed with carbamidomethylation as a fixed modification, and protein N-terminal acetylation, methionine oxidation, and STY phosphorylation as variable modifications. A maximum of two missed cleavage was allowed while requiring strict trypsin specificity. Peptides with a minimum sequence length of seven were considered for further data analysis. Peptides and proteins were identified with a false discovery rate (FDR) of 1 %.

2. The Evidence file is one of the output files generated by MaxQuant. The Evidence file combines all the available information about the identified peptides. When the column "Reverse" or "Contaminant" is marked with "+", this particular "evidence" should be removed before further data analysis. After removing "Reverse" or "Contaminant," the evidence file in our dataset contained 67,075 entries, which corresponds to 37,769 peptides, 21,355 phosphorylation sites, and 7586 protein groups (Table 3).

3. The Phospho (STY) Sites file contains information on the identified phosphorylation sites in the processed raw files. When the column "Reverse" or "Contaminant" is marked with "+", these phosphorylation sites should be removed before further data analysis. Localization probability is the

Table 3
A summary of number of evidence spectra, peptides, phosphorylated sites, and protein groups identified in the dataset

Evidence	Peptides	Phosphorylated sites	Protein groups
67,075	37,769	21,355	7586

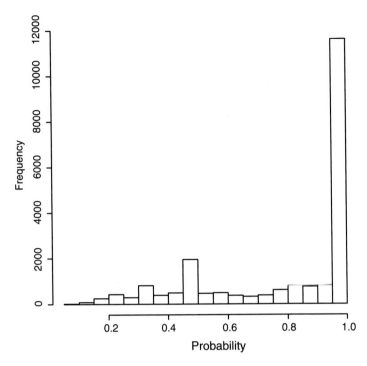

Fig. 1 Distribution of phosphorylation site localization probabilities. MaxQuant software was used to perform phosphorylation site localization. Phosphorylation site localization probabilities in the Phospho (STY) Sites file were plotted in the histogram. The majority of phosphorylation sites have a localization probability of around 100 %

confidence level of assignment of the phosphorylation sites by MaxQuant. When we plot the distribution of localization probabilities of identified phosphorylation sites as shown in Fig. 1, the majority of phosphorylation sites have a localization probability of around 100 %.

4. Figure 2 showed the number of phospho- and nonphospho-peptides in each fraction after TiO_2 enrichment. Phosphopeptides dominated in later fractions (9–18) while more nonphosphopeptides were identified in early fractions (4–8).

3.7.2 Functional Annotation and Pathway Analysis Using DAVID Bioinformatics Resources

1. Based on the Phospho(STY)Sites table, we created a gene list for phosphorylated proteins (*see* **Note 14**).

2. Submit the gene list to DAVID at http://david.abcc.ncifcrf.gov and click on "Start Analysis" on the header [14]. Use the gene list manager panel and perform the following steps:

    (a) Copy and paste the list of gene IDs into box A.

    (b) Select the "OFFICIAL_GENE_SYMBOL" as gene identifier type.

    (c) Indicate the list to be submitted as a gene list (*see* **Note 15**).

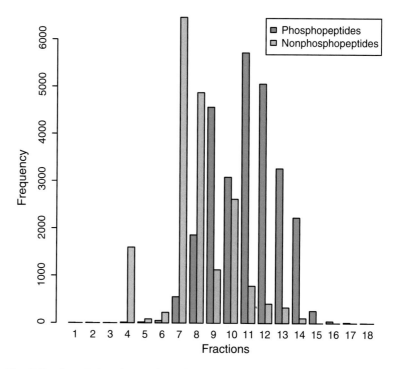

Fig. 2 Number of phospho- and nonphosphopeptides in each fraction after TiO_2 enrichment. Eighteen fractions were collected after HILIC chromatography and phosphopeptides were subsequently enriched from each fraction by TiO_2 Stage-Tips. Phosphopeptides dominated in later fractions (9–18) while more nonphosphopeptides were identified in early fractions (4–8)

(d) Click on the "Submit List" button.

(e) Under Gene List Manager, select one or more species to limit annotations and click on "Select Species" (*see* **Note 16**).

(f) Under Population Manager, select the same species as background and click on "Use."

3. Analysis Wizard page will indicate whether the gene list is submitted successfully.

4. If so, click on the "Functional Annotation Tool" link, and the Annotation Summary Results page will appear.

5. Click on "Functional Annotation Chart." You should expect the term "phosphoprotein" to appear on top of the chart with the highest count and lowest P-Value (Fig. 3a) [15, 16].

6. Expand "Pathways" box and select "Chart" behind KEGG_-PATHWAY. A list of pathways will appear (Fig. 3b).

7. Click on "ErbB signaling pathway" to view genes in the ErbB signaling pathway picture. Genes with blinking red stars are in the gene list submitted to the DAVID server (Fig. 4a).

Functional Annotation Chart

Current Gene List: List_1
Current Background: Mus musculus
2567 DAVID IDs

a **1704 chart records**

Category		Term	RT	Genes	Count	%	P-Value	Benjamini
SP_PIR_KEYWORDS	phosphoprotein		**RT**	▬▬▬▬▬▬▬▬	2044	79.6	0.0E0	0.0E0
SP_PIR_KEYWORDS	acetylation		**RT**	▬▬▬	769	30.0	9.1E-157	2.1E-154
SP_PIR_KEYWORDS	nucleus		**RT**	▬▬▬	900	35.1	6.0E-87	9.2E-85
SP_PIR_KEYWORDS	cytoplasm		**RT**	▬▬▬	759	29.6	1.0E-82	1.2E-80
GOTERM_CC_FAT	non-membrane-bounded organelle		**RT**	▬▬	485	18.9	1.2E-76	6.5E-74

b **54 chart records**

Category		Term	RT	Genes	Count	%	P-Value	Benjamini
KEGG_PATHWAY	Insulin signaling pathway		**RT**	▪	49	1.9	1.7E-12	2.6E-10
KEGG_PATHWAY	Spliceosome		**RT**	▪	45	1.8	6.7E-12	5.3E-10
KEGG_PATHWAY	ErbB signaling pathway		**RT**	▪	36	1.4	1.7E-11	8.9E-10
KEGG_PATHWAY	Neurotrophin signaling pathway		**RT**	▪	45	1.8	4.2E-11	1.6E-9
KEGG_PATHWAY	Chronic myeloid leukemia		**RT**	▪	30	1.2	4.1E-9	1.3E-7

Fig. 3 Layout of a DAVID annotation chart and a KEGG_PATHWAY category. A gene list was created from the Phospho(STY)Sites table and submitted to DAVID for functional annotation. (**a**) "Phosphoprotein" was the top enriched functional annotation term associated with the genes. (**b**) ErbB signaling pathway and neurotrophin signaling pathway were among the top five pathways in KEGG_PATHWAY category

8. Click on "Neurotrophin signaling pathway" to view genes in the neurotrophin signaling pathway picture. Genes with blinking red stars are in the gene list submitted to the DAVID server (Fig. 4b).

4   Notes

1. Residual PBS can dilute the lysis buffer, resulting in inefficient cell lysis. Tilting the culture plate surface helps to remove residual PBS.

2. The choice of lysis buffer volume is a balancing act. Too little lysis buffer will not cover the entire surface area. Too much will dilute the protein concentration in the lysate. Diluted lysate is difficult to precipitate.

3. We recognize that residual PBS cannot be removed completely. Also, cells contribute to the volume. These two factors lead to a dilution of lysis buffer and decrease its lysis efficiency. We find that adjusting Triton X-100 to a final concentration of 1 % (v/v) increases protein extraction efficiency.

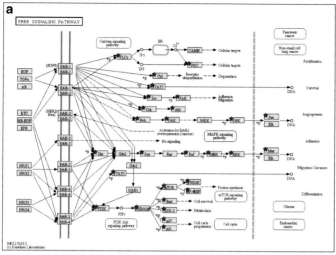

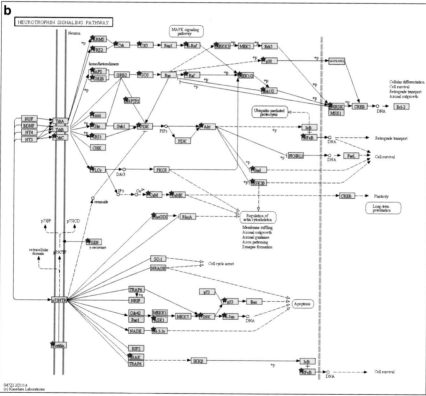

Fig. 4 Phosphoproteins identified in ErbB and neurotrophin signaling pathways. A detailed view of phosphoproteins covered in ErbB and Neurotrophin signaling pathways. Phosphoproteins identified in the ErbB signaling pathway (**a**) and neurotrophin signaling pathway (**b**) were marked with *red stars*. In the ErbB signaling pathway, 28 out of 36 known downstream effector proteins were identified. In the neurotrophin signaling pathway, 35 out of 68 known downstream effector proteins were identified

4. For each 1.5 mL microcentrifuge tube, the volume should not exceed 300 μL. Six hundred microgram lysate per tube has been tested and can be dissolved by urea buffer.

5. Higher speed and longer time will result in a more compact pellet, which is more difficult to dissolve later.

6. If the urea buffer aliquots are thawed on ice, insoluble particles will appear and cannot dissolve. We recommend thawing it at room temperature.

7. Do not incubate at a temperature higher than 60 °C or for too long, as urea-based carbamylation of lysines and protein N-termini can occur.

8. Do not let the sorbent dry. Dried sorbent will decrease binding efficiency.

9. Using sample buffer to equilibrate the sample loop will help the subsequent HILIC fractionation. We recommend equilibrating the system for at least 20 min before loading your sample.

10. Do not apply too much force when filling in the disks into the tip, otherwise it will require a higher speed and longer time for the liquid to pass through.

11. Keep the flow-through of the samples. They may be used to analyze the phosphopeptide enrichment efficiency or nonphosphopeptides.

12. Dry the combined eluent immediately after the elution. Peptides tend to be unstable in alkaline solution. Dried peptides can be stored at −80 °C.

13. It is important to load samples to the bottom of the vial and avoid generating bubbles.

14. Because DAVID does not allow lists of >3000 symbols to be uploaded, we chose phosphorylation sites with Andromeda score ≥131. We ended up with 2991 unique genes. DAVID recognized 2567 of them and converted them to DAVID IDs.

15. DAVID has an automatic procedure to "guess" the background as the global set of genes in the genome on the basis of the uploaded gene list. In our case, DAVID selected the *Mus musculus* genome as a population background. Users also have the freedom to choose their own "gene population background" for enrichment analysis in the population manager.

16. We chose *Mus musculus* as the species because 2567 proteins are annotated in *Mus musculus* compared to in Rattus norvegicus (2396).

Acknowledgments

We would like to acknowledge support from NIH grants P30NS050276 and S10RR027990 to T.A.N.

References

1. Wilson KE (2004) Functional genomics and proteomics: application in neurosciences. J Neurol Neurosurg Psychiatry 75:529–538

2. Witze E, Old W, Resing K, Ahn N (2007) Mapping protein post-translational modifications with mass spectrometry. Nat Methods 4

3. Blackburn K, Goshe MB (2009) Challenges and strategies for targeted phosphorylation site identification and quantification using mass spectrometry analysis. Brief Funct Genomic Proteomic 8:90–103

4. McNulty DE, Annan RS (2008) Hydrophilic interaction chromatography reduces the complexity of the phosphoproteome and improves global phosphopeptide isolation and detection. Mol Cell Proteomics 7:971–980

5. Di Palma S, Mohammed S, Heck AJR (2012) ZIC-cHILIC as a fractionation method for sensitive and powerful shotgun proteomics. Nat Protoc 7:2041–2055

6. Mcnulty DE, Annan RS (2009) Hydrophilic interaction chromatography for fractionation and enrichment of the phosphoproteome. Humana, Totowa, NJ

7. Villén J, Gygi SP (2008) The SCX/IMAC enrichment approach for global phosphorylation analysis by mass spectrometry. Nat Protoc 3:1630–1638

8. Thingholm TE, Jørgensen TJD, Jensen ON, Larsen MR (2006) Highly selective enrichment of phosphorylated peptides using titanium dioxide. Nat Protoc 1:1929–1935

9. Zhang G, Neubert TA (2011) Comparison of three quantitative phosphoproteomic strategies to study receptor tyrosine kinase signaling. J Proteome Res 10:5454–5462

10. Rappsilber J, Mann M, Ishihama Y (2007) Protocol for micro-purification, enrichment, prefractionation and storage of peptides for proteomics using StageTips. Nat Protoc 2:1896–1906

11. Michalski A, Damoc E, Hauschild J-P, Lange O, Wieghaus A, Makarov A, Nagaraj N, Cox J, Mann M, Horning S (2011) Mass spectrometry-based proteomics using Q Exactive, a high-performance benchtop quadrupole Orbitrap mass spectrometer. Mol Cell Proteomics 10:M111.011015

12. Cox J, Mann M (2008) MaxQuant enables high peptide identification rates, individualized ppb-range mass accuracies and proteome-wide protein quantification. Nat Biotechnol 26:1367–1372

13. Cox J, Neuhauser N, Michalski A, Scheltema R, Olsen JV, Mann M (2011) Andromeda: a peptide search engine integrated into the MaxQuant environment. J Proteome Res 10:1794–1805

14. Huang DW, Sherman BT, Lempicki R (2009) Systematic and integrative analysis of large gene lists using DAVID bioinformatics resources. Nat Protoc 4:44–57

15. Kanehisa M, Goto S, Sato Y, Kawashima M, Furumichi M, Tanabe M (2014) Data, information, knowledge and principle: back to metabolism in KEGG. Nucl. Acids Res. 42:D199–D205 16. Kanehisa M, Goto S (2000) KEGG: Kyoto Encyclopedia of Genes and Genomes. Nucl. Acids Res. 28:27–30

16. Kanehisa M, Goto S (2000) KEGG: Kyoto Encyclopedia of Genes and Genomes. Nucl. Acids Res. 28:27–30

Neuromethods (2016) 114: 143–154
DOI 10.1007/7657_2015_83
© Springer Science+Business Media New York 2016
Published online: 14 April 2016

Phosphoproteomics of Tyrosine Kinases in the Nervous System

Robert J. Chalkley and Ralph A. Bradshaw

Abstract

Protein phosphorylation represents one of the most prevalent posttranslational modifications (PTMs) in mammalian cellular processes and is most easily identified/measured by mass spectrometry or with specific recognition molecules, such as antibodies, usually in an array format. Phosphorylation of the phenolic group of tyrosine occurs to a much lesser extent than on the hydroxyl side chains of serine and threonine, but because of its principal involvement in the initiation of intracellular signaling events, its determination is of singular importance. Generally, it can be most easily identified in bottom-up proteomic analyses in which the germane peptides bearing the pTyr residue(s) have been first concentrated with anti-pTyr sera. Levels of pTyr in resting cells are quite low, so it is most readily measured in samples following a bolus stimulation, such as by a growth factor like nerve growth factor (NGF), after a period of 1–2 min. The identification of the modified residue and its location in the sequence of the peptide can be determined following collision induced dissociation of the precursor peptide in the mass spectrometer and the observation of fragment ions that frame the modification site. Quantification can be accomplished using either isotope or non-isotope dependent methodology. Targeted proteomics of individual sites, once they have been ascertained, can allow for detailed analyses of the involvement of individual sites in the biological processes under investigation.

Keywords: Mass spectrometry, Signal transduction, Neurotrophic factors, Receptor tyrosine kinases, Soluble tyrosine kinases, Protein tyrosine phosphatases, SH2 and PTB domains, Bottom-up proteomics, Immunoprecipitation, Database search engines

1 Introduction

Posttranslational modifications (PTMs) are the principal mechanism for perpetrating signals that originate in the extracellular environment and are ultimately manifested in modulations in gene expression leading to phenotypic change [1]. While such changes require new protein synthesis, short term stimuli are dependent on the activation of pre-existing proteins either by covalent modification, by protein-protein interactions, or both. There are a large number of PTMs that are involved in signal transduction, some of which are transient (reversible) and some of which are basically permanent and require the turnover of the protein itself to eliminate the effect imparted by the modification.

At our present level of understanding, phosphorylation (Ser, Thr, Tyr), GlcNAcylation (Ser, Thr), *N*-acetylation (Lys), and ubiquitination (Lys) are the most important of the reversible modifications, although it is clear that the full range of their effects and the complete scope of their involvement in cellular processes has not been elucidated. Historically, phosphorylation was probably the earliest reversible PTM to be identified [2] and was considered for a number of years as being primarily important as a mechanism for controlling enzyme activity through the alteration of specific Ser or Thr residues. The discovery of tyrosine phosphorylation in viral [3] and then signaling systems [4], rapidly expanded the appreciation of the important role it played in cell signaling. The advent of large-scale (often called shotgun) proteomic experiments, made possible by the introduction of mass spectrometric ionization methodologies suitable for fragile biomolecules (ESI and MALDI) [5, 6], revealed the enormous extent to which phosphorylation occurs, even in unstimulated cells [7]. It is now known that the phosphoproteome of any given cell type may run to tens of thousands of sites and may involve most of the expressed proteome of a cell [7–9]. Although the most detailed studies have been in transformed cell lines, the phosphoproteome of nervous tissue is anticipated to be as broad, although there are no detailed demonstrations of this yet [10]. These large-scale mass spectrometric analyses were also responsible for establishing the wide spread distribution of the other modifications [11].

Large-scale phosphoproteomic experiments are generally based either on static (unstimulated) or dynamic (stimulated) samples, and this is well illustrated by experiments conducted with nervous tissue. An example of the former is analyses of synaptosomal preparations [12, 13]. These measurements show the basal levels of phosphorylation. However, samples from different brain regions can be compared, or the effects of drugs or other treatments on the modification patterns compared to untreated controls can provide activity-related data [13]. In contrast, the acute responses of target cells to external signaling molecules, usually inducing their effects through plasma membrane-bound receptors, can be measured kinetically and thus related to intracellular signaling (and ultimately transcriptomic and phenotypic) responses. The profile of the phosphoproteome changes dramatically during acute stimulation because of the transient nature of the phosphorylation modification and the abundance of protein phosphatases. Thus, experiments usually focus on significant changes relative to controls (both increases and decreases), which substantially reduces the number of proteins identified as regulated. This is particularly true with respect to the number and quantity of the phosphotyrosine peptides detected. In samples from unstimulated tissues the phosphoproteome is dominated by pSer and pThr sites, whereas pTyr modifications are generally about 1 % or less of the total identifications [3, 7]. In stimulated samples, the levels of pTyr are highest in the early time points (usually 1–2 min after the response is initiated) and samples are normally treated with phosphotyrosine phosphatases inhibitors, such as orthovanadate, before analysis to preserve these readily reversed modifications. Samples from stimulated cells taken at longer time points (5 min and beyond) are, like unstimulated samples, largely composed of pSer and pThr peptides [7].

The domination of the phosphoproteome by hydroxyl modifications in mammalian cells is also reflected in the kinome, i.e., the distribution and expression of protein kinases. There are over 500 protein kinases annotated from the human genome, of which only 90 are specific for tyrosine [14]. These are divided between the transmembrane receptor tyrosine kinases (RTKs) (58 members) and the soluble (or non-receptor) group (32 members). Both are involved in signal transduction. The former group provides the loci for many of the neurotrophic factors that are responsible for neuronal growth, differentiation and survival (or lack thereof), while the latter group interacts with a variety of other signaling entities to perpetrate intracellular signals. In one case (the JAKs), they are the principal mediators of a class of receptors that do not have kinase domains as an integral part of their structure. Thus they play a comparable role to the tyrosine kinase domains of the RTKs. Some members of this cytokine class of receptors have important neurotrophic ligands [15].

Neurotrophic factors are generally defined as humoral agents that stimulate neuronal targets in either the peripheral or central nervous systems (or both). Nerve growth factor (NGF) was the first of this class of substances to be identified as such and it is generally held to be the patriarch of the family [16]. It is one of a group of four neurotrophins that utilize one of the RTK subfamilies, i.e., the Trks [17, 18]. However, many of the other subfamilies are also expressed in various parts of the nervous system, making their ligands neurotrophic, at least in part, as well. In fact, most, if not all of the neurotrophic factors identified to date have both neuronal and non-neuronal targets. Thus it is more accurate to refer to these substances as having neurotrophic activity as opposed to being specifically neurotrophic. The importance of this distinction is that studies of signaling responses of substances such as insulin or epidermal growth factor (EGF) in non-neuronal targets may have relevance to neuronal responses as well.

The unbiased determination of phosphorylated residues in tissue samples and extracts, including from neuronal paradigms, is usually carried out by mass spectrometric analysis of protein proteolytic digests that have undergone a phosphopeptide enrichment step to concentrate phosphorylated peptides. In samples that have been stimulated (which includes most pTyr measurements), quantification is also required. This can be accomplished by spectral counting (or other non-isotope-based methods) or by the introduction of isotopic labels. SILAC (stable isotope labeling of amino acids in culture) [19] and chemical tagging approaches such as iTRAQ [20], where isobaric tags are introduced into the peptides after proteolysis of the sample but prior to fractionation, are among the most commonly employed. The choice of the method of quantification utilized is governed by the nature of the sample (e.g., metabolic labeling is generally not possible in primary tissue), the number of samples/conditions that are sought be compared, and resources available.

The method of concentration of the phosphopeptides represents the only major distinction between studies that are designed to determine pTyr modifications and those that mainly identify pSer- and pThr-labeled sites. The most commonly utilized enrichment methods for phosphopeptides involve metal (or metal oxide) affinity chromatography (IMAC; TiO_2; Ti^{4+}[21–23]). However, for pTyr-labeled peptides the method of choice is to use immunoprecipitation (IP). pTyr clearly provides a better epitope than do pSer and pThr: antibodies against these epitopes all show sequence specificity and have not proven to be particularly useful in IP-based protocols. There are several antibodies that have excellent specificity for the modified phenol but are not particularly sequence specific and these have been shown to be effective in precipitating a broad range of pTyr containing peptides. As already noted, most protocols usually also incorporate phosphatase inhibitors. Details of the IP protocol (which distinguishes it from phosphoproteomic experiments to measure pSer and PThr) are given in the next section.

It should be noted that once information has been obtained regarding a site of modification, targeted mass spectrometry measurements or the development of site-specific antibodies can be utilized in place of broad-scale experiments, and these approaches generally have the capacity to screen a larger number of samples in a quantitative fashion.

2 Materials

See **Note 1**

2.1 Cell Lysis and Protein Digestion

1. Lysis Buffer = 8 M urea, 1 mM sodium orthovanadate, 25 mM ammonium bicarbonate pH 8 (or 20 mM HEPES pH 8). Make fresh immediately before use (*see* **Note 2**).

2. 50 mM Tris (2-carboxyethyl)phosphine (TCEP) or 5 mM dithiothreitol (DTT) dissolved in 25 mM ammonium bicarbonate pH 8 (or 20 mM HEPES, pH 8).

3. 100 mM iodoacetamide dissolved in 25 mM ammonium bicarbonate, pH 8 (or 20 mM HEPES, pH 8). Make solution fresh: light sensitive.

4. 25 mM ammonium bicarbonate.

5. TPCK-modified trypsin (e.g., sequencing-grade modified trypsin from Promega, Madison, WI) dissolved in 25 mM ammonium bicarbonate.

6. 20 % TFA in water.

2.2 Peptide Purification

1. Sep-Pak C18 columns (Waters, Milford, MA).

2. 0.1 % TFA in water.

3. 50 % acetonitrile, 0.1 % TFA.

2.3 Immunoprecipitation (IP) of Phosphotyrosine-Containing Peptides

1. Phosphotyrosine mAb P-Tyr-100 beads (Cell Signaling Technology, Danvers, MA).

2. IP Buffer: 50 mM 3-(N-morpholino)propanesulfonic acid (MOPS), pH 7.2, 10 mM sodium phosphate, 50 mM NaCl (store at 2–8 °C).

3. 0.15 % TFA in water.

2.4 Phosphopeptide Desalting

1. C18 Tips: either ZipTips (Millipore, Billerica, MA) or OMIX tips (Agilent, Santa Clara, CA).

2. 20 % TFA in water.

3. 0.1 % TFA in water.

4. 50 % acetonitrile, 0.1 % TFA.

2.5 Liquid Chromatography: Mass Spectrometry

See **Note 3**.

1. Nanoacquity (Waters, Milford, MA).

2. BEH130 C18 75 μm ID × 150 mm UPLC column (Waters, Milford, MA).

3. 0.1 % formic acid in water.

4. 0.1 % formic acid in acetonitrile.

5. Q-Exactive Plus mass spectrometer (Thermo, San Jose, CA).

2.6 MSMS Data Analysis

1. Proteowizard (proteowizard.sourceforge.net).

2. Protein Prospector (prospector2.ucsf.edu).

3. Excel (Microsoft, Redmond, WA) or alternative spreadsheet software.

3 Methods

See **Note 4**.

3.1 Cell Lysis and Protein Digestion

1. Lyse cells by homogenizing then sonicating in 2 ml lysis buffer.

2. Add 1/10 volume 50 mM TCEP (or DTT; *see* **Note 5**). Mix well, and then incubate at 55°C for 20 min.

3. Allow solution to cool to room temperature, and then add equal volume of 100 mM iodoacetamide solution as TCEP solution added in **step 2**. Incubate in the dark for 40 min.

4. Dilute solution fourfold with 25 mM ammonium bicarbonate to a final concentration of less than 2 M urea, then add 1/50 w/w TPCK-trypsin dissolved in 25 mM ammonium bicarbonate (e.g., if starting with 10 mg protein, then add 200 μg trypsin) and digest overnight at 37°C (*see* **Note 6**).

5. Stop the digestion and acidify the solution by adding 1/20 volume 20 % TFA (to give final concentration of 1 % TFA).

3.2 Peptide Purification

See **Note 7**.

1. Centrifuge the acidified sample solution for 5 min at 1800 × *g* to pellet any precipitate.

2. Connect 10 ml plastic syringe (with plunger removed) to the shorter end of the Sep-Pak column.
   See **Note 8**.

3. Pre-wet column with 5 ml 50 % ACN, 1 % TFA.

4. Equilibrate column with 2 × 3 ml 0.1 % TFA.

5. Load supernatant from acidified sample solution.

6. Wash with 2 × 3 ml 0.1 % TFA.

7. Elute peptides with 2 × 2 ml 50 % ACN, 0.1 % TFA, collecting eluate.

8. Lyophilize/vacuum centrifuge to dryness.
   See **Note 9**.

3.3 Immunoprecipitation of Phosphotyrosine-Containing Peptides

1. Resuspend peptides in 1.4 ml IP buffer. Briefly sonicate to assist resuspension. Allow 5 min for peptides to redissolve under gentle shaking.

2. Cool sample on ice.

3. Transfer sample into microfuge tube containing phosphotyrosine antibody beads (80 μl slurry), and then incubate at 4°C for 30 min on a rotator.
   See **Note 10**.

4. Centrifuge at 1500 × *g* for 1 min and remove supernatant.
   See **Note 11**.

5. Wash beads for 1 min with 1 ml IP buffer by inverting the tube multiple times, and then centrifuge at 1500 × *g*. Remove supernatant.

6. Repeat **step 5**.

7. Wash beads with 1 ml water by inverting the tube multiple times. Centrifuge at $1500 \times g$ and get rid of the supernatant.

8. Repeat **step 7**.

9. Add 60 µl 0.15 % TFA to the beads, tap the tube on the bench a couple of times to ensure mixing, and then incubate at room temperature for 10 min.

10. Remove and collect the supernatant.

11. Repeat **steps 9** and **10**, combining the supernatants.

3.4 Phosphopeptide Desalting

See **Note 12**.

1. Pre-wet column with 50 µl 50 % ACN, 1 % TFA.

2. Equilibrate column with 50 µl 0.1 % TFA.

3. Load sample.

4. Wash with 50 µl 0.1 % TFA, and then repeat this step (i.e., 2×50 µl washes).

5. Elute peptides with 2×20 µl 50 % ACN, 0.1 % TFA, collecting eluate.

3.5 Liquid Chromatography: Mass Spectrometry

See **Note 13**.

1. Attach BEH130 C18 75 µm ID $\times$ 150 mm UPLC column to NanoAcquity.

2. Make solvent A = 0.1 % formic acid in water; solvent B = 0.1 % formic acid in acetonitrile.

3. Set flow rate at 300 nl/min, and starting condition as 2 % solvent B. Allow 15 min for column to equilibrate at starting conditions.

4. Run sample using gradient:

Time (min)	% Solvent B
0	2
100	30
105	50
106	2
120	2

5. Mass spectrometry data should be acquired using the QExactive Plus in a data-dependent analysis approach where the top 10 precursors are automatically selected for MS/MS analysis.

6. MS data should be acquired at a resolution of 70,000 in profile mode, whereas MS/MS should be acquired at a resolution of 17,500 in centroid mode.

7. Target AGC for MS scans should be 2×10^6 and for MS/MS 5×10^4.

8. Use a normalized collision energy of 25. Dynamic exclusion should be set for 30 s.

3.6 MS/MS Data Analysis

See **Note 14**.

1. Create mgf file from raw data file using MSConvert. *See* **Note 15**.

2. Go to the Protein Prospector website (prospector2.ucsf.edu) and select the program Batch-Tag Web.

3. It will ask for a username and password. If you already have an account then log in; if you do not then click "add user" and create an account.

4. Fill in appropriate search parameters. For protein database select a SwissProt.xxx.random.concat database (*see* **Note 16**). For QExactive Plus data select "Instrument" as ESI-Q-high-res; Parent Tol 10 ppm and Fragment Tol 20 ppm. In the Variable Mods field add Phospho (Y) as an option.

5. Browse and select the mgf file created in **step 1**, and then Start Search. *See* **Notes 17** and **18**.

6. From the Search Compare page, select Report Type as "False Positive Rate." Hit "Compare Searches" and from the resulting plot note the E-value threshold to employ for a 1 % FDR at the spectrum level. *See* **Note 19**.

7. Set the Max Expect Value for protein and peptide to the value that corresponds to a 1×10^{-2} (1 %) FDR rate according to the False Positive Rate plot.

8. Set Report Type to "peptide" and Format to "Tab-delimited text," then "Compare Searches."

9. Copy and Paste Special (as text) results into a spreadsheet package such as Microsoft Excel.

4   Notes

1. Many of the materials required for lysis, digestion, and tyrosine phosphopeptide purification can be purchased together as part of the PhosphoScan P-Tyr-100 kit (Cell Signaling Technologies, Danvers, MA). All solvents used should be of HPLC grade.

2. Urea forms an equilibrium with ammonium cyanate in solution, which can react with primary amino groups on proteins forming carbamylation products. To minimize this, urea solutions should either be made fresh before use or passed over mixed bed resin before use.

3. Any combination of a nanoflow HPLC system and a high-mass-accuracy mass spectrometer (time-of-flight or Orbitrap) will be effective. An instrument that can perform beam-type collision-induced dissociation (as opposed to resonant excitation in an ion trap) is preferable, as the presence of the pTyr immonium ion at m/z 216.04 is useful confirmation that the precursor is a phosphopeptide, and an ion trap will not be able to trap this fragment ion for some precursors. Example instrumentation is provided here.

4. As tyrosine phosphorylation is a low stoichiometry modification, it is recommended to start with at least 1 mg of protein; preferably 10–20 mg. The cell lysis, protein digestion, and peptide purification steps are identical to those used for global peptide analysis, with the exception of the addition of orthovanadate to the lysis buffer, which is a tyrosine phosphatase inhibitor.

5. Both TCEP and DTT are effective at reducing disulfide bonds to free sulfhydryls. This allows the protein to unfold more effectively, giving a proteolytic enzyme such as trypsin better access, leading to more efficient digestion. The advantage of TCEP is that it is active over a wider pH range, whereas if DTT is used then the solution needs to be around pH 7–8.

6. Trypsin is not active in urea concentrations above 2 M. TPCK modification of trypsin is generally used to reduce trypsin autolysis so that in the subsequent peptide analysis there are fewer peptides derived from trypsin itself. For a solution digest such as here, where a 50-fold excess of protein to trypsin is used, this is less of an issue. However, the TPCK modification also increases the stability of the trypsin, making it more tolerant of the urea present in the solution.

7. Sep-Pak cartridges have a maximum capacity of about 20 mg and a bed volume of a little under 1 ml. They can be used more than once, so if you may have more sample than 20 mg, then it is possible to repeat this protocol multiple times with aliquots of the sample.

8. The Sep-Pak steps may be possible to complete using gravity alone, but the plunger can be used to speed up the process. However, flow rates should not exceed 1 ml per 10 s.

9. Sample can be stored at this point at −80 °C for months, if necessary.

10. For the washing steps of the immunoprecipitate, it does not matter if there are still a few microliters of solution left after any step, but all solution should be removed during the final elution step.

11. This supernatant will contain unmodified peptides, so depending on other interests, may be worth keeping.

12. This protocol is analogous to 3.2, except it is performed on a smaller scale, because there will be much less sample at this point (tyrosine phosphorylated peptides only). Several companies sell C18 tips in two different sizes: 10 µl tips that typically have a capacity of about 0.8 µg, and 100 µl tips that have a capacity of about 8 µg. In most cases the smaller tips will have enough capacity to bind all tyrosine phosphorylated peptides, but it is permissible to use the larger tips if you want to make sure that you purify all peptides, and using the larger tips means one does not need to perform an extra concentration step at the end of 3.3. The volumes listed in the protocol are assuming use of 100 µl tips. If 10 µl tips are used, then divide all volumes by five.

    C18 tips can be used by pipetting solutions up and down through the tip. However, best results are generally achieved if they are used like a column, loading the sample on the top of the tip and then re-attaching the pipettor to push solution through.

13. Electrospray (ESI) mass spectrometry is a concentration-sensitive process, so for the highest sensitivity chromatography is performed at sub-microliter flow rates.

14. In order to analyze the data using a database search engine it is necessary to convert the raw data into peak list files. A free, open-source package, Proteowizard includes a file conversion tool called MSConvert. There are multiple output peak list formats, but the most popular and convenient for database searching is the mgf format.

15. There are many database search engines that can be used to search the data. The section here describes the use of Protein Prospector. This is free web-based software that is unusual in having modification site localization scoring built-in, which is important for PTM analysis. There are video tutorials that explain the use of this software in more detail than can be provided here (https://vimeo.com/channels/194363/videos).

16. Searching against a SwissProt database that has a random database concatenated to the end is generally most appropriate, as SwissProt databases are the best curated, and the addition of random sequences allows estimation of a false discovery rate for your results [24]. Filtering by taxonomy can be appropriate, providing the species is well represented in the database (e.g.,

mouse or human). For a slightly less well-cataloged species, such as rat, then a higher level taxonomy filter, such as rodent, or even mammals may be more effective at identifying homologous sequence in related species.

17. Updates on search progress will be provided, and when the search finishes it will automatically go to the Search Compare page to set parameters for viewing results. One does not need to keep this window open during searching. When a search is finished it can be accessed at any time by selecting Search Compare from the Protein Prospector homepage, and then selecting the relevant project, which will have the same name as the mgf file that was created in **step 1** of the Section 3.6.

18. Search Compare allows reporting of different report types, different thresholding and choices of columns in output. For interactive viewing of results leave "Format" as HTML; for exporting results to a table or spreadsheet, select "Format" as Tab-delimited text.

Site Localization Scoring in Peptide (SLIP) scoring is automatically reported, and the SLIP threshold is set at 6 as a default. This corresponds to all site assignment results have a score greater than 95 % confidence (and typically global site localization results are between 98 and 99 % correct) [25].

19. It is up to the user to decide the reliability threshold at which they wish to report their results. This plot reports an estimated false discovery rate (FDR) at the spectrum level. The FDR at the spectrum level is nearly always lower than at the unique peptide level and protein level, because correct peptides are more likely to be observed more than once, and incorrect peptide identifications are commonly the only identification to a particular protein, whereas correct identifications are usually to proteins identified by other peptides.

References

1. Bradshaw RA and Dennis EA (Editors) (2009) Handbook of Cell Signaling. Elsevier Academic Press, San Diego CA, Vol 1-3 (2nd edition)

2. Krebs EG, Fischer EH (1964) Phosphorylase and related enzymes of glycogen metabolism. Vitam Horm 22:399–410

3. Hunter T, Sefton BM (1980) Transforming gene product of Rous sarcoma virus phosphorylates tyrosine. Proc Natl Acad Sci U S A 77:1311–1315

4. Ushiro H, Cohen S (1980) Identification of phosphotyrosine as a product of epidermal growth factor- activated protein kinase in A-431 cell membranes. J Biol Chem 255:8363–8365

5. Fenn JB (2003) Electrospray wings for molecular elephants. Angewandte Chemie 42:3871–3894

6. Tanaka K (2003) The origin of macromolecule ionization by laser irradiation. Angew Chem Int Ed Engl 42:3860–3870

7. Sharma K, D'Souza RCJ, Tyanova S et al (2014) Ultradeep human phosphoproteome reveals a distinct regulatory nature of Tyr and Ser/Thr-based signaling. Cell Rep 8:1583–1594

8. Junger MA, Aebersold R (2013) Mass spectrometry-driven phosphoproteomics:

patterning the systems biology mosaic. Wiley Interdiscip Rev Dev Biol 3:83–112

9. Olsen JV, Mann M (2013) Status of large-scale analysis of post-translational modifications by mass spectrometry. Mol Cell Proteomics 12:3444–3452

10. Biarc J, Chalkley RJ, Burlingame AL et al (2013) Dissecting the roles of tyrosines 490 and 785 of TrkA in the induction of downstream protein phosphorylation using chimeric receptors. J Biol Chem 288:16606–16618

11. Minguez P, Letunic I, Parca L et al (2013) PTMcode: a database of known and predicted functional associations between post-translational modifications in proteins. Nucleic Acids Res 41:D306–D311

12. Trinidad JC, Thalhammer A, Specht CG et al (2008) Quantitative analysis of synaptic phosphorylation and protein expression. Mol Cell Proteomics 7:684–696

13. Munton RP, Tweedie-Cullen R, Livingstone-Zatchej M et al (2007) Qualitative and quantitative analyses of protein phosphorylation in naive and stimulated mouse synaptosomal preparations. Mol Cell Proteomics 6:283–293

14. Blume-Jensen P, Hunter T (2001) Oncogenic kinase signaling. Nature 411:355–365

15. Deister C, Schmidt CE (2006) Optimizing neurotrophic factor combinations for neurite outgrowth. J Neural Eng 3:172–179

16. Levi-Montalcini R (1987) The nerve growth factor 35 years later. Science 237:1154–1162

17. Reichardt LF (2006) Neurotrophin-regulated signalling pathways. Philos Trans R Soc Lond B Biol Sci 361:1545–1564

18. Hondermarck H (2012) Neurotrophins and their receptors in breast cancer. Cytokine Growth Factor Rev 23:357–365

19. Ong SE, Blagoev B, Kratchmarova I et al (2002) Stable isotope labeling by amino acids in cell culture, SILAC, as a simple and accurate approach to expression proteomics. Mol Cell Proteomics 1:376–386

20. Ross PL, Huang YN, Marchese JN et al (2004) Multiplexed protein quantitation in Saccharomyces cerevisiae using amine-reactive isobaric tagging reagents. Mol Cell Proteomics 3:1154–1169

21. Li S, Dass C (1999) Iron(III)-immobilized metal ion affinity chromatography and mass spectrometry for the purification and characterization of synthetic phosphopeptides. Anal Biochem 270:9–14

22. Thingholm TE, Jørgensen TJD, Jensen ON et al (2006) Highly selective enrichment of phosphorylated peptides using titanium dioxide. Nat Protoc 1:1929–1935

23. de Graaf EL, Giansanti P, Altelaar AFM et al (2014) Single-step enrichment by Ti4 + -IMAC and label-free quantitation enables in-depth monitoring of phosphorylation dynamics with high reproducibility and temporal resolution. Mol Cell Proteomics 13:2426–2434

24. Elias JE, Gygi SP (2007) Target-decoy search strategy for increased confidence in large-scale protein identifications by mass spectrometry. Nat Methods 3:207–214

25. Baker PR, Trinidad JC, Chalkley RJ (2011) Modification site localization scoring integrated into a search engine. Mol Cell Proteomics 10:M111.008078

Neuromethods (2016) 114: 155–170
DOI 10.1007/7657_2015_85
© Springer Science+Business Media New York 2015
Published online: 18 November 2015

Analysis of PINK1 and CaMKII Substrates Using Mass Spectrometry-Based Proteomics

Yan Li, Lesley A. Kane, Michael A. Bemben, and Katherine W. Roche

Abstract

Mass spectrometry is a powerful tool for protein phosphorylation analysis. Collision-induced dissociation (CID) is a widely applied fragmentation method. Complementary fragmentation techniques such as electron transfer dissociation (ETD) and higher-energy C-trap dissociation (HCD) enhance the accurate elucidation of phosphorylation sites. Here we present proteomic approaches used for identifying phosphorylation sites of in vitro-phosphorylated neuroligin-1 (NL-1) and for identifying PINK1 substrates from outer mitochondrial membrane proteins. Technical details on how to identify phosphorylation sites using CID, ETD, and HCD fragmentation are described, including sample preparation, in-gel and in-solution protein digestion, peptide separation, and data acquisition.

Keywords: Mass spectrometry, CID, ETD, HCD, Phosphorylation, CaMKII, Neuroligin, PINK1, Ubiquitin

1 Introduction

Protein phosphorylation is an important posttranslational modification (PTM) that controls many cellular processes and plays a crucial role in the function of both the nervous and immune systems [1]. Methods capable of characterizing protein phosphorylation are in high demand. Since the early 1980s, when techniques such as electrospray ionization [2] became available, coupling liquid chromatography (LC) to tandem mass spectrometry (MS/MS) [3–6] has greatly advanced the study of proteins by mass spectrometry. The development of computer algorithms to search MS/MS spectra against protein databases offers the ability to match the spectra to peptide sequences [7, 8]. LC-MS/MS workflows allow site-specific assignment of posttranslational modifications at the level of the individual amino acid, becoming a powerful tool to identify and locate the residues that are phosphorylated [9, 10].

Supported by the National Institute of Neurological Disorders and Stroke Intramural Research Program

Studies performed on LTQ Orbitrap indicate that CID, ETD, and HCD are complementary ion fragmentation methods; each has its own advantages and disadvantages [11–15]. Collision-induced dissociation (CID) is commonly employed to generate MS/MS fragments. On triple quadrupole and quadrupole-TOF instruments, a beam-type CID approach is used; while on ion trap instruments, an ion trap CID approach is employed [16]. It has been shown that CID works less effectively for large peptides and highly charged peptides, and the number of structurally informative ions generated is peptide sequence dependent [17–19]. CID spectra of phosphopeptides, for example, usually generate a dominant peak corresponding to the precursor ion with a neutral loss of the phosphate group (HPO_3 or H_3PO_4), and yet few structurally significant fragments are obtained [20]. Compared to the CID process, during which mainly b and y ions are produced, two relatively new fragmentation techniques, electron capture dissociation (ECD) [21] and electron transfer dissociation (ETD) [22], generally produce c and z ions. ECD and ETD have quickly become important complementary fragmentation techniques to CID. They allow sequencing of larger peptides (even proteins), fragment highly charged peptides more efficiently, and preserve labile modifications, such as phosphorylation [20, 23, 24]. Recently, a higher-energy C-trap dissociation (HCD) fragmentation has been developed for the LTQ Orbitrap mass spectrometer [11]. HCD is a beam-type CID, which can generate more backbone amide bond cleavage compared to ion trap CID [16]. Since fragments are generated in a multiple collision cell and analyzed in the Orbitrap, high resolution MS/MS data are acquired with no low mass cut-off. However, compared with ion trap CID, Orbitrap HCD scanning requires a larger time investment.

Phosphoproteomics discovery benefits from the ability to generate large datasets. However, fast acquisition speed often comes at the sacrifice of the high quality MS/MS data needed to localize phosphorylation sites. Building specific data acquisition methods optimized for the experimental goal and the sample condition facilitates a better balance between spectral quality and speed. Here we present mass spectrometry-based methods used successfully in two projects: (1) Mapping phosphorylation sites on neuroligin-1 (NL-1) [25] and (2) Identification and analysis of PINK1 substrates [26]. A brief outline of each study is provided for context below, followed by detailed protocols for the pertinent MS experiments performed in each.

1. Mapping phosphorylation sites on NL-1
Ca^{2+}/calmodulin kinase II (CaMKII) is an activity-dependent kinase that plays a critical role in synaptic plasticity at excitatory synapses. Neuroligins are cell adhesion molecules that are localized in the postsynapse [27–30]. NL-1 is predominantly

expressed at excitatory synapses with its ability to induce synaptogenesis dependent on CaMKII [25, 31]. To determine if CaMKII can phosphorylate NL-1, we produced GST fusion proteins of the intracellular tail of NL-1 and performed in vitro kinase assays on the fusion proteins. Following the reaction, we purified GST-NL-1 by SDS-PAGE, and NL-1 protein bands were digested with chymotrypsin. The extracted peptides were desalted and then separated with a 15 cm nano-column. The LC-MS/MS data were acquired on an Orbitrap Elite in a data-dependent decision tree (DDDT) fashion [32]. A phosphopeptide was detected on GST-NL-1. The CID and ETD spectra were obtained for the triply and quadruply charged phosphopeptide, respectively. The phosphorylation site on the peptide $_{730}$RRCSPQRTTpTNDLTHAPEEEIM$_{751}$ was identified using the ETD data (Fig. 1). The analysis showed that amino acid T739 in the intracellular c-tail of NL-1 was phosphorylated when incubated with CaMKII. When NL-1 was treated either without enzyme or alternatively with other kinases such as cyclic AMP (cAMP)-dependent protein kinase A (PKA) or Protein Kinase C (PKC), phosphorylation of T739 was not detected (Fig. 1). Hence, NL-1 is a direct substrate of CaMKII [25].

2. Identification and analysis of PINK1 substrates by CID, ETD, and HCD

PINK1 is a kinase that is imported into the mitochondrion, processed by the inner membrane protease PARL, and then normally degraded by the proteasome [33–36]. Upon loss of mitochondrial membrane potential or accumulation of unfolded protein, stabilized PINK1 recruits the E3 ubiquitin ligase Parkin to the outer mitochondrial membrane (OMM) and causes its activation [37–40]. This results in the selective autophagy of damaged mitochondria. PINK1 phosphorylates Parkin at serine 65 (S65) [41, 42], but Parkin is still recruited in a PINK1-dependent manner to mitochondria despite mutation of S65 [26], which indicates that another PINK1 substrate is essential for Parkin translocation. A proteomics approach was designed to identify PINK1 substrates using mass spectrometry. Mitochondria were isolated from WT and PINK1 KO cells after carbonyl cyanide 3-chlorophenylhydrazone (CCCP) treatment. Isolated mitochondria were then treated with trypsin to proteolyze exposed OMM proteins and the remaining intact mitochondria were removed by centrifugation. The supernatant containing the OMM peptides was alkylated. To ensure complete digestion, this preparation was again incubated with trypsin. Peptides were separated with a 25 cm nano-column using a 4 h gradient. To maximize the number of ions being scanned, only CID fragmentation was performed. MS/MS data were acquired in the ion trap of an Orbitrap Elite

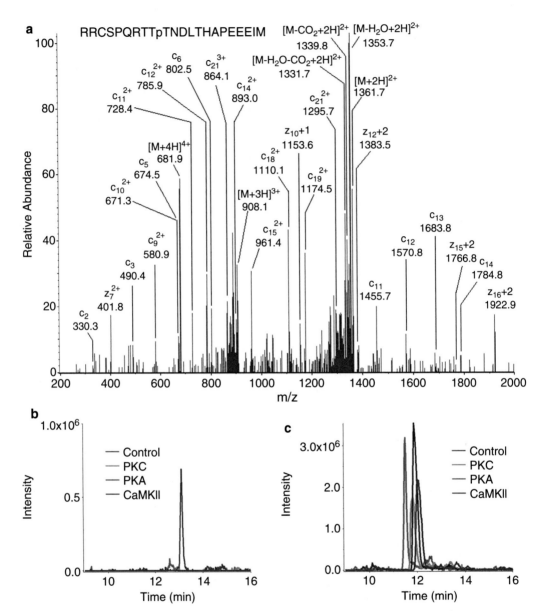

Fig. 1 T739 in the intracellular c-tail of NL-1 is phosphorylated when incubated with CaMKII. (**a**) ETD MS/MS spectrum of the phosphorylated NL-1 peptide $_{730}$RRCSPQRTTpTNDLTHAPEEEIM$_{751}$ found only in GST–NL-1 fusion proteins incubated with ATP and purified CaMKII and not those incubated with PKA or PKC. (**b**) Extracted ion chromatogram of a quadruply charged ion at m/z 681.30, which corresponds to the phosphorylated NL-1 730–751 peptide, as shown in **a**, for GST–NL-1 without enzyme (*red*), with PKA (*gray*), with PKC (*green*), or with CaMKII (*blue*). (**c**) Extracted ion chromatogram of a quadruply charged ion at m/z 661.31, which corresponds to the nonphosphorylated NL-1 730–751 peptide in GST-NL-1 without enzyme (*red*), with PKA (*gray*), with PKC (*green*), or with CaMKII (*blue*). ©Bemben et al., 2014. Originally published in Nature Neuroscience. doi:10.1038/nn.3601

mass spectrometer. A unique phosphopeptide (TLSDYNIQ-KEpSTLHLVLR) corresponding to phosphorylated ubiquitin was found only in the PINK1 WT mitochondrial sample. To confirm, recombinant His-ubiquitin (His-Ub) was incubated in vitro with mitochondria isolated from control (no PINK1) or CCCP-treated (with PINK1) cells. His-Ub samples exposed to mitochondria from CCCP-treated cells contained an identical phosphopeptide, but the untreated controls did not. A second LC/MS/MS run was then performed where CID, ETD, and HCD spectra were acquired for the peptide. CID, ETD, and HCD spectra obtained for the phosphopeptide from PINK1 WT mitochondria samples were almost identical to those from His-Ub CCCP-treated samples. The CID spectra contained a dominant neutral loss-associated peak. The ETD and HCD spectra clearly showed that S11 of $_{55}$TLSDYNIQKEpSTLHLVLR$_{72}$, which corresponds to S65 of ubiquitin, was phosphorylated (Fig. 2). This analysis demonstrates that PINK1 phosphorylates ubiquitin. This is an important cellular process, as phosphorylated ubiquitin activates Parkin E3 ubiquitin ligase [26, 44, 45].

2   Materials and Methods

2.1   GST-NL-1 Protein Purification

1. Transform pGEX-NL-1 constructs in BL21 cells (+amp).

2. Grow *E. coli* from single colony in 10 ml LB (+amp) overnight at 37 °C.

3. Transfer 10 ml culture to a 2 l flask containing 500 ml LB (+amp).

4. Grow culture at 37 °C (250 RPM shake) until OD600 = 1.1–1.2.

5. Add IPTG (0.25 μM final concentration) and shake overnight at 16 °C.

6. Centrifuge (9000 × *g*) for 20 min.

7. Resuspend pellet in 40 ml TBS supplemented with 100 μg/ml lysozyme, four tablets of protease inhibitors (Roche), 10 mM ethylenediaminetetraacetic acid (EDTA), 15 mM dithiothreitol (DTT), and 1.5 % sarkosyl.

8. Incubate on ice for 10–15 min.

9. Sonicate the 40 ml lysate and centrifuge for 30 min (12,000 × *g*) (*see* **Note 1**).

10. Add TritonX-100 to a final concentration of 4 % (neutralization step).

11. Incubate lysate with a 10:1 ratio of glutathione-Sepharose 4B (GE Healthcare) for 1 h at 4 °C.

12. Wash beads thoroughly in TBS-based buffer, then resuspend beads in an optimal buffer.

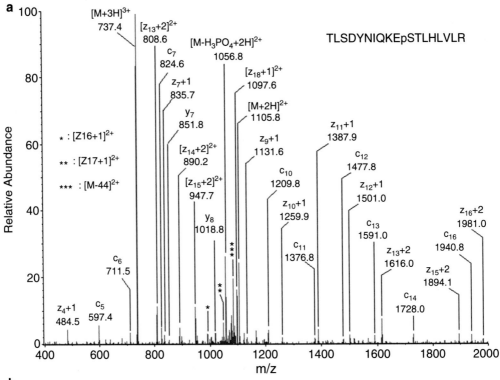

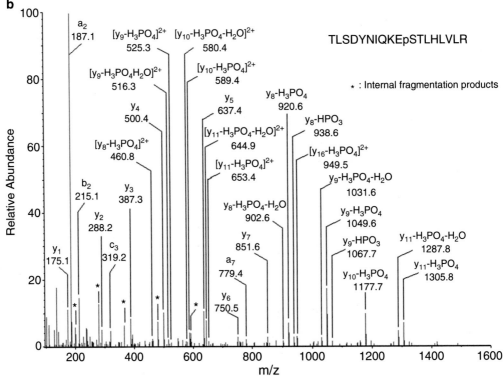

Fig. 2 His-Ub is phosphorylated after exposed to CCCP-treated mitochondria. Both ETD (**a**) and HCD (**b**) MS/MS spectra of the triply charged Ub phosphopeptide TLSDYNIQKEpSTLHLVLR indicate that S65 of Ub is

2.2 GST-NL-1 In Vitro Kinase Assay (CaMKII)

1. 5–10 % of isolated protein is incubated in 20 mM Tris–HCl (pH 7.5), 10 mM MgCl₂, 0.5 mM DTT, 0.1 mM EDTA, 2.4 µM calmodulin, 2 mM CaCl₂, 100 µM adenosine triphosphate (ATP), with 25 ng of recombinant CaMKIIα (Calbiochem) at 30 °C for 30 min.

2. Reaction is halted with addition of SDS-PAGE sample buffer and incubation at 65 °C for 5 min.

3. Proteins are resolved by SDS-PAGE and identified by Coomassie blue staining.

2.3 Mitochondrial Isolation and Outer Mitochondrial Membrane Proteolysis

1. Prepare Isolation Buffer (IB); 220 mM mannitol, 20 mM 4-(2-hydroxyethyl)-1-piperazineethanesulfonic acid (HEPES) pH 7.6, 70 mM sucrose.

2. Add 2 mg/ml BSA to an aliquot of IB; keep ice cold.

3. Harvest three confluent 15 cm plates of cells: Rinse 2 × 10 ml ice-cold PBS, scrape into 5 ml of ice-cold PBS, and spin down in 15 ml tube.

4. Resuspend cell pellet in 7 ml IB. Leave cells on ice to swell for 10 min.

5. Transfer cell suspension to a Teflon/glass homogenizer and homogenize cells with ~30 strokes.

6. Transfer homogenate back into 15 ml tube and spin at 850 × *g* for 10 min.

7. Transfer supernatant to 1.5 ml tubes and spin at 10,000 × *g* for 20 min.

8. Discard supernatant and combine pellets, then spin at 10,000 × *g* for 20 min.

9. Discard supernatant. Resuspend mitochondrial pellet in 200 µl of IB. Remove a 2 µl aliquot to lyse, then measure protein concentration.

10. Take 100 µg mitochondria and transfer to a clean tube and add 0.5 µg trypsin for 2 h at 24 °C to proteolyze the outer membrane proteins.

Fig. 2 (continued) phosphorylated. In the ETD spectrum the phosphorylated residue remains intact. In the HCD spectrum, loss of HPO_3 or H_3PO_4 groups occurs for fragments containing the phosphorylated residue. Many low mass product ions including immonium ions and several internal fragment ions are detected in HCD scan. This exact phosphopeptide, TLSDYNIQKEpSTLHLVLR, is also found in Ub from endogenous PINK1 WT samples with ETD and HCD spectra almost identical to what shown in **a** and **b**. Phosphorylation of S65 is not detected on Ub from PINK1 KO or His-Ub exposed to mitochondria from untreated cell. ©Kane et al., 2014. Originally published in *J. Cell Biol.* doi: 10.1083/jcb.201402104

11. Spin at 20,000 × g to remove mitochondria and harvest supernatant as the mitochondrial outer membrane proteome for further digestion with trypsin.

2.4 In-Gel Digestion of GST-NL-1 Samples with Chymotrypsin

1. Cut NL-1 protein band (contains ~1 μg protein) into approximately 1 × 1 mm pieces and place into a 0.5 ml LoBind Eppendorf tube (*see* **Note 2**).

2. Add 400 μl of 10 % acetic acid (Sigma, St Louis, MO) in 50 % methanol (MeOH) (Honeywell B&J, Muskegon, MI) into the sample tube, vortex on Thermomixer 5436 (Eppendorf, Hauppauge, NY) with medium speed for 1 h. Discard the supernatant. Repeat this step twice (*see* **Note 3**).

3. Dehydrate the gel pieces for 5 min with 400 μl HPLC grade acetonitrile (ACN) (Honeywell B&J, Muskegon, MI). Discard the supernatant.

4. Dry the gel pieces in the Savant SpeedVac Concentrators (Thermo Scientific) for 30 min.

5. Add 30 μl of 10 mM DTT (Sigma, St Louis, MO) in 50 mM ammonium bicarbonate (NH_4HCO_3) (Fluka, Germany). Incubate at room temperature (RT) for 45 min (*see* **Note 4**). Discard the supernatant.

6. Add 30 μl of 100 mM iodoacetamide (IAM) (Sigma, St Louis, MO) in 50 mM NH_4HCO_3. Incubate at room temperature in the dark for 30 min (*see* **Note 5**). Discard the supernatant.

7. Add 400 μl of HPLC grade H_2O (Honeywell B&J, Muskegon, MI). Incubate at RT for 10 min. Discard the supernatant (*see* **Note 6**).

8. Dehydrate the gel pieces for 5 min with 400 μl ACN. Discard the supernatant.

9. Dry the gel pieces in the SpeedVac for 30 min.

10. Make fresh 5 ng/μl chymotrypsin (sequencing grade, Roche, Mannheim, Germany) solution in 5 mM NH_4HCO_3.

11. Add 10 μl of chymotrypsin solution to each sample (*see* **Note 7**). Incubate at 4 °C for 30 min.

12. Add 30 μl of 5 mM NH_4HCO_3 to cover the gel pieces. Carry out the digestion at 25 °C overnight in Mini Incubator (Labnet, Edison, NJ).

13. Add 50 μl of 5 % trifluoroacetic acid (TFA) (Thermo Scientific Pierce, Rockford, IL) in 50 % ACN. Vortex with low speed for 20 min. Collect and transfer the supernatant in a new 0.5 ml LoBind Eppendorf tube. Repeat this step once. Combine the supernatants together.

14. Add 50 µl of 0.5 % TFA in 80 % ACN. Vortex with medium speed for 10 min. Collect the supernatant. Combine with the supernatant from **step 13**.

15. Dry the combined supernatants in the SpeedVac.

2.5 In-Solution Digestion of PINK1-WT, PINK1-KO, and His-Ub Samples with Trypsin

1. Samples are in 220 mM Mannitol, 70 mM Sucrose, 20 mM HEPES (pH 7.6). PINK1-WT and PINK1-KO samples contain ~5 µg protein in 30 µl solution. His-Ub samples contain ~0.1 µg protein in 10 µl solution.

2. Make 100 mM HEPES (Sigma, St Louis, MO) aqueous solution; adjust the pH to 8.2 with NaOH (Sigma, St Louis, MO).

3. Add 30 µl of 100 mM HEPES into each PINK1 KO/WT sample. For His-Ub samples, 10 µl of 100 mM HEPES is added.

4. Add 1 µl of 50 mM tris-(2-carboxyethyl) phosphine (TCEP) (AB Sciex, Framingham, MA). Vortex to mix, and briefly spin to bring solution to the bottom of the tube. Incubate at room temperature for 45 min (*see* **Note 8**).

5. Add 0.5 µl of 200 mM methyl methanethiosulfonate (MMTS) in isopropanol (AB Sciex, Framingham, MA) (*see* **Note 9**). Vortex and briefly spin. Incubate at room temperature for 30 min.

6. Make fresh trypsin (Mass Spectrometry Grade, Promega, Madison, WI) solution with a concentration of 10 ng/µl in 50 mM NH_4HCO_3.

7. Add 20 µl of trypsin solution into each PINK1 KO/WT sample. For His-Ub samples, 1 µl of trypsin solution is used for each sample. Incubate at 37 °C overnight in Mini Incubator.

8. Add 5 % TFA into the solution to bring the pH to 3. Vortex and briefly spin. Store the digests in −20 °C freezer if not immediately used for desalting.

2.6 Peptide Desalting

1. For in-gel samples, dissolve the extracted peptides in 100 µl of 0.1 % TFA (*see* **Note 10**). Vortex for 5 min and briefly spin.

2. For in-solution samples, check the pH of the solution; use 5 % TFA to adjust the pH to 3 if needed.

3. Mount the HLB µElution plate (Waters, Milford, MA) on the Extraction Plate Minifold (Waters, Milford, MA) and condition the sorbent in each well (*see* **Note 11**) with 0.5 ml MeOH. Equilibrate with 0.5 ml 0.1 % TFA. Adjust the vacuum level to 15 In Hg. Discard the flow-through.

4. Load the sample solution. Adjust the vacuum level at 3–5 In Hg, let the sample solution pass through the sorbent slowly. Discard the flow-through (*see* **Note 12**).

5. Wash with 0.5 ml 0.1 % TFA. Adjust the vacuum level to 15 In Hg. Discard the flow-through.

6. Elute with 100 µl HPLC grade MeOH. Adjust the vacuum level to 3–5 In Hg. Collect the flow-through in a clean 0.5 ml LoBind Eppendorf tube.

7. Elute with 100 µl 65 % ACN, 35 % MeOH. Adjust the vacuum level to 3–5 In Hg. Combine the flow-through with that collected in **step 7**.

8. Dry the eluent in the SpeedVac.

2.7 LC-MS/MS Analysis

An Ultimate 3000 RSLCnano HPLC system with an RS autosampler (Thermo-Dionex) is coupled to an Orbitrap Elite Mass Spectrometer (Thermo Scientific, Bremen, Germany). Xcalibur 2.2 sp 1.48 software provides instrument control for Orbitrap Elite mass spectrometer. Xcalibur also controls Ultimate 3000 HPLC system through Dionex Chromatography Mass Spectrometry Link (DCMSLink) 2.11.

2.7.1 LC-MS/MS Method for NL-1 Samples

1. Samples are loaded into the LC-MS system using the autosampler with a 5 µl sample loop. The temperature of the sampler tray is kept at 4 °C. Dissolve the desalted and dried protein digests in 15 µl 1 % formic acid (FA) (Fluka, Germany) in 2 % ACN. Vortex for 3 min, briefly spin. Transfer 7.5 µl of the sample solution into sampler vial. For each LC-MS/MS run, ~6.4 µl of sample solution is taken from the vial, and ~4 µl of it is injected to the column directly (*see* **Note 13**). Store remainder solution at −20 °C. A user-defined injection method is used.

2. The peptides are separated at a flow rate of 300 nl/min using an ES800 nano-LC column (15 cm × 75 µm ID, packed with 3 µm PepMap C18 particles, Thermo Scientific). A gradient of mobile phase A (MPA) and mobile phase B (MPB) is employed: 2–27 % MPB for 23 min, 27–80 % MPB for 7 min, and 80 % MPB for 15 min (*see* **Note 14**).

   MPA contains 0.1 % FA in 2 % LC-MS grade ACN (J. T. Baker, Philipsburg, NJ), 97.9 % LC-MS grade H_2O (J. T. Baker, Philipsburg, NJ).

   MPB contains 0.1 % FA in 98 % LC-MS grade ACN, 1.9 % LC-MS grade H_2O.

3. The Orbitrap Elite is operated in a DDDT fashion (*see* **Note 15**). The precursor ion scan (MS scan) is performed in the Orbitrap with a resolution of 60 K at m/z 400. The m/z range for the survey scans was 300–2000. The automatic gain control (AGC) target value for MS scan is 1×10^6. The fragment ion scan (MS/MS scan) is performed in the linear ion trap with an AGC value of 1×10^4. The maximum ion injection time for MS scan is 10 ms, and it is 100 ms for MS/MS

scan. Minimum signal threshold for MS/MS scan is 3×10^4, and up to ten MS/MS scans are performed after each MS scan. The dynamic exclusion duration is 9 s. "Charge state screening," "monoisotopic precursor selection," "charge state dependent ETD time," and "FT master scan preview mode" functions are activated. The micro scan count is 1 for both MS and MS/MS scans. The isolation width is 1.9 Da. The value for normalized collision energy is 35, and the default charge state is 2. Singly charged ions and ions with unassigned charge states are rejected from MS/MS scan. The default decision tree logic embedded within the Xcalibur method is used except for the value for "charge state 3 mass." All doubly charged ions are fragmented with CID and all ions with charge state more than 5 are fragmented with ETD. For triply charged ions, the default rule is that ions are fragmented with ETD if the m/z value is less than 650. In our experiments the m/z value was set to 750 (*see* **Note 16**). As a result, triply charged ions with m/z 750 or less are fragmented with ETD. The default values of m/z 900 and m/z 950 are used for ions with charge state 4 and 5, respectively.

2.7.2 LC-MS/MS Method for PINK1 KO and WT, and His-Ub Samples

1. Samples are injected with the same method described in Section 2.7.1.

2. For PINK KO and WT samples, an ES802 nano-LC column (25 cm $\times$ 75 μm ID, packed with 2 μm PepMap C18 particles, Thermo Scientific) is used. A 4 h gradient is employed: 2–24 % MPB for 170 min, 24–50 % MPB for 35 min, 50–80 % MPB for 15 min, and 80 % MPB for 20 min (*see* **Note 14**).

3. For His-Ub samples, an ES800 nano-LC column is used. A 75 min gradient is employed: 2–24 % MPB for 38 min, 24–50 % MPB for 12 min, 50–80 % MPB for 5 min, and 80 % MPB for 20 min (*see* **Note 14**).

4. For the first LC/MS/MS run, the Orbitrap Elite is operated in a data-dependent acquisition (DDA) mode (*see* **Note 17**). The m/z range for survey scans is 300–1600. A narrower m/z range helps to reduce the MS scan time. The dynamic exclusion duration is 30 s for PINK KO and WT samples, while it is 12 s for His-Ub samples. The decision tree procedure is not activated. Settings for MS scan and CID MS/MS scan are the same as those discussed in Section 2.7.1.

5. Based on the data analysis results of the first LC/MS/MS run, the m/z values of the potentially phosphorylated peptides are included in the "Parent mass list" when building the data acquisition method for the second LC/MS/MS run. Three MS/MS scan events are set for each precursor ion from the parent mass list: CID, ETD, and HCD. The parameters for MS

scan, CID and ETD MS/MS scan are the same as those discussed in Section 2.7.1. The HCD MS/MS scan is performed in the Orbitrap with a resolution of 15 K. Minimum signal threshold for MS/MS scan is 3×10^4. The isolation width is 2.0 Da. The value for normalized collision energy is 35, and the default charge state is 2.

2.8 Data Analysis

1. LC-MS/MS raw data are processed with Mascot Distiller 2.4.3.3 (Matrix Science Inc., http://www.matrixscience.com). The MS peak picking and processing method is the same for data acquired with different methods: Minimum signal to noise: 2; Expected peak width: 0.02 Da; Data points per Da: 200; Maximum charge: 7. For MS/MS peak picking and processing, the time domain is on with maximum intermediate time 30 s. For data acquired in DDA mode, the expected MS/MS peak width is 0.2 Da and default charge range is 2–3. For data acquired in DDDT mode, the expected MS/MS peak width is 0.1 Da and default charge range is 2–7. For data acquired in HCD mode, the expected MS/MS peak width is 0.05 Da and default charge range is 2–7.

2. The processed peaks are combined into Mascot generic format files (MGF). Mascot Daemon 2.4.0 is used to submit the MGF files to Mascot Server 2.4 (Matrix Science Inc., http://www.matrixscience.com) for database search. The following parameters are used for all searches: Monoisotopic peptide mass; Peptide tolerance: $\pm$ 10 ppm; MS/MS tolerance: 0.2 Da; Instrument type: CID + ETD. Parameters only for NL-1 samples: Searching against a house-built database which contains the sequences of NCBI human database and the GST-NL-1 sequence; Enzyme: None (*see* **Note 18**); Variable modifications: carbamidomethyl (C), oxidation (M), phospho (ST), phospho (Y). Parameters only for PINK WT/KO and His-UB samples: Searching against the NCBI human database; Trypsin as enzyme with up to two missed cleavages; Variable modifications: MMTS (C), oxidation (M), phospho (ST), phospho (Y).

3. Manually check the MS and MS/MS spectra of phosphopeptides matched in the database search. The spectra shown in Figs. 1 and 2 are individual scans without any merging.

3 Notes

1. The lysate should be a clear yellow color. If cloudy you will need to re-sonicate or spin down the lysate at $100,000 \times g$ for 30 min

2. Sample loss can be severe due to adsorption to the plastic surface, especially when sample amount is very limited. Using LoBind Eppendorf tubes helps increasing sample recovery.

3. The washing step can be carried out overnight. If the gel pieces are still blue, perform another wash step with 400 μl of 25 mM NH_4HCO_3 in 50 % MeOH for 20 min.

4. TCEP can be used as the reducing agent to break disulfide bonds.

5. IAM is light-sensitive. The reaction should be carried out in the dark. Iodoacetic acid or MMTS can be used to block the cysteine residues.

6. IAM can react with histidine and lysine residues at high pH, and react with methionine residues at low pH. This wash step helps to remove the extra IAM after the free cysteine residues are blocked. Don't incubate for longer than 20 min to minimize sample loss.

7. For PTM analysis it is critical to achieve high sequence coverage of the protein. Trypsin is the most commonly used enzyme for protein digestion, but is not always the best choice. Due to the sequence of the GST-NL-1 protein, a higher sequence coverage can be obtained when chymotrypsin is used instead of trypsin to digest the protein.

8. When using TCEP, thiols are not introduced to the sample mixture, unlike use of DTT. The effective pH range for TCEP is 1.5–8.5, much wider than that of DTT. TCEP is also more stable when pH > 7.5 and is more effective when pH < 8 [43]. The solution is acidic when dissolving TCEP in water. For trypsin digestion, it is better to adjust the pH of TCEP solution to 7–8.

9. MMTS is a reversible cysteine blocker. This feature is very important for experiments that need to selectively isolate cysteine-containing peptides after the digestion.

10. When using the μElution plate to desalt the sample, a typical sample volume is 100 μl. Although each well of the collection plate can hold up to 600 μl solution, to avoid cross-contamination, the volume of the solution used to elute the peptides should be less than 300 μl.

11. The amount of sorbent in μElution plate is 2 mg/well, with a maximum mass capacity estimated to be ~50 μg (see Waters Oasis Brochure youngwha.com/front/file/download.do?seq=1097). We use the μElution plate to desalt samples containing up to 5 μg peptides/proteins. For sample with larger amount, the plate with 5 mg/well sorbent or cartridges should be considered.

12. To minimize the possibility of sample loss, collect the flow-through from the sample loading and wash steps. Store at −20 °C. Discard it only after the LC/MS/MS data confirm the experiment is successful.

13. Alternatively, samples can be loaded using a microliter Pick-Up method with minimum sample loss. Since the microliter Pick-Up method requires a minimum of 10 μl sample loop, samples are usually loaded on a trapping column first at a flow rate of 10–30 μl/min, then flushed to the nano-column at a flow rate of 200–300 nl/min. Loading the sample onto a trapping column using the microliter Pick-Up method minimizes, and removes the need to desalt the digests before LC-MS/MS injection. However, it requires a loading pump and an extra column switching valve. In addition, because the same trapping column is used for different samples, the possibility of sample carryover increases.

14. At the time of performing the LC-MS/MS runs on the samples, the HPLC system delivers more MPB than indicated in the gradient, even right after viscosity calibration. To find out the right condition for the samples under the circumstance, LC-MS/MS data of BSA digests are acquired using different gradients. The gradient offering better separation is used for real samples.

15. Coon and coworkers introduced a DDDT logic [32] that makes "on-the-fly" decision on whether to use CID or ETD for a certain precursor ion, based on the peptide charge state z and mass-to-charge ratio (m/z). DDDT mode increases the number of successful MS/MS events that can be achieved in a single run. Otherwise, to obtain the complementary CID and ETD spectra without sacrificing the number of ions being scanned, two separate runs would be necessary.

16. In our hands, it is beneficial to change the value for "charge state 3 mass" from $m/z\,650$ to $m/z\,750$. For phosphopeptides with molecular weight ranging from 1950 to 2250 Da, it is very rare to see the quadruply charged form of the peptides. Mainly doubly and triply charged ions are detected. Using the default value embedded in the decision tree method, only CID spectra are obtained for those peptides. With the new setting, doubly charged ions of the peptides are fragmented with CID, and triply charged forms are fragmented with ETD. Since CID and ETD data often offer complementary information, the new setting helps increase the chance of localizing the phosphorylation site.

17. The purpose of using the DDA method instead of the DDDT method is to perform the MS/MS scan as fast as possible,

which increases the likelihood of finding phosphopeptides within a complex mixture.

18. Chymotrypsin often produces peptides that are cleaved non-specifically at one end or even both ends. Choosing "None" as the enzyme makes it possible to detect those nonspecific peptides. However, doing so largely increases the search time and the identity threshold.

References

1. Manning G et al (2002) The protein kinase complement of the human genome. Science 298(5600):1912–1934

2. Yamashita M, Fenn JB (1984) Electrospray ion-source—another variation on the free-jet theme. J Phys Chem 88(20):4451–4459

3. Hunt DF et al (1986) Protein sequencing by tandem mass-spectrometry. Proc Natl Acad Sci U S A 83(17):6233–6237

4. Mcluckey SA et al (1991) Ion spray liquid-chromatography ion trap mass-spectrometry determination of biomolecules. Anal Chem 63(4):375–383

5. Biemann K (1992) Mass-spectrometry of peptides and proteins. Annu Rev Biochem 61:977–1010

6. Arnott D, Shabanowitz J, Hunt DF (1993) Mass-spectrometry of proteins and peptides—sensitive and accurate mass measurement and sequence-analysis. Clin Chem 39(9):2005–2010

7. Eng JK, Mccormack AL, Yates JR (1994) An approach to correlate tandem mass-spectral data of peptides with amino-acid-sequences in a protein database. J Am Soc Mass Spectrom 5(11):976–989

8. Perkins DN et al (1999) Probability-based protein identification by searching sequence databases using mass spectrometry data. Electrophoresis 20(18):3551–3567

9. McLachlin DT, Chait BT (2001) Analysis of phosphorylated proteins and peptides by mass spectrometry. Curr Opin Chem Biol 5(5):591–602

10. Mann M et al (2002) Analysis of protein phosphorylation using mass spectrometry: deciphering the phosphoproteome. Trends Biotechnol 20(6):261–268

11. Olsen JV et al (2007) Higher-energy C-trap dissociation for peptide modification analysis. Nat Methods 4(9):709–712

12. Nagaraj N et al (2010) Feasibility of large-scale phosphoproteomics with higher energy collisional dissociation fragmentation. J Proteome Res 9(12):6786–6794

13. Frese CK et al (2011) Improved peptide identification by targeted fragmentation using CID, HCD and ETD on an LTQ-orbitrap velos. J Proteome Res 10(5):2377–2388

14. Jedrychowski MP et al (2011) Evaluation of HCD- and CID-type fragmentation within their respective detection platforms for murine phosphoproteomics. Mol Cell Proteomics 10(12):M111.009910

15. Shen YF et al (2011) Effectiveness of CID, HCD, and ETD with FT MS/MS for degradomic-peptidomic analysis: comparison of peptide identification methods. J Proteome Res 10(9):3929–3943

16. Xia Y, Liang XR, McLuckey SA (2006) Ion trap versus low-energy beam-type collision-induced dissociation of protonated ubiquitin ions. Anal Chem 78(4):1218–1227

17. Dongre AR et al (1996) Influence of peptide composition, gas-phase basicity, and chemical modification on fragmentation efficiency: evidence for the mobile proton model. J Am Chem Soc 118(35):8365–8374

18. Huang YY et al (2005) Statistical characterization of the charge state and residue dependence of low-energy CID peptide dissociation patterns. Anal Chem 77(18):5800–5813

19. Sleno L, Volmer DA (2004) Ion activation methods for tandem mass spectrometry. J Mass Spectrom 39(10):1091–1112

20. Boersema PJ, Mohammed S, Heck AJR (2009) Phosphopeptide fragmentation and analysis by mass spectrometry. J Mass Spectrom 44(6):861–878

21. Zubarev RA, Kelleher NL, McLafferty FW (1998) Electron capture dissociation of multiply charged protein cations. A nonergodic process. J Am Chem Soc 120(13):3265–3266

22. Syka JEP et al (2004) Peptide and protein sequence analysis by electron transfer dissociation mass spectrometry. Proc Natl Acad Sci U S A 101(26):9528–9533

23. Good DM et al (2007) Performance characteristics of electron transfer dissociation mass spectrometry. Mol Cell Proteomics 6 (11):1942–1951

24. Molina H et al (2008) Comprehensive comparison of collision induced dissociation and electron transfer dissociation. Anal Chem 80 (13):4825–4835

25. Bemben MA et al (2014) CaMKII phosphorylation of neuroligin-1 regulates excitatory synapses. Nat Neurosci 17(1):56–64

26. Kane LA et al (2014) PINK1 phosphorylates ubiquitin to activate Parkin E3 ubiquitin ligase activity. J Cell Biol 205(2):143–153

27. Koyano F, Okatsu K, Kosako H, Tamura Y, Go E, Kimura M, Kimura Y, Tsuchiya H, Yoshihara H, Hirokawa T, Endo T, Fon EA, Trempe JF, Saeki Y, Tanaka K, Matsuda N. Nature. 2014 Jun 5;510(7503):162–6. doi:10.1038/nature13392. Epub 2014 Jun 4.

28. Kazlauskaite A, Kondapalli C, Gourlay R, Campbell DG, Ritorto MS, Hofmann K, Alessi DR, Knebel A, Trost M, Muqit MM. Biochem J. 2014 May 15;460(1):127–39. doi:10.1042/BJ20140334

29. Ichtchenko K et al (1995) Neuroligin-1—a splice site-specific ligand for beta-neurexins. Cell 81(3):435–443

30. Iida J et al (2004) Synaptic scaffolding molecule is involved in the synaptic clustering of neuroligin. Mol Cell Neurosci 27(4):497–508

31. Sudhof TC (2008) Neuroligins and neurexins link synaptic function to cognitive disease. Nature 455(7215):903–911

32. Bemben MA, Shipman SL, Nicoll RA, Roche KW. Trends Neurosci. 2015 Jul 21. pii: S0166-2236(15)00149-6. doi:10.1016/j.tins.2015.06.004. [Epub ahead of print] Review. PMID:26209464

33 Chubykin AA et al (2007) Activity-dependent validation of excitatory versus inhibitory synapses by neuroligin-1 versus neuroligin-2. Neuron 54(6):919–931

34. Swaney DL, McAlister GC, Coon JJ (2008) Decision tree-driven tandem mass spectrometry for shotgun proteomics. Nat Methods 5(11):959–964

35. Lin W, Kang UJ (2008) Characterization of PINK1 processing, stability, and subcellular localization. J Neurochem 106(1):464–474

36. Jin SM et al (2010) Mitochondrial membrane potential regulates PINK1 import and proteolytic destabilization by PARL. J Cell Biol 191 (5):933–942

37. Deas E et al (2011) PINK1 cleavage at position A103 by the mitochondrial protease PARL. Hum Mol Genet 20(5):867–879

38. Yamano K, Youle RJ (2013) PINK1 is degraded through the N-end rule pathway. Autophagy 9(11):1758–1769

39. Matsuda N et al (2010) PINK1 stabilized by mitochondrial depolarization recruits Parkin to damaged mitochondria and activates latent Parkin for mitophagy. J Cell Biol 189 (2):211–221

40. Geisler S et al (2010) PINK1/Parkin-mediated mitophagy is dependent on VDAC1 and p62/SQSTM1. Nat Cell Biol 12(2):119–131

41. Narendra DP et al (2010) PINK1 is selectively stabilized on impaired mitochondria to activate parkin. PLoS Biol 8(1):e1000298

42. Vives-Bauza C et al (2010) PINK1-dependent recruitment of Parkin to mitochondria in mitophagy. Proc Natl Acad Sci U S A 107 (1):378–383

43. Kondapalli C et al (2012) PINK1 is activated by mitochondrial membrane potential depolarization and stimulates Parkin E3 ligase activity by phosphorylating Serine 65. Open Biol 2

44. Shiba-Fukushima K et al (2012) PINK1-mediated phosphorylation of the Parkin ubiquitin-like domain primes mitochondrial translocation of Parkin and regulates mitophagy. Sci Rep 2:1002

45. Han JC, Han GY (1994) A procedure for quantitative determination of tris(2-carboxyethyl)phosphine, an odorless reducing agent more stable and effective than dithiothreitol. Anal Biochem 220(1):5–10

Neuromethods (2016) 114: 171–185
DOI 10.1007/7657_2015_84
© Springer Science+Business Media New York 2015
Published online: 08 April 2016

Liquid Chromatography Tandem Mass Spectrometry Analysis of Tau Phosphorylation

Jhoana Mendoza, Georgia Dolios, and Rong Wang

Abstract

Hyperphosphorylation, aggregation, and formation of neurofibrillary tangles of the microtubule-associated protein tau have been implicated in the pathogenesis of Alzheimer's disease (AD) and other tauopathies. Phosphorylation and dephosphorylation of tau regulate its attachment or detachment from microtubules (Yoshiyama et al., J Neurol Neurosurg Psychiatry 84(7):784–795, 2013). The abnormal hyperphosphorylation of tau, however, disrupts its proper binding to microtubules and induces microtubule disassembly. Accumulation of unbound tau, then, results to aggregate and neurofibrillary tangle formation. Knowing the mechanism behind the abnormal phosphorylation of tau, therefore, is important to understanding the pathogenesis of neurodegenerative diseases. A protocol that employs nanospray liquid chromatography tandem mass spectrometry (nanoLC-MS^3) for the analysis of tau phosphorylation by the checkpoint kinases, Chk1 and Chk2, in vitro, is provided here. Technical details on phosphorylation, protein digestion, peptide desalting, reversed-phase liquid chromatography (RPLC), mass spectrometry, and proteomics data analysis are discussed.

Keywords: Tau phosphorylation, Liquid chromatography, Mass spectrometry, LC-MS, LC-MS/MS

1 Introduction

Tau protein promotes the assembly and stability of neuronal microtubules under normal physiological conditions [1, 2]. There are six isoforms in the adult human brain which have been found to differ from each other in having either three (3R) or four (4R) microtubule-binding repeats and from zero (0N) to two (2N) amino- or N-terminal inserts [3, 4]. However, modifications such as hyperphosphorylation under pathological conditions could impair the binding of tau to microtubules [1, 4]. The accumulation of unbound tau results to the formation of aggregates such as those found in neurofibrillary tangles in Alzheimer's disease (AD). Knowing the mechanism behind the abnormal phosphorylation of tau in AD, therefore, is important to understanding the disease's pathogenesis as well as those of other neurodegenerative disorders.

The longest isoform of tau consists of 441 amino acid residues and has 85 potential phosphorylation sites (45 serine, 35 threonine, and 5 tyrosine). To date, 45 sites have been identified to be phosphorylated in AD brains [5]. Phosphorylation in the microtubule-binding domain (residues 244–368) is believed to be important in regulating microtubule stability. Phosphorylation at other sites, such as Ser214 and Thr231, has also been found to maintain cytoskeletal stability [6]. The interaction of a peptidyl-prolyl isomerase, Pin1, with phosphorylated Thr231, has been reported to restore binding of tau to microtubules indicating a possible role of this site in AD pathogenesis [7]. Indeed, the relative importance of phosphorylation at other sites to tau function and/ or toxicity remains to be studied and established, more importantly in in vivo systems.

Several kinases have been shown to phosphorylate tau in vitro and in cells [4, 8]. These are classified into proline-directed or non-proline directed protein kinases. The DNA damage-activated cell cycle checkpoint kinases, Chk1 and Chk2, have also been shown to phosphorylate tau at AD-related sites and enhance tau toxicity [8, 9]. On the other hand, protein phosphatase PP1, PP2A, PP2B, and PP5 are reported to dephosphorylate tau [10]. The binding and unbinding of tau to microtubules is actually coordinated by the balance of phosphorylation and dephosphorylation of these enzymes.

Mass spectrometric analysis [5, 8], Edman degradation of phosphopeptides [5, 11], and use of phospho-tau-specific antibodies [5, 8] have been employed to identify tau phosphorylation sites in control and AD human brain and in in vitro kinase studies. However, the low abundance and lower ionization efficiency of phosphopeptides compared to non-modified peptides impose challenges to their detection [12, 13]. Advances in MS instrumentation and methodology have provided increased sensitivity and specificity to the detection of protein phosphorylation and other posttranslational modifications (PTMs; [14, 15]). High-pressure liquid chromatography (HPLC) in combination with tandem mass spectrometry has been used to resolve and simplify complex peptide mixtures from protein digests for improved sequence coverage. Trypsin has been the protease of choice for PTM analysis because of its high cleavage specificity [16] and the remaining positively charged amino acid residues, Lys and Arg, at the C-terminals of the resulting peptides. In cases where tryptic peptides may be too large for LC-MS/MS analysis or too hydrophilic to be retained by reversed-phase C_{18} material, as in the case of many phosphorylated tryptic peptides, substitution of trypsin for another enzyme having different cleavage specificity should be considered [14]. Asp-N, chymotrypsin, or Glu-C are popular choices. When tryptic peptides are too large, it is often desirable to add the second enzyme to the trypsin digest. The use of phosphopeptide enrichment strategies such as metal affinity chromatography [17, 18] and/or ion-exchange chromatography [19, 20] further provides enhanced sensitivity and efficient MS characterization of phosphopeptides. Another difficulty often encountered in phosphopeptide analysis by mass spectrometry is the predominance of phosphorylation-specific neutral losses in the MS/MS spectra. This yields lower intensities of b- and y-ions resulting from peptide backbone fragmentations which are important for phosphopeptide identification and precise phosphorylation site determination. Data-dependent neutral loss (DDNL) MS^3 methods and multistage activation (MSA) protocols are among the strategies developed for generating efficient phosphopeptide fragmentation in ion trap instruments [14]. In DDNL MS^3 methods, additional fragmentation of the product of the precursor neutral loss in the form of an MS^3 scan was initiated when a dominant NL-associated peak was detected in the MS/MS spectrum [21, 22]. The MSA approach, on the other hand, uses consistent supplemental activation of neutral loss product ions and records all the fragments from both the precursor and NL product activation in the same MS/MS spectrum (pseudo-MS^3) [23].

In this chapter, we provide specifics on how we perform these experiments. We employ nanospray LC-MS^3 to identify the sites in tau protein phosphorylated by Chk1 and Chk2 in vitro. A data-dependent neutral loss MS^3 method is used to improve fragmentation of phosphopeptides (Fig. 1). As illustrated in Fig. 2, Phosphorylation site assignment from database search was further evaluated with A-scores using Scaffold PTM [24]. A total of 25 Ser/Thr residues was identified as Chk1- or Chk2-target sites. Many of these sites are located within the microtubule-binding domain and C-terminal domain (Fig. 3), phosphorylation of which has been shown to reduce tau binding to microtubules and/or has been implicated in tau toxicity [8].

2 Materials

2.1 Tau Phosphorylation

1. Recombinant human tau 0N4R (variant 3, NM_016834.3, T9825, Sigma, St. Louis, MO).

2. Recombinant active human GST-tagged Chk1 (C0870, activity 169–229 nmol/min mg, Sigma, St. Louis, MO).

3. Recombinant active human GST-tagged Chk2 (C0995, activity 654–884 nmol/min mg, Sigma, St. Louis, MO).

4. Reaction buffer: 5 mM MOPS (pH 7.2), 2.5 mM glycerol 2-phosphate, 5 mM $MgCl_2$, 1 mM EGTA, 0.4 mM EDTA, 0.05 mM DTT and 5 µM ATP (Reagents are purchased separately from Sigma-Aldrich, St. Louis, MO).

2.2 Reduction, Alkylation, and Enzyme Digestion

1. Ammonium bicarbonate, NH_4HCO_3 (Sigma-Aldrich, St. Louis, MO).

2. Reducing reagent: Tris(2-carboxyethyl)-phosphine, TCEP (Pierce, Thermo Fisher Scientific, Waltham, MA).

3. Alkylating reagent: Iodoacetamide, IAM (Sigma-Aldrich, St. Louis, MO).

4. Calcium chloride, $CaCl_2$ (Sigma-Aldrich, St. Louis, MO).

5. Formic acid (analytical/chromatography grade, Sigma-Aldrich, St. Louis, MO).

6. Trifluoroacetic acid, TFA (LC-MS grade, Pierce, Thermo Fisher Scientific, Waltham, MA).

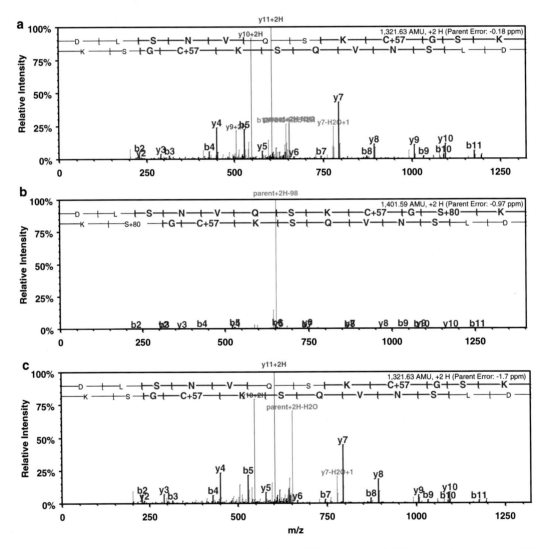

Fig. 1 MS/MS spectra of a tau peptide from (**a**) control, (**b**) Chk1-treated, (**c**) Chk2-treated samples and (**d**) MS^3 spectrum showing that Ser293 was phosphorylated by Chk1 only. The detection of the corresponding non-phosphopeptide in the sample treated with Chk2 suggests that phosphorylation at residue Ser293 is unique to the Chk1-treated sample. Reprinted with permission from Mendoza et al. [8]. Copyright 2013 American Chemical Society

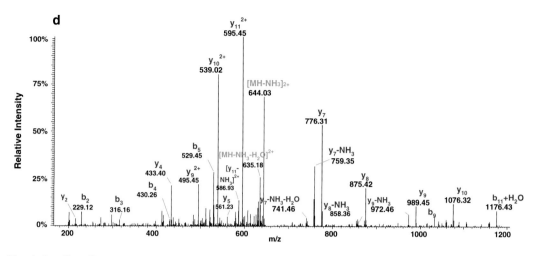

Fig. 1 (continued)

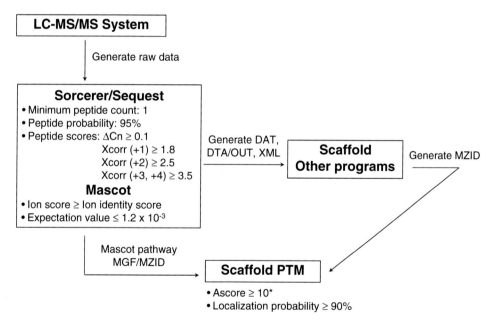

Fig. 2 Data flow in phosphorylation analysis using Scaffold and/or Mascot and Scaffold PTM. Scaffold PTM requires a peak list in MGF format and a list of proteins and peptides in MZIdentML format. The scores/criteria used for phosphopeptide identification and phosphorylation site assignment are included. (*Star*) Typically, *A*-score ≥ 15 is used but the high localization probability (99 %) and good MS/MS quality of the phosphopeptide with *A*-score ≥ 10 in this study led us to include it

7. Sequencing-grade modified trypsin (Porcine; Promega, Madison, WI).

8. Endoproteinase Asp-N sequencing grade (Roche Applied Science, Indianapolis, IN).

```
1     MAEPRQEFEV  MEDHAGTYGL  GDRKDQGGYT  MHQDQEGDTD  AGLKESPLQT  50

51    PTEDGSEEPG  SETSDAKSTP  TAEDVTAPLV  DEGAPGKQAA  AQPHTEIPEG  100

101   TTAEEAGIGD  TPSLEDEAAG  HVTQARMVSK  SKDGTGSDDK  KAKGADGKTK  150

151   IATPRGAAPP  GQKGQANATR  IPAKTPPAPK  TPPSSGEPPK  SGDRSGYSSP  200

201   GSPGTPGSRS  RTPSLPTPPT  REPKKVAVVR  TPPKSPSSAK  SRLQTAPVPM  250

251   PDLKNVKSKI  GSTENLKHQP  GGGKVQIINK  KLDLSNVQSK  CGSKDNIKHV  300

301   PGGGSVQIVY  KPVDLSKVTS  KCGSLGNIHH  KPGGGQVEVK  SEKLDFKDRV  350

351   QSKIGSLDNI  THVPGGGNKK  IETHKLTFRE  NAKAKTDHGA  EIVYKSPVVS  400

401   GDTSPRHLSN  VSSTGSIDMV  DSPQLATLAD  EVSASLAKQG  L
```

Fig. 3 Phosphorylation sites on tau by Chk1 and Chk2. The residues phosphorylated by Chk1 but not Chk2 are shown in *open boxes*, the residues phosphorylated by Chk2 but not Chk1 are shown in *black boxes*, and the residues phosphorylated by both Chk1 and Chk2 are shown in *gray boxes*. Stars indicate the residues phosphorylated in AD brains, and an *open circle* indicates the residues possibly phosphorylated in AD brains. The amino acid sequence of the longest isoform of human brain tau (2N4R) is shown. Residues corresponding to N-terminal inserts are *italicized*. The microtubule-binding domain is *underlined*. Phosphorylation at Ser320 and Ser409 is detected with Western blot analysis using anti-pSer320 tau and anti-pSer409 tau. Reprinted with permission from Mendoza et al. [8]. Copyright 2013 American Chemical Society

9. Water (Optima LC/MS, Thermo Fisher Scientific, Waltham, MA).

10. Thermomixer (Eppendorf, Hauppage, NY).

11. pH paper (Whatman Inc., Piscataway, NJ).

12. 1.5 mL Microcentrifuge tubes (Thermo Fisher Scientific, Waltham, MA).

13. Speed vacuum concentrator (Savant™, Thermo Fisher Scientific, Waltham, MA).

2.3 Peptide Desalting

1. ZipTip C$_{18}$ (Millipore, Billerica, MA).

2. POROS R2 beads (Applied Biosystems, Foster City, CA).

3. Ethanol (Thermo Fisher Scientific, Waltham, MA).

4. Water (Optima LC/MS, Thermo Fisher Scientific, Waltham, MA).

5. Trifluoroacetic acid, TFA (LC-MS grade, Pierce, Thermo Fisher Scientific, Waltham, MA).

6. Activation solution: 100 % acetonitrile, ACN (Optima LC/MS, Thermo Fisher Scientific, Waltham, MA).

7. Equilibration solution: 0.1 % TFA in water.

8. Wash solution: 0.1 % TFA in water.

9. Elution solution: 0.1 % TFA in 50:50 ACN:H$_2$O.

10. Reconstitution solvent: 0.1 % formic acid in 2:98 ACN:H$_2$O.

11. 50 mL Conical centrifuge tubes (Falcon™, Thermo Fisher Scientific, Waltham, MA).

12. 1.5 mL Microcentrifuge tubes (Thermo Fisher Scientific, Waltham, MA).

2.4 Reversed-Phase Liquid Chromatography

1. Solvent A: 0.1 % formic acid in LC-MS-grade water.

2. Solvent B: 0.1 % formic acid in LC-MS-grade acetonitrile.

3. NanoAcquity Ultrahigh Pressure Liquid Chromatography (UPLC) system (Waters, Milford, MA).

4. C_{18} trap column (Waters Symmetry®, 180 μm id × 20 mm length, 5 μm particle size, Waters, Milford, MA).

5. C_{18} capillary column (BEH130, 100 μm i.d × 100 mm length, 1.7 μm particle size, Waters, Milford, MA).

6. 300 μL Autosampler microvials with pre-slit caps (National Scientific™, Thermo Fisher Scientific, Waltham, MA).

2.5 Mass Spectrometry

1. LTQ-Orbitrap mass spectrometer equipped with a nanospray ionization source (Thermo Scientific, San Jose, CA).

2. LTQ calibration solution for positive ion mode: 200 μL of 1.0 mg/mL caffeine in 100 % methanol, 100 μL of 166.7 μM L-methionyl-arginyl-phenylalanyl-alanine acetate × H_2O (MRFA) in 50:50 methanol:water, 100 μL of 0.1 % Ultramark 1621 in 100 % acetonitrile and 100 μL glacial acetic acid, 5 mL acetonitrile, and 4.5 mL 50:50 methanol:water (all reagents except solvents are purchased separately from Sigma-Aldrich, St. Louis, MO).

3. LTQ Orbitrap calibration solution for positive and negative ion mode: 10 mL of LTQ positive ion mode calibration solution, 100 μL of 1.0 mM sodium dodecyl sulfate (SDS) in 50:50 methanol:water, 100 μL of 1.0 mM sodium taurocholate in 50:50 methanol:water and 200 μL of 166.7 μM of MRFA in 50:50 methanol:water (reagents are purchased separately from Sigma-Aldrich, St. Louis, MO).

4. Angiotensin 5-peptide mix (Michrom Bioresources Inc., Pleasanton, CA).

5. Water (Optima LC/MS, Thermo Fisher Scientific, Waltham, MA).

6. Methanol (Optima LC/MS, Thermo Fisher Scientific, Waltham, MA).

7. Acetonitrile (Optima LC/MS, Thermo Fisher Scientific, Waltham, MA).

2.6 Data Analysis

1. Sorcerer/Sequest (Sage-N Research, Milpitas, CA).

2. Mascot (Matrix Science Inc., Boston, MA).

3. Scaffold (Proteome Software Inc., Portland, OR).

4. Scaffold PTM (Proteome Software Inc., Portland, OR).

3   Methods

3.1   Tau Phosphorylation

1. Dilute the recombinant active human GST-tagged Chk1 to 1:2 in the reaction buffer, mix with 2 μg of recombinant human tau 0N4R, and incubate at 30 °C for 3 h.

2. Similarly, dilute the recombinant active human GST-tagged Chk2 to 1:5 in the reaction buffer, mix with 2 μg of recombinant human tau 0N4R, and incubate at 30 °C for 3 h.

3.2   Reduction, Alkylation, and Enzyme Digestion

1. Prepare all reagent solutions using LC-MS-grade water.

2. Adjust the pH of tau protein samples with or without kinases to pH 8.5 with 100 mM NH_4HCO_3—Reduce sample volumes first to approximately 70 % of the original using a speed vacuum. Titrate solution pH to 8.5 with 100 mM NH_4HCO_3. During pH titration, vortex-mix the samples, centrifuge briefly, and check the pH with pH paper. Keep adding NH_4HCO_3 in small increments until desired pH is achieved and restore samples to the original volumes (*see* **Note 1**).

3. Reduce the proteins with 1 M TCEP to a final concentration of 5 mM. Vortex-mix the samples and centrifuge briefly. Place the sample tubes in a Thermomixer set to 37 °C for 20 min, with mixing.

4. Centrifuge the samples briefly and cover the tubes with foil. Add the appropriate volume of 500 mM iodoacetamide to achieve a final concentration of 10 mM in solution. Vortex-mix the samples and centrifuge briefly. Allow the alkylation process to proceed for 30 min in the dark at room temperature (*see* **Note 2**).

5. Add 1 M $CaCl_2$ to the samples to a final concentration of 1 mM, vortex-mix, and centrifuge briefly (*see* **Note 3**).

6. Reconstitute a vial (20 μg/vial) of sequencing-grade trypsin with 40 μL of resuspension buffer (50 mM acetic acid) provided by the manufacturer. Vortex-mix the enzyme solution for 30 s and centrifuge briefly. Divide the trypsin solution into 5 μL aliquots, set aside the amount needed for digestion, and keep the tubes on ice; store the rest in a −30 °C freezer for future use.

7. Check if the pH of sample solutions is at 8.3–8.5, using a pH paper. If pH is less than the optimal value, adjust with 100 mM NH_4HCO_3 (*see* **Note 3**).

8. Prepare a $10\times$ dilution of the stock trypsin solution. Add 0.05 µg/µL trypsin at an enzyme to substrate ratio of 1:50. Incubate the sample mixture overnight in a Thermomixer set to 37 °C in the dark while mixing.

9. Quench the digestion by adding an appropriate volume of 10 % $TFA_{(aq)}$ to achieve a pH of 2–4.

10. Reconstitute a vial (2 µg/vial) of sequencing-grade endoproteinase Asp-N with 50 µL of LC/MS-grade water. Vortex-mix the enzyme solution for 30 s and centrifuge briefly. Divide the solution into 2.5 µL aliquots, dry down in a speed vacuum, set aside the amount needed for digestion and keep the tubes on ice; store the rest in a -30 °C freezer for future use.

11. Follow the same procedure for digestion with 0.04 µg/µL Asp-N (reconstituted in 100 mM NH_4HCO_3) except the addition of $CaCl_2$ (*see* **Note 4**).

3.3 Peptide Desalting

1. For POROS R2 beads conditioning, add 40 mL of acetonitrile into the original bottle, mix and transfer 20 mL into a 50 mL conical tube. Vortex-mix vigorously for 1 min, centrifuge for 10 min at $1700 \times g$ at room temperature, and discard the supernatant. Repeat this wash procedure with 20 mL of 0.1 % $TFA_{(aq)}$ and with 20 mL water (twice). Add 13.5 mL of 10 % ethanol for an approximate 100 µg/µL bead slurry preparation.

2. Place ZipTip C_{18} tips in holders positioned in 1.5 mL microcentrifuge tubes. Load a 50 µL aliquot of 100 µg/µL POROS R2 bead slurry in 10 % ethanol onto ZipTip C_{18} tips. Centrifuge at $240 \times g$ for 2 min at room temperature. Continue centrifugation until almost all solution has been eluted.

3. Activate the POROS R2 slurry and C_{18} resin by adding 50 µL of 100 % acetonitrile and centrifuging at $240 \times g$ for 2 min. Continue centrifugation until almost all solution has been eluted.

4. Equilibrate by adding 50 µL of 0.1 % $TFA_{(aq)}$ and centrifuging at 1800 rpm for 2 min or until all solution has been eluted. Minimize column drying from this point on.

5. Check the sample pH to make sure it is less than 4 (*see* **Note 5**).

6. Replace the microcentrifuge tubes with clean ones. Load the samples onto the ZipTips with POROS R2 beads and centrifuge at $340 \times g$ for 2 min. Continue centrifugation until all solution has been eluted. Collect the flow-through and repeat sample loading two more times (*see* **Note 6**).

7. Replace the microcentrifuge tubes with the ones used to collect the eluate from activation and equilibration of the resin. Wash off unbound salts by adding 50 µL of 0.1 % $TFA_{(aq)}$ onto the

ZipTip columns and centrifuging at $340 \times g$ for 2 min or until all solution has been eluted.

8. Add 25 µL of 0.1 % TFA in 50:50 ACN:H_2O to elute the peptides of interest and collect in clean microcentrifuge tubes. Centrifuge at $340 \times g$ for 2 min and continue centrifugation until all solution has been eluted.

9. Dry down the eluate in a speed vacuum (*see* **Note 7**).

3.4 Reversed-Phase Liquid Chromatography

1. Reconstitute the peptide residue in 25 µL of solvent A. Vortex-mix for 15 s and centrifuge briefly. Mix, centrifuge again, and make sure that all residue has been dissolved.

2. Transfer the sample into an autosampler microvial and check for bubbles.

3. Equilibrate the RPLC column for at least 10 min with 1 % solvent B at 0.5 µL/min (*see* **Note 8**).

4. Load 5–10 µL of the reconstituted peptides onto a C_{18} trap column with 0.5 % solvent B at a flow rate of 15 µL/min (*see* **Note 9**).

5. Resolve or separate the bound peptides using a C_{18} capillary column at a flow rate of 0.5 µL/min using the following gradient:

Time (min)	% Solvent A	% Solvent B
0	99	1
3	99	1
33	50	50
43	15	85
53	15	85
55	99	1
70	99	1

3.5 Mass Spectrometry

1. Prior to sample run, calibrate the mass spectrometer using a standard calibration solution to ensure the mass accuracy of the instrument. Switch to using an electrospray ionization source and directly infuse the calibration solution into the mass spectrometer at a flow rate of 3–5 µL/min. Please refer to the LTQ Orbitrap Getting Started manual for detailed instructions on how to prepare the calibration solution (*see* **Note 10**).

2. For optimum ion transmission conditions, tune the instrument with 100 nM angiotensin 5-peptide mix prepared in 50:50 ACN:H_2O with 0.1 % formic acid. Directly infuse the tune solution into the mass spectrometer at a flow rate of 3 µL/min.

Specifically, the $[M + 3H]^{3+}$ ion of angiotensin-1 at m/z 433 is used for automatic tuning (*see* **Note 11**).

3. Introduce peptides eluted from the reversed-phase capillary column into the mass spectrometer using a nanospray ionization source (*see* **Note 12**).

4. Operate the mass spectrometer in positive mode with spray voltage at 2.1 kV, ion transfer tube voltage at 49 V, and ion transfer tube temperature at 170 °C. Set the sheath and auxiliary gases to zero (*see* **Note 13**).

5. For a full-scan mass spectrum acquisition, set the target value at 1×10^6 ions with resolution (R) of 60,000 at m/z 400. Enable the lock mass option, using the polydimethylcyclosiloxane ion (PCM; protonated $(Si(CH3)_2O)_6$) at m/z 445.120025, for accurate mass measurement [25].

6. Set an ion signal threshold of 1000 for MS/MS. Use a normalized collision energy of 35 %, an activation of $q = 0.25$, and activation time of 30 ms for MS/MS acquisitions.

7. Employ data-dependent acquisition with automatic switching between MS and MS/MS modes. Select the top eight most intense ions for fragmentation in the LTQ. Use a collision-induced dissociation (CID) target value of 10,000 ions for fragmentation.

   Apply the following dynamic exclusion settings to precursor ions chosen for MS/MS analysis: repeat count—1; repeat duration—30 s; and exclusion duration—120 s.

8. Carry out a neutral loss experiment where data-dependent settings were chosen to trigger an MS^3 scan when a neutral loss of 97.97, 48.99 or 32.66 m/z units (relative to the singly, doubly, or triply charged phosphorylated precursor ion, respectively), was detected among the eight most intense product ions to improve fragmentation of phosphopeptides.

9. To ensure reliable mass spectrometric identification of phosphorylation sites, repeat the phosphorylation reaction, sample processing, and mass spectrometric analysis at least two more times.

3.6 Data Analysis

The following database and search engines are used for the search: human component of the NCBI non-redundant database (11/01/2010 version; 113,484 entries); Sequest (Ver.27, Rev. 11); and Mascot (Ver. 2.3.01).

1. Search the MS, MS/MS and MS^3 spectra against the human component of the NCBI non-redundant database using Sequest and Mascot algorithms.

2. Perform searches with full tryptic specificity (two missed cleavages); carbamidomethylated cysteine residues (+57.0340 Da) as static modification and oxidized methionine, histidine, and

tryptophan (+15.9949 Da); deamidated asparagine and gluta-mine (+0.9840 Da); phosphorylated serine, threonine, and tyrosine (+79.9663 Da); and dehydroalanine and dehydroami-nobutyric acid (−18.0106 Da) as differential modifications.

3. Use a precursor mass error tolerance of 10 ppm and the default product ion mass error tolerance of the above searching algorithms.

4. Use Scaffold to view Sequest search results. Set the minimum protein probability to 95 %, minimum number of peptides to 1, minimum peptide probability to 95 %, and the following pep-tide scores: $\Delta Cn \geq 0.1$, XCorr ≥ 1.8 for +1, XCorr ≥ 2.5 for +2, and XCorr ≥ 3.5 for +3 and +4.

5. Consider the ion score, ion identity score, and expectation value in Mascot search results for evaluating the phosphopep-tides identified.

6. Inspect the tandem mass spectra and product ion lists manually to ensure quality of the phosphorylation site identification (*see* **Note 14**).

7. Evaluate further the phosphorylation site assignment from database search with *A*-scores using Scaffold PTM. Generate MZID and MGF files of the raw data with Scaffold. Export the MZID and corresponding MGF files in Scaffold. Create a new experiment and load the MZID and MGF files of the data that needs to be analyzed into Scaffold PTM. Scaffold PTM then calculates the *A*-score and localization probability for the mod-ified peptides (*see* **Note 15**).

4   Notes

1. Separation of tau samples on polyacrylamide gels and staining with colloidal Coomassie Blue G could also be carried out for further sample purification [5].

2. The alkylating reagent, iodoacetamide, is light sensitive, so minimize light exposure during sample alkylation. It is also preferable to prepare it fresh before use.

3. For optimal trypsin activity, concentrations of salts and other reagents (e.g., reducing and alkylating agents, SDS, urea and acetonitrile) must be within tolerable limits; otherwise dilution with 50–100 mM NH_4HCO_3 is necessary. Addition of calcium ions and adjustment of sample pH are also important to ensure effective digestion. Trypsin works best at pH 7–9 [26, 27].

4. Carrying out the sample digestion twice with enzymes of dif-ferent cleavage specificity helps improve proteolytic efficiency. Proteolysis by Asp-N provided access to additional cleavage

sites that resulted in smaller peptide size for regions of the tau protein that could not be well digested by trypsin.

5. Follow the user guide for reversed-phase ZipTip (Millipore).

6. Loading the sample repeatedly onto the ZipTip column allows for maximum binding of peptides to the packing material.

7. Phosphopeptide enrichment of the sample using TiO_2, ZrO_2, and/or Fe^{3+} and Ga^{3+} immobilized beads usually in conjunction with ion-exchange chromatography such as strong cation exchange (SCX), hydrophilic interaction chromatography (HILIC), and electrostatic repulsion—hydrophilic interaction chromatography (ERLIC) could be employed prior to LC-MS^3 analysis for improved phosphopeptide detection.

8. Another important consideration is to evaluate and optimize the biocompatibility of the UPLC system to minimize phosphopeptide loss during analysis.

9. Loading capacity of columns to be used should be considered to prevent overloading and promote efficient resolution of peptide mixture.

10. It is recommended to designate and use an ion transfer tube for calibration purposes only to minimize carry-over of calibrant ions in sample runs. Use a different ion transfer tube for tuning and sample runs.

11. Ideally, the corresponding compound of interest (e.g., peptide, small molecule) is used for tuning the instrument parameters. However, for instances when the sample is complex such as a protein digest, a peptide or mixture of peptides with masses that fall within the mass range of interest can be used for tuning.

12. Nanospray ionization offers better sensitivity for samples of limited quantities [28, 29].

13. Sheath and auxiliary gases are not used with nanospray. For experiments conducted at higher LC flow rates (>10 μL/min), optimizing gas flow settings contributes to MS signal enhancement.

14. In cases where an MS/MS spectrum resulted to phosphopeptide identification and has a corresponding MS^3 scan, provided that the MS/MS spectrum shows a good signal-to-noise quality and relatively high number of signal peaks, the mass list of b- and y-ions is consulted and detection of the corresponding fragment ions especially the phosphorylated residue is confirmed. If the phosphorylated residue and several other b- and y-ions are detected, the phosphopeptide is considered for *A*-score evaluation and possible inclusion in the list of modified peptides identified.

15. *A*-score measures the probability or likelihood that a site is modified by chance by examining the product ion peaks

between the first and second most likely site. The localization probability is calculated based on the P value used in *A*-score calculation. Scaffold PTM "extends the *A*-score algorithm to consider overlapping data from several peptides simultaneously to improve confidence in specific site assignments" [24, 30].

References

1. Yoshiyama Y, Lee VM, Trojanowski JQ (2013) Therapeutic strategies for tau mediated neurodegeneration. J Neurol Neurosurg Psychiatry 84(7):784–795

2. Avila J et al (2004) Role of tau protein in both physiological and pathological conditions. Physiol Rev 84(2):361–384

3. Goedert M et al (1989) Multiple isoforms of human microtubule-associated protein tau: sequences and localization in neurofibrillary tangles of Alzheimer's disease. Neuron 3 (4):519–526

4. Hanger DP, Anderton BH, Noble W (2009) Tau phosphorylation: the therapeutic challenge for neurodegenerative disease. Trends Mol Med 15(3):112–119

5. Hanger DP et al (2007) Novel phosphorylation sites in tau from Alzheimer brain support a role for casein kinase 1 in disease pathogenesis. J Biol Chem 282(32):23645–23654

6. Cho JH, Johnson GV (2003) Glycogen synthase kinase 3beta phosphorylates tau at both primed and unprimed sites. Differential impact on microtubule binding. J Biol Chem 278(1):187–193

7. Buerger K et al (2006) CSF phosphorylated tau protein correlates with neocortical neurofibrillary pathology in Alzheimer's disease. Brain 129(Pt 11):3035–3041

8. Mendoza J et al (2013) Global analysis of phosphorylation of tau by the checkpoint kinases Chk1 and Chk2 in vitro. J Proteome Res 12 (6):2654–2665

9. Iijima-Ando K et al (2010) A DNA damage-activated checkpoint kinase phosphorylates tau and enhances tau-induced neurodegeneration. Hum Mol Genet 19(10):1930–1938

10. Liu F et al (2005) Contributions of protein phosphatases PP1, PP2A, PP2B and PP5 to the regulation of tau phosphorylation. Eur J Neurosci 22(8):1942–1950

11. Hasegawa M et al (1992) Protein sequence and mass spectrometric analyses of tau in the Alzheimer's disease brain. J Biol Chem 267 (24):17047–17054

12. Mann M, Jensen ON (2003) Proteomic analysis of post-translational modifications. Nat Biotechnol 21(3):255–261

13. Thingholm TE, Jensen ON, Larsen MR (2009) Analytical strategies for phosphoproteomics. Proteomics 9(6):1451–1468

14. Olsen JV, Mann M (2013) Status of large-scale analysis of post-translational modifications by mass spectrometry. Mol Cell Proteomics 12 (12):3444–3452

15. Roux PP, Thibault P (2013) The coming of age of phosphoproteomics—from large data sets to inference of protein functions. Mol Cell Proteomics 12(12):3453–3464

16. Olsen JV, Ong SE, Mann M (2004) Trypsin cleaves exclusively C-terminal to arginine and lysine residues. Mol Cell Proteomics 3 (6):608–614

17. Aryal UK, Ross AR (2010) Enrichment and analysis of phosphopeptides under different experimental conditions using titanium dioxide affinity chromatography and mass spectrometry. Rapid Commun Mass Spectrom 24 (2):219–231

18. Pinkse MW et al (2004) Selective isolation at the femtomole level of phosphopeptides from proteolytic digests using 2D-NanoLC-ESI-MS/MS and titanium oxide precolumns. Anal Chem 76(14):3935–3943

19. Villen J, Gygi SP (2008) The SCX/IMAC enrichment approach for global phosphorylation analysis by mass spectrometry. Nat Protoc 3(10):1630–1638

20. Zarei M et al (2011) Comparison of ERLIC-TiO$_2$, HILIC-TiO$_2$, and SCX-TiO$_2$ for global phosphoproteomics approaches. J Proteome Res 10(8):3474–3483

21. Olsen JV, Mann M (2004) Improved peptide identification in proteomics by two consecutive stages of mass spectrometric fragmentation. Proc Natl Acad Sci U S A 101(37):13417–13422

22. Villen J, Beausoleil SA, Gygi SP (2008) Evaluation of the utility of neutral-loss-dependent MS3 strategies in large-scale phosphorylation analysis. Proteomics 8(21):4444–4452

23. Schroeder MJ et al (2004) A neutral loss activation method for improved phosphopeptide sequence analysis by quadrupole ion trap mass spectrometry. Anal Chem 76(13):3590–3598

24. Beausoleil SA et al (2006) A probability-based approach for high-throughput protein phosphorylation analysis and site localization. Nat Biotechnol 24(10):1285–1292

25. Olsen JV et al (2005) Parts per million mass accuracy on an Orbitrap mass spectrometer via lock mass injection into a C-trap. Mol Cell Proteomics 4(12):2010–2021

26. Beynon RJ, Bond JS (1989) Proteolytic enzymes: a practical approach, The Practical approach series. IRL Press at Oxford University Press, Oxford, New York, xviii, 259 p

27. Sipos T, Merkel JR (1970) An effect of calcium ions on the activity, heat stability, and structure of trypsin. Biochemistry 9 (14):2766–2775

28. Ramanathan R et al (2007) Response normalized liquid chromatography nanospray ionization mass spectrometry. J Am Soc Mass Spectrom 18(10):1891–1899

29. Wilm M, Mann M (1996) Analytical properties of the nanoelectrospray ion source. Anal Chem 68(1):1–8

30. Turner M, Searle BC (2011) Probabilistically assigning sites of protein modification with scaffold PTM (poster), in association of biomolecular resource facilities conference, San Antonio, TX

Neuromethods (2016) 114: 187–207
DOI 10.1007/7657_2015_92
© Springer Science+Business Media New York 2015
Published online: 08 November 2015

Identification of Protease Substrates in Complex Proteomes by iTRAQ-TAILS on a Thermo Q Exactive Instrument

Tobias Kockmann, Nathalie Carte, Samu Melkko, and Ulrich auf dem Keller

Abstract

The human genome encodes more than 560 proteases, but only for few of them the substrate proteins are known. This is mainly due to high numbers of potential substrate targets for any given protease that cannot be comprehensively explored by conventional candidate approaches. In this chapter, we describe a proteomics protocol for the reliable identification of protease substrates on a proteome-wide scale. Notably, this method termed iTRAQ-Terminal Amine Isotopic Labeling of Substrates (TAILS) does not require any prior knowledge on candidate proteins. Instead, it is used as an unbiased discovery approach to identify protease-substrate relations in complex biological samples. In addition, iTRAQ-TAILS not only identifies substrate proteins but also maps cleavage sites with amino acid precision. Knowing the cleavage site enables the researcher to perform specific downstream analyses and eliminates the need for laborious follow-up experiments like Edman sequencing. iTRAQ-TAILS acquires this rich information by exploiting the power of latest generation mass spectrometers as the Thermo Q Exactive instrument. Through quantitative assessment of protein N-termini in protease-exposed and control samples and a robust data analysis pipeline, iTRAQ-TAILS can systematically screen proteome-wide substrate spaces for proteolytic events exerted by proteases of interest.

Keywords: Proteolysis, Protease substrate screening, N-terminomics, TAILS, iTRAQ, Thermo Q Exactive

---

1 Introduction

Recently, the first drafts of the human proteome have been published [1, 2]. These drafts include about 300,000 peptides mapping to approximately 30,000 proteins expressed by cultured cell lines and adult/fetal human tissues. Overall, these peptides constitute the basis of a zoomable proteome map [3]. Peptides belonging to the first two levels of this map indicate the expressed proteome and its splice isoforms. The third and by far most complex level encompasses peptides that arise from posttranslational protein processing and modifications. Protein processing—sometimes also referred to as maturation—is mainly driven by proteases. In total, more than 560 proteases have been identified in the human genome so far [4, 5]. Well-known forms of processing are the release of the initiator methionine by aminopeptidases or the cleavage of signal peptides by signal peptidases. Thus, proteases can be perceived as the final shapers of the functional proteome. Since protein maturation is irreversible, most third level peptides are static (or precisely in steady state). In analogy to genomic sciences, proteases catalyzing such reactions can be viewed as house-keeping proteases.

In addition, proteases drive highly dynamic processing events with direct impact on protein activity and function [6]. Here, protease activity is tightly regulated in time and space, thereby steering the level of cleavage products. The cleavage products act as effector molecules that propagate, restrict, amplify, or dampen signals [7]. A well-known example from neurosciences is the γ-secretase complex, which is implicated in cell-surface receptor signaling via the Delta/Notch pathway [8]. The enzymatic subunit of γ-secretase called presenilin cleaves the Notch receptor upon ligand binding, thereby releasing its intracellular domain (Notch ICD) into the nucleus, where it acts as transcriptional regulator.

Not surprisingly, malfunctions of proteases can lead to severe diseases. For instance, mutations in presenilin are positively correlated with Alzheimer's disease (AD) [9]. In addition, brains of AD patients show accumulations (plaques) of amyloid beta (Aβ) protein that is generated by γ-secretase cleavage of the amyloid precursor protein (APP) [10]. Another striking example relating to neurodegenerative disorders is the pathological processing of huntingtin in Huntington's disease [11].

These examples show that proteases are integral parts of biological systems and need to be studied in more detail. In this chapter, we present a versatile technique termed iTRAQ-Terminal Amine Isotopic Labeling of Substrates (TAILS) that facilitates studying protease activity on a proteome-wide scale [12–14]. Depending on the exact experimental setup, this technique enables answering different protease-related questions. Here, we will focus on a 2plex-design (two-group comparison without replication) that tries to solve the question, which substrates are cleaved by a specific protease. This substrate-screening setup will help explaining the basic principles of iTRAQ-TAILS and can be readily expanded to more complicated designs (e.g., two-group comparison with replication using 4plex or 8plex chemistry, multi-group comparisons, time-series studies) [15–17]. A typical iTRAQ-TAILS experiment is subdivided into three major parts, as outlined below (Fig. 1):

1. In the first part, a proteome of choice (substrate space) is incubated under two experimental conditions (+, −). In the + condition the potential substrates are exposed to an active protease, which we call the test protease. The minus condition lacks this proteolytic activity and serves as control/baseline for the experiment. This can happen in a test tube (in vitro), e.g., by incubating a purified protease with a complex protein

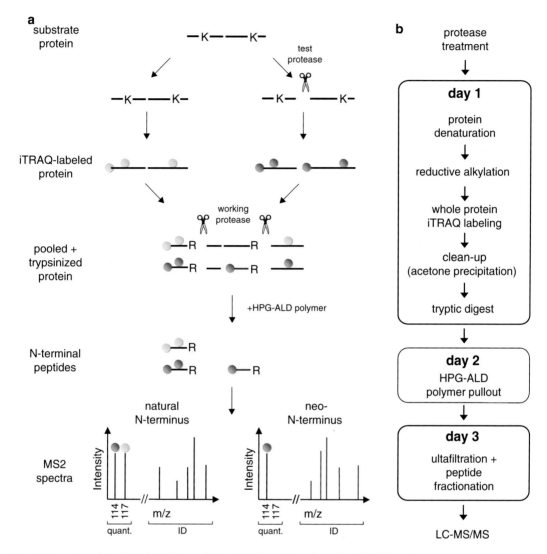

Fig. 1 Overview of the iTRAQ-TAILS workflow. (**a**) Illustration of a 2plex iTRAQ-TAILS experiment for protease substrate screening. For simplicity only a single polypeptide is indicated by *colored line* segments. *Red colored* segments are peptides generated by activity of the test protease. *Colored spheres* denote tandem mass tags (iTRAQ labels) attached to protein N-termini and lysine residues (K). *Scissor symbols* indicate proteases. (**b**) Experimental time line for the iTRAQ-TAILS procedure

mixture of choice, or in vivo under conditions with inherently different proteolytic activity (e.g., tissue from wild type and protease-deficient animals) [16, 18]. In general, the goal is to obtain samples containing differentially abundant cleavage products that can be identified and quantified by iTRAQ-TAILS.

2. In the second part, all C-terminal cleavage products are chemically labeled by an isobaric mass tag (in this protocol we use iTRAQ labels), which become attached to the newly exposed N-terminus of a released fragment. These tags come in

different flavors that can be resolved by mass spectrometry (MS). Technically, they allow multiplex MS measurements, since proteolysis products originating from different samples/conditions can be analyzed simultaneously. Label attachment also happens at free protein N-termini and all lysine side chains. In order to generate peptides that can be analyzed by MS, thus revealing the identity of the iTRAQ-labeled protein fragments, labeled proteins need to be digested by a second protease with known specificity (termed the working protease). We usually apply trypsin for this task, which cleaves C-terminal to arginine (Arg) and lysine (Lys), but notably skips labeled residues. Therefore, in the context of an iTRAQ-TAILS experiment trypsin cleaves proteins following ArgC specificity. Every trypsin cleavage creates a novel reactive N-terminus. This feature is exploited by our method to segregate the resulting peptides into two populations. The majority of peptides—harboring a free N-terminus created by the working protease—can form covalent bonds with an amine-reactive polymer and are thereby eliminated from the sample. The smaller portion (about 10 %) is protected by their N-terminal mass tag and therefore stays in the mixture. In summary, iTRAQ-TAILS effectively reduces the complexity of the peptide mixture to the N-terminome, consisting of natural mature protein N-termini, N-terminal peptides derived from active proteases other than the test protease in both samples (background proteolysis), or neo-N-termini that have been created by the test protease and thus released from its substrates.

3. In the third part, the remaining peptides are analyzed by liquid chromatography tandem mass spectrometry (LC-MS/MS). The mass spectrometer is operated in data-dependent analysis (DDA) mode, meaning that peptide ions get selected for fragmentation and recording of fragment ion spectra (MS2) based on their intensity in the precursor (MS1) scan. This is also referred to as shotgun proteomics. How many peptides can be fragmented during MS analysis primarily depends on instrument speed. During fragmentation the so-called reporter ions are released from the tandem mass tags. These reporter ions are visible as peaks in the low m/z range of the MS2 spectrum. Reporter ion intensity is proportional to the relative quantitative contribution of each sample to the analyzed peptide population. Their extraction from the MS2 spectrum indicates the relative abundance of each N-terminal peptide in the multiplexed sample. The remaining fragment ion peaks in the spectrum originate from the common peptide backbone and can be used to identify the corresponding peptide. Based on the quantitative information attached to the peptide, it can be

determined if this peptide represents the natural mature protein N-terminus of a protein, was derived from background proteolysis in the sample, or was generated by the test protease and thus released from a substrate protein. While natural mature protein N-termini and those generated by background proteolysis are equal in abundance in both conditions, peptides released from substrates are significantly more abundant in samples derived from the test protease (the + condition). In statistical terms the peptide abundance depends on the experimental condition. After removing systematic biases from the quantitative data (normalization), standard statistical methods are applied to guide the decision, if a neo-N-terminus was generated by the test protease or not (*see* Section 3.9 for details).

Taken together, the final result of an iTRAQ-TAILS experiment is a list of protein N-termini observed in multiplexed samples with high confidence. Peptides that are statistically higher in abundance in the protease-treated compared to the control sample are potential direct substrates of the test protease.

It should be noted that iTRAQ-TAILS only identifies potential substrate candidates. Since a complex mixture of proteins is used as substrate space, the test protease might have acted indirectly on an identified substrate, e.g., by activating a downstream protease or by inactivating a protease inhibitor [15, 16]. Thus, direct protease-substrate relationships can only be demonstrated by independent validation experiments using purified components. These experiments may be difficult to perform, because not all proteases can be easily obtained in their active form. Some may also need unknown cofactors that have not been identified by the iTRAQ-TAILS procedure.

Moreover, iTRAQ-TAILS is bound to the limits of bottom-up proteomics. Proteolysis by a combination of test and working protease may release peptides whose physical and chemical properties prevent detection by LC-MS/MS. Possible constraints are, e.g., the length of peptides, their hydrophobicity, or ionization properties. Even if peptides ionize well, their fragmentation behavior may result in MS2 spectra that cannot support peptide identification. This is a well-described problem in targeted MS, but often overlooked in peptide-centric shotgun MS. Since cleavages by the test protease are fixed in position, the only strategy to tackle this problem is to perform multiple iTRAQ-TAILS experiments with different working proteases (e.g., trypsin and endoproteinase GluC). This increases the chances of generating an N-terminal peptide with favorable LC-MS/MS properties, but does not necessarily guarantee its detection by MS.

2 Materials

2.1 Protein Preparation and Test Protease Treatment

1. Starting material: tissue, cells, conditioned growth medium.

2. TAILS sample buffer: 2.5 M Guanidine hydrochloride (GnHCl), 250 mM 4-(2-hydroxyethyl)-1-piperazineethane-sulfonic acid (HEPES), pH 7.8.

3. Acetone. HPLC-grade acetone is available from Sigma-Aldrich (34850).

4. Amicon Ultra 0.5 ml centrifugal filter units, 5 kDa (Millipore).

2.2 Whole Protein iTRAQ Labeling

1. Reducing agent: 350 mM Tris(2-carboxyethyl)phosphine hydrochloride (TCEP). Store aliquots at −20 °C. Molecular biology grade TCEP is available from Sigma-Aldrich (C4706).

2. Cysteine alkylating agent: 250 mM Iodoacetamide (IAA). Should always be prepared fresh by dissolving x mg of IAA in $x * 21.6$ μl of ddH$_2$O. Molecular biology grade IAA is available from Sigma-Aldrich (A3221).

3. Labeling reagents. iTRAQ® Reagents Methods Development Kit, AB Sciex (4352160) Labeling reagents should be stored at −20 °C and used before expiry date.

4. Dimethyl sulfoxide (DMSO). Molecular biology grade DMSO is available from Sigma-Aldrich (D8418).

5. 1 M ammonium bicarbonate solution. Store ammonium bicarbonate stock solution at room temperature in airtight vessels. Molecular biology grade ammonium bicarbonate is available from Sigma-Aldrich (09830).

2.3 Protein Precipitation and Digestion

1. Acetone. HPLC-grade acetone is available from Sigma-Aldrich (34850).

2. Methanol (MeOH). HPLC-grade MeOH is available from Sigma-Aldrich (34860).

3. 100 mM sodium hydroxide (NaOH) solution.

4. Mass spectrometry grade trypsin. Trypsin should be reconstituted in 50 mM acetic acid at 1 μg/μl, aliquoted, and stored at −20 °C until usage. Minimize the number of freeze/thaw cycles to maintain maximum activity. Mass spectrometry grade trypsin (Trypsin Gold) is available from Promega (V5280).

2.4 HPG-ALD Polymer Pullout

1. 1 M Hydrochloric acid (HCl).

2. HPG-ALD (hyperbranched polyglycerol-aldehydes) polymer. Store polymer aliquots under argon at −80 °C. Polymer is available without commercial or company restriction from Flintbox Innovation Network, The Global Intellectual Exchange and Innovation Network (http://www.flintbox.com/public/project/1948/).

3. 1 M ALD coupling solution (Sterogene, cat. no. 9704-01).

4. Amicon Ultra 0.5 ml centrifugal filter units, 30 kDa (Millipore).

5. 0.1 M ammonium bicarbonate solution.

2.5 Strong Cation Exchange Chromatography (SCX)

1. 50 % phosphoric acid.

2. Agilent 1200 Series High Pressure Liquid Chromatography (HPLC) system incl. UV detector and fraction collector.

3. Column: PolySULFOETHYL A column, 200 × 2.1 mm, 5 µm 300-Å (PolyLC Inc.).

4. SCX buffer A: 10 mM KH_2PO_4 pH 2.7, 25 % ACN.

5. SCX buffer B: 10 mM KH_2PO_4 pH 2.7, 0.5 M KCl, 25 % ACN.

2.6 Peptide Cleanup

1. SpeedVac.

2. OMIX C18 pipette tips, 10–100 µl (Agilent).

3. Peptide wash buffer: 3 % ACN, 0.1 % Trifluoroacetic acid (TFA).

4. Peptide elution buffer: 60 % ACN, 0.1 % Trifluoroacetic acid (TFA).

5. 100 % ACN.

6. MS sample buffer: 3 % Acetonitrile, 0.1 % Formic acid (FA).

2.7 LC-MS/MS Analysis

1. Thermo Q Exactive™ Hybrid Quadrupole-Orbitrap Mass Spectrometer.

2. PicoTip Emitter/SilicaTip, Tip 10 µm (New Objective, FS360-20-10-N-20-C12).

3. Reprosil-Pur 120 C18-AQ, 1.9 µm (Dr. Maisch GmbH).

2.8 Data Analysis Software

1. Xcalibur™, the data acquisition and analysis software for Thermo Scientific™ mass spectrometers.

2. Trans-Proteomic Pipeline (TPP) 4.7.1, a collection of integrated tools for MS/MS proteomics, developed at the Seattle Proteome Center (SPC) [19]. Current binaries for Windows and Linux can be obtained via http://tools.proteomecenter.org/wiki/index.php?title=Software:TPP.

3. Mascot, a commercial search engine developed by Matrix Science. Further information can be obtained via http://www.matrixscience.com.

4. CLIPPER, a Perl script for processing of TAILS data [20]. The current release can be obtained via http://clipserve.clip.ubc.ca/tails and needs to be installed on top of a working TPP installation.

3 Methods

3.1 Protein Preparation and Test Protease Treatment

Choices of source material for substrate preparation are very flexible. In the past, iTRAQ-TAILS has been successfully applied to conditioned cell culture media and mammalian tissues [14–17, 21]. Ideally, the substrate proteome is derived from "naïve" cells or tissues that are deficient for the test protease [13]. The most appropriate extraction procedure primarily depends on the source material and protein fraction of interest (e.g., membrane or nuclear compartment). Suitable protocols for mass spectrometry compatible protein extraction procedures from cells and tissues are beyond this chapter and should be taken from the comprehensive literature.

1. Transfer substrate proteins to a suitable digestion buffer that fits your protease of interest (*see* **Note 1**). This can be achieved, e.g., by ultrafiltration with 5 kDa cut-off membranes or acetone precipitation (*see* **Note 2**) and subsequent resuspension in the desired buffer.

2. Split your substrate mixture into test (+) and control (−) samples and incubate the + sample at a fixed protease/substrate ratio (e.g., $1/100$ w/w) for several hours at 37 °C. The control (−) sample is incubated in parallel without addition of the test protease.

Before proceeding to the following section, we recommend concentrating proteins in TAILS sample buffer (250 mM HEPES pH 7.8, 2.5 M GnHCl) using acetone precipitation. Alternatively, proteins may be directly digested in HEPES buffer and adjusted to final TAILS sample buffer conditions after completing the digest.

3.2 Whole Protein iTRAQ Labeling

The following steps assume: + and − sample at $(1\ \mu g/\mu l)$ in TAILS sample buffer.

1. Transfer 250 μl protein solution from + and − sample (equal to 250 μg total protein each) to fresh Eppendorf tubes (*see* **Note 3**).

2. Denature proteins by incubating sample for 15 min at 65 °C.

3. Reduce proteins by adding 2.5 μl of 350 mM TCEP solution and continue incubating for 45 min at 65 °C.

4. Cool samples down to room temperature (RT).

5. Alkylate proteins by adding 5 μl of 250 mM IAA solution and incubate for 30 min in the dark.

6. Bring iTRAQ labeling reagents (114 and 117) to RT and pulse-spin to collect reagents at the bottom of the tubes (*see* **Note 4**).

7. Add 250 μl DMSO to each reagent tube (*see* **Note 5**), mix by pipetting, and transfer diluted reagents to sample tubes (114 to protease (+), 117 to control (−)).

8. Incubate labeling reaction for 30 min at RT.

9. Quench excess iTRAQ reagent by adding 50 µl of 1 M ammonium bicarbonate solution and incubate for 30 min at RT.

3.3 Protein Precipitation and Digestion

1. Combine iTRAQ-labeled samples in 50 ml falcon tube and mix by vortexing.

2. Precipitate proteins by adding 32 ml acetone and 4 ml MeOH (both −20 °C cold). Subsequently, incubate at −80 °C for 1–2 h.

3. Pellet proteins by centrifugation at max rpm and 4 °C for 20 min.

4. Discard supernatant by decanting and wash protein pellet in 15 ml ice-cold MeOH.

5. Collect precipitate by centrifugation at max rpm and 4 °C for 10 min.

6. Discard supernatant by decanting and invert falcon tubes on tissue paper.

7. Resuspend protein pellet in 200 µl of 0.1 M NaOH solution by pipetting and transfer dissolved proteins to fresh Eppendorf tube.

8. Add 700 µl of ddH$_2$O, 100 µl of 1 M HEPES, pH 7.8 and mix by vortexing.

9. Digest labeled proteins by adding 20 µg of MS-grade Trypsin and incubating at 37 °C overnight.

10. (Optional) If analysis of sample without N-terminal enrichment is desired, remove 100 µl (10 %) from digested sample prior to proceeding to the following step.

3.4 HPG-ALD Polymer Pullout

1. Adjust pH of peptide solution to 6.5–7 by adding 20–30 µl of 1 M HCl (*see* **Note 6**).

2. Add 5 mg (140 µl) of HPG-ALD polymer (*see* **Note 7**) and 50 µl of 1 M ALD coupling solution. Mix by gentle pipetting and incubate at 37 °C overnight.

3. Condition a 30 kDa Amicon with 350 µl of ddH$_2$O.

4. Add one third (~340–350 µl) of polymer containing sample to filtration device and collect free peptides by centrifugation according to the manufacturer's instructions. Repeat for a total of three times and combine filtrates.

5. Wash polymer by adding 100 µl of 0.1 M ammonium bicarbonate solution to filtration device and gentle pipetting.

6. Collect flow-through by centrifugation and combine with filtrates from **step 4**.

7. Store peptides at −20 °C in case you do not directly proceed to peptide fractionation or C18 cleanup.

3.5 Strong Cation Exchange Chromatography (SCX)

1. Adjust pH of sample to ≤2.7 by adding 20 μl of 50 % phosphoric acid.

2. Fractionate peptides using strong cation exchange chromatography (*see* **Note 8** for details).

3.6 Peptide Cleanup

1. Dry peptide fractions in SpeedVac and resuspend in 100 μl peptide wash buffer.

2. Clean and optionally pool peptides on C18 OMIX tips (*see* **Note 9** for details).

3. Dry cleaned peptides in SpeedVac and resuspend in 20 μl MS sample buffer.

3.7 LC-MS/MS Analysis

Analyze peptide fractions using liquid chromatography coupled to tandem mass spectrometry (LC-MS/MS, *see* **Note 10** for details). We recommend injecting 10–20 % (2–4 μl) of your total sample volume as a first try. The optimal injection volume depends on sample amount/complexity and should be chosen empirically, based on the results of this first test injection (see next section).

3.8 Inspection of Mass Spectrometry Raw Data

Raw data quality should be rigorously confirmed before moving to the data analysis section. The most basic parameter that needs to be inspected is the total ion count (TIC) and its distribution over retention time (RT). A TIC much smaller than the machine specific optimal value indicates technical problems or that too little material was injected. This might be caused by sample loss during preparation or incomplete sample pickup from the auto sampler (AS) vial. To exclude pickup problems, weigh sample tubes before and after injection and compare weight differences to expected volume reduction. After excluding pickup problems (see also next paragraph) re-inject larger volumes of analyte, bearing in mind that nanoLC columns have a limited loading capacity. Unequal signal distribution along the RT dimension often indicates chromatography/ionization problems. Inspect peptide signals by computing extracted ion chromatograms (XICs) across injections. XICs are calculated from raw data files using the MS instrument vendor software (for Thermo instruments: Xcalibur™). Another attribute that can be judged on the raw data scale is sample complexity. The number of dependent scans (MS2 scans) per analysis cycle (MS1 scan) serves as a good first proxy. Given a sensible MS2 triggering threshold, high dependent scan numbers over time indicate complex peptide mixtures. Re-injection of these samples may provide additional data that has been missed during the first injection due to systematic undersampling. Dependent scan numbers and peak picking behavior are inspected using the instrument vendor software. Apart from re-injecting complex samples with the same chromatography protocol, longer linear gradient should be considered (*see* **Note 10**). Flatter gradients at constant cycle times provide higher chances to select specific peptides for fragmentation, because corresponding MS1 peaks will cover more analysis cycles. However, adjusting the gradient slope needs to be harmonized with dynamic exclusion settings. Otherwise, additional MS2 spectra will cover mainly peaks that have already been annotated in previous cycles, thereby resulting in redundant information.

3.9 Data Analysis

Interpretation of mass spectrometry data will depend on the type of mass spectrometer used and how spectra have been acquired. Here, we describe an analysis workflow for data obtained on a Thermo Q Exactive™ instrument (*see* **Note 11**) (Fig. 2). In the first part of the analysis, MS2 spectra are matched probabilistically against a reference proteome (*see* Section 3.9.1). This process infers the most likely origin (explanation) of each spectrum given a collection of proteins potentially expressed by the experimental system under study. In order to control the total number of false negative and false positive peptide identifications, these search results need to be merged and confirmed using statistical error models (*see* Section 3.9.2). The result is a list of peptides that have been observed by the experimenter with known confidence. Finally, the user assigns functional annotations to these confirmed peptides by exploiting information on protein maturation/processing (Section 3.9.3). This step divides the peptides into functional categories like N-termini of unprocessed precursors or processed mature

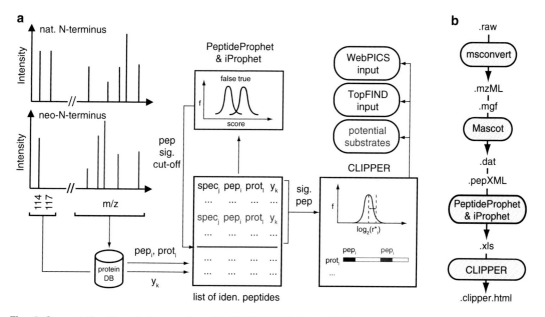

Fig. 2 Computational analysis overview for iTRAQ-TAILS data. (**a**) Illustration outlines analysis principles exemplified for the two N-terminal peptides shown in Fig. 1. *Red colored* entities are related to a neo-N-terminus generated by the test protease. (**b**) The flowchart shows which data files/formats are processed by which software modules during the analysis workflow

proteins (e.g., upon signal peptide removal). In addition, the quantitative information obtained from the iTRAQ reporter ion region is evaluated in order to decide if the abundance of a given peptide is dependent on one of the experimental conditions. For two-group comparisons between protease-treated and control samples, this evaluation translates into finding protease substrate candidates and defining cleavage site specificities.

3.9.1 Data Conversion and Peptide Identification

As an analysis platform for iTRAQ-TAILS proteomics data, we use the Trans-Proteomic Pipeline (TPP) in combination with the Mascot search engine. These perform well in peptide-centric approaches, and with pep.xml the TPP provides a common output format for raw data from different vendors.

1. Open the TPP graphical user interface *Petunia* by directing browser to http://www.localhost/tpp-bin/tpp_gui.pl. Log in with user name "guest" and password "guest" and set analysis pipeline to "Mascot".

2. Within *Petunia* ("Utilities -> Browse Files") create a directory for your project and upload Thermo RAW files to the TPP data tree. Details on the general use of TPP can be found here: http://tools.proteomecenter.org/wiki/index.php?title=TPP_Tutorial.

3. Convert Thermo RAW data files for each sample (e.g., SCX fraction) to mzML format using ProteoWizard's msconvert tool that is automatically installed with TPP v4.7.1 ("Analysis Pipeline (Mascot) -> mzML/mzXML"). Check options for centroiding, compression of peak lists, and output as gzipped file.

4. Convert mzML.gz output files to Mascot Generic Format (mgf) applying TPP default settings ("mzXML Utils -> Convert mz[X]ML Files"). Make sure to select the correct activation method (HCD).

5. Perform database (*see* **Note 12**) search with Mascot search engine and following peptide modifications: fixed: iTRAQ4-plex (K), Carbamidomethyl (C); variable: iTRAQ4plex (N-term), Acetyl (N-term), Oxidation (M). Additional variable N-terminal modifications, e.g., pyro-Glu formation, might be added. Enzyme should be set to ArgC and semi-specificity to allow for identification of peptides with nonspecific N-termini. Set precursor and fragment ion mass tolerances to 10 ppm and 0.02 Da, respectively (*see* **Note 13**).

6. Retrieve search result files (dat) from Mascot server. If the server is accessible from within *Petunia*, "Intermediate Files" can be downloaded via "Analysis Pipeline (Mascot) -> Database Search". If not, locate search results in Mascot search log and save "Intermediate file" under the same name as mzML.gz and mgf files to the same directory in the TPP data tree.

7. Convert dat files to pep.xml ("Analysis Pipeline (Mascot) -> pepXML") choosing a local copy of the corresponding Mascot search database (*see* **Note 12**) and "argc - Semi" as enzyme.

3.9.2 Sample Merging, Secondary Validation, and iTRAQ Quantification

Since iTRAQ-TAILS relies on identification of protein N-termini and thus in many cases on a single peptide per protein, very stringent criteria for assignment of mass spectra to peptides are required. Therefore, we apply several steps for secondary validation of Mascot search hits by integrating multiple statistical models (PeptidePro- phet, iProphet, ProteinProphet) [22–24].

1. Generate "condition.xml" file for iTRAQ quantification by Libra ("Utilities -> Libra Conditions"). Check "Use Reagent m/z" for 114.1 and 117.1 and leave correction values at "0.0" (*see* **Note 14**). Save "condition.xml" to TPP project directory.

2. Run "xinteract" ("Analysis Pipeline (Mascot) -> Analyze Pep- tides") to combine pep.xml search result files from multiple SCX fractions, secondary validate search results by PeptidePro- phet and iProphet, and quantify iTRAQ reporter ions by Libra. Select pep.xml files for all SCX fractions and apply default "Output File and Filter Options" and following parameters: PeptideProphet Options -> RUN PeptideProphet, Use accu- rate mass binning (using PPM), Do not use the NTT model; InterProphet Options -> RUN InterProphet; Libra Quantifi- cation Options -> RUN Libra.

3. (Optional) Run "ProteinProphet" ("Analysis Pipeline (Mas- cot) -> Analyze Proteins") to generate a list of proteins to be used for calculation of isoform assignment scores (IAS). Select "interact.ipro.pep.xml" and apply following settings: Protein- Prophet Parameters -> Input is from iProphet; Advanced Pro- teinProphet Options -> Do not assemble protein groups.

3.9.3 Peptide Filtering, Positional Annotation, and Statistical Evaluation of Protease Cleavage Events

In the final part of data analysis high confidence peptides are extracted, annotated for their position in the corresponding protein and quantitatively compared in protease-treated and control sam- ple. This allows distinguishing natural mature protein N-termini and background proteolysis events from neo-N-termini generated by the test protease (*see* **Note 15**). For all these steps the CLIPPER analysis pipeline was developed that can be easily installed on top of an existing TPP installation [20].

1. Open [PepXML] link to "interact.ipro.pep.xml" in "Output Files" of "xinteract" analysis (*see* Section 3.9.2). In "PepXML- Viewer" go to "Filtering Options" and set "min iProphet prob- ability" to 0.95 (*see* **Note 16**). Next, go to "Pick Columns", add "All>>" to "columns to display" and "Update Page". Make sure all information is displayed, go to "Other Actions" and export "interact.ipro.pep.xls" with "Export Spreadsheet".

2. (Optional) Open [ProtXML] link to "interact.prot.xml" in "Output Files" of "ProteinProphet" analysis (*see* Section 3.9.2). In "ProteinProphet protXML Viewer" set "min probability" to 0.95 (*see* **Note 16**). Set "protein groups", "annotation", and "peptides" to "hide". Check "export to excel" and export to "interact.prot.xls" with "Filter/Sort/Discard checked entries".

3. In *Petunia* open "CLIPPER" tab and select "interact.ipro.pep.xls" as "PepXMLViewer Excel export file to analyze". Select the local copy of the protein sequence database that had been used in Section 3.9.1 for the Mascot search in "Specify database to use". Apply the following "Output Options": Merge multiple spectra for same peptide, Annotate peptides, Calculate cutoff from natural N-termini (select appropriate channels for protease and control) (*see* **Note 17**). (Optional) To calculate IAS values enter "interact.prot.xls" in "Proteinprophet Excel export after enrichment of N termini". Run CLIPPER.

4. Optional parameters. If checked "Generate output file for upload to TopFIND" will automatically generate a file containing information on identified protease-substrate relations that can be directly uploaded to the TopFIND database (http://clipserve.clip.ubc.ca/topfind/contribute) [25, 26] (strongly recommended). "Generate output file for analysis with Web-PICS" provides data for analysis of protease specificity [27]. "Generate pdf file for all spectra" outputs a single pdf for all spectra assigned to peptides included in CLIPPER analysis.

5. Open "interact.ipro_merge.clipper.html" in "clipper" subdirectory for inspection of results. All additional files for each step of analysis will also be saved to this directory.

3.9.4 CLIPPER Output Files

In addition to the integrated HTML frameset "interact.ipro_merge.clipper.html", CLIPPER will generate output files for each step of analysis in comma-separated values (*.csv) format that can be opened in spreadsheet applications (Excel etc.) for further inspection. Moreover, histograms for distributions of $\log_2$(protease/control) ratios before (nterm_hist.png) and after (nterm_norm_hist.png) normalization are provided. Additional detailed information can be found in the CLIPPER Manual, which is freely available at http://clipserve.clip.ubc.ca/tails/ [20].

1. interact.clip.csv: lists extracted data from interact.pep.xls filtered for peptides with N-terminal modification (iTRAQ, acetylation etc.) and tryptic C termini.

2. interact_merge.clip.csv: lists peptides from "interact.clip.csv" upon merging of multiple spectra and modifications (e.g., nonoxidized, oxidized) for the same peptide.

3. interact_merge.annotate.clip.csv: lists all data from "interact. ipro_merge.clipper.html" but in csv format. This is the main working file for downstream data interpretation. Neo-N-termini from substrate candidates can be extracted by filtering for p-value cut-offs given in column "substr_p_1".

4. nterm_hist.png and nterm_norm_hist.png: histograms should be inspected for number of contributing N-termini (n), distribution fit, and normalization factor. For valid results n should be ≥ 30, the fitted curve should closely follow the histogram, and the maximum of "Density" should not extensively deviate from 0 in nterm_hist.png.

4 Notes

1. Proteases have very different needs when it comes to conditions like pH, buffering agent, ions (e.g., Mg^{2+}, Ca^{2+}, Zn^{2+}), and other additives. As a starting point, use a digestion buffer that has already been described in the literature. Be careful: Many extraction buffers contain protease inhibitors and detergents. Both may inhibit your test protease if carried over into the digest.

2. Standard acetone precipitation is performed by adding six volumes of ice-cold (-20 °C) acetone to the sample tube and incubating for 30 min to 4 h at -20 °C. The protein flocculent is pelleted by spinning the sample tube at $6000 \times g$ for 10 min. Be careful not to overdry the protein pellet after decanting the acetone. Inverting the sample tubes on tissue paper for 5–10 min is usually sufficient. Some protein pellets are difficult to redissolve in TAILS sample buffer. Therefore, we recommend first dissolving the protein pellets in 8 M GnHCl by gentle pipetting. Once the pellets have been completely dissolved, add 1 M HEPES pH 7.8 and ddH$_2$O to match final TAILS sample buffer concentrations. If you wish to deviate from the TAILS sample buffer, consider the following: GnHCl assures complete protein denaturation prior to protein labeling. We have not used buffers without chaotropic salt for TAILS and do not recommend this. In case you replace HEPES by a different buffering agent, make sure that your buffer of choice is devoid of free amine group, since these would interfere with the labeling reaction, thereby leading to partial labeling.

3. Do not use coated Eppendorf tubes at any step of the protocol, since these can release polymers into the sample that might interfere with MS analysis.

4. iTRAQ labeling reagent is supplied in a minimal volume of organic solvent.

5. iTRAQ makes use of NHS chemistry to form covalent bonds between the labeling reagent and free amine groups. Efficient labeling requires ≥50 % organic solvent. Therefore, care should be taken to add sufficient DMSO to the labeling reaction.

6. Before adding polymer, the pH should be quickly checked by applying 10 µl of peptide solution to a pH paper. If needed, the pH can be further adjusted with NaOH or HCl.

7. This amount is based on the actual binding capacity of a 100 kDa HPG-ALD polymer (35 mg/ml) but may vary with the specific batch. Information is provided on the package insert.

8. We perform strong cation exchange chromatography of peptides using the SCX buffers and column listed in Section 2 and the gradient table given below. If you choose to use a different column, make sure that it is suited for peptide separation and that its loading capacity exceeds 2 mg of input material. We favor the use of a matching precolumn to protect the preparative column and to ease column cleaning. Test column/instrument performance prior to sample fractionation by separating a standard mixture of synthetic peptides. The optimal number of fractions depends on sample complexity and desired MS instrument time. We routinely divide the gradient into 27 fractions and pool low intensity fractions, judged by UV signal, on C18 OMIX tips prior to MS analysis (*see* **Note 9**). General instructions regarding the operation of HPLC systems and SCX can be found elsewhere.

Step	Time [min]	%B	Flow [ml/min]
1	0	0	0.2
2	60	0	
3	65	5	
4	100	35	
5	110	100	
6	115	100	
7	115.01	0	
8	140	0	

9. Wet and equilibrate C18 OMIX tips by 5 aspiration/infusion cycles in 100 µl of 100 % ACN, elution buffer and wash buffer. Bind and optionally pool peptides on C18 resin by 10 cycles in redissolved HPLC fractions. Finally, elute peptides by five cycles in 100 µl peptide elution buffer.

10. For inline peptide separation we use custom-made 15 cm × 75 μm (ID) frit columns, packed with C18 chromatography medium of 1.9 μm particle size. Columns are operated at 50 °C in connection with a 10 μm microTip emitter. The chromatographic elution time of a single peptide on such a column will be in the 20–30 s range. A typical nanoLC gradient table suited for DDA experiments is given below. For longer gradients simply adjust **step 2**.

Step	Time [min]	%B	Flow rate [nl/min]
1	0	2	300
2	60	35	
3	63	95	
4	70	95	

Beam- and trapping-type MS instruments are suitable for iTRAQ-TAILS analysis. We recommend using a high resolution, high mass accuracy model providing higher-energy collisional dissociation (HCD) fragmentation of precursor ions. The instrument of choice should be operated in data-dependent analysis (DDA) mode. In DDA, also called shotgun mode, the instrument sequentially dissociates the most intense (top n) ions from a precursor scan (MS1) and records corresponding fragment ion spectra (MS2). Post acquisition, these MS2 spectra are probabilistically matched against a proteome database for peptide identification. We routinely use top15 to top30 DDA methods, depending on the instrument speed, with cycle times around 3 s (necessary time to perform MS1 and all depending MS2 scans). This corresponds to fragmentation frequencies of 5–10 Hz. The following table lists MS instrument settings that we have used successfully for iTRAQ-TAILS analysis on the Thermo Q Exactive™ instrument:

MS1	Resolution: 70,000 AGC target: 1e6 Max. IT: 250 ms Scan range: 400–2000 m/z
MS2	Resolution: 17,500 AGC target: 2e5 Max. IT: 120 ms topN: 15 Isolation window: 2 m/z Fixed first mass: 100 m/z NCE: 30
Data-dependent settings	Underfill ratio: 5 % Charge state exclusion: undefined, +1, >5+ Dynamic exclusion[*]: 10 s

In theory, single MS2 spectra/precursor can be sufficient for correct peptide identification and iTRAQ quantification. Still, we do not recommend using very strict dynamic exclusion settings for iTRAQ-TAILS analysis (e.g., 1 min exclusion after first appearance). Generally, single spectra recordings/precursor bear the risk of failed peptide assignments and inaccurate quantification due to low quality spectra. In protein centric DDA experiments, matching other peptides from the same protein usually compensates for this. In contrast, iTRAQ-TAILS analysis is more likely to be affected by such missed identifications, since detection of proteolytic events implies the correct assignment and quantification of a single peptide. Thus, we prefer to sample each precursor multiple times.

11. To test the functionality of the installed data analysis pipeline, a preprocessed test dataset recorded on a Thermo Q Exactive™ instrument might be downloaded from http://clipserve.clip.ubc.ca/tails/. These data were obtained by labeling two aliquots of 0.5 ng of a commercially available tryptic digest of bovine serum albumin (MassPREP Digestion Standard Kit, Waters, 186002329, SwissProt P02769) with iTRAQ reagents 114 and 117, respectively, following the protocol provided with the iTRAQ® Reagents Methods Development Kit, AB Sciex (4352160). Mass spectrometry was performed applying parameters described in **Note 10**. To see expected results unpack "BSA_test.zip" and move folder to "C:\Inetpub\wwwroot\ISB\data". To test your own pipeline create a new folder "C:\Inetpub\wwwroot\ISB\data\BSA_test_2" and copy files "bsa_digest.raw" and "uniprot-organism-taxid-9913-bostaurus.fasta" to this location. Follow Section 3.9 for processing and compare results to preprocessed test output.

12. To allow for positional annotation and compatibility with CLIPPER (*see* Section 3.9.3), we recommend using organism-specific UniprotKB fasta databases. These can be downloaded from ftp://ftp.uniprot.org/pub/databases/uniprot/current_release/knowledgebase/proteomes/. As an example a database for Bos Taurus is provided with the test dataset (*see* **Note 11**). For later use in TPP, the same database has to be uploaded to the TPP data tree.

13. These are standard values for a Thermo Q Exactive™ mass spectrometer. They might differ on your specific instrument.

14. Due to their differences in masses of reporter ions released from iTRAQ reagents 114 and 117, no purity corrections have to be entered. If reagents generating reporter ions with mass differences of less than 3 Da are used, appropriate correction factors have to be entered. If provided, these may be obtained from package inserts with individual iTRAQ reagent kits. Otherwise, commonly used values suggested by TPP's Libra condition.xml tool should be applied.

15. CLIPPER exploits the basic assumption that the levels of natural N-termini are not affected by the test protease treatment and thus equal in both conditions. Accordingly, the $\log_2$-transformed peptide abundance ratio follows a Normal distribution:

$$r_i = \log_2\left(\frac{\text{pep}_{i,\,\text{treated}}}{\text{pep}_{i,\text{control}}}\right) \sim \mathcal{N}\left(\mu, \sigma^2\right)$$

In the absence of any experimental bias, the mean parameter (μ) of the model should be zero. CLIPPER normalizes r_i by subtracting the mean ratio over all natural N-termini (this re-centers the ratio distribution on zero). The prefix of the normalized $\log_2$ abundance ratio ($r_i{}^*$) now indicates the direction of regulation relative to the control level. The variance parameter of the model (σ^2) can be considered as the total measurement noise of a given iTRAQ-TAILS experiment. According to the cumulative distribution function (cdf) of the Normal distribution, ratios more than x *sigma* away from the mean are unlikely to occur by chance. Thus, the standard deviation of r_i over all natural protein N-termini can be used to calculate p-values for statistical significance of substrate cleavage events applying the Gaussian error function. CLIPPER estimates the peptide abundances as $\text{pep}_{i,\,k} = \sum_j y$ with y being the reporter ion intensity for condition k of a spectrum j matched to a stripped peptide i.

16. In PeptideProphet, iProphet, and ProteinProphet analyses, the probability translates into an error rate for spectrum to peptide assignments or protein inferences, respectively, calculated from the actual statistical model [22–24]. In our experience, in iTRAQ-TAILS experiments iProphet and ProteinProphet probabilities of ≥ 0.95 correspond to error rates of <1 % on the peptide and <2 % on the protein level [16] and thus represent stringent cut-offs. Models for your specific experiment and "Sens/Error Tables" are accessible via "interact.ipro. pep-MODELS.html" and "interact.prot-MODELS.html" in the TPP project directory. Based on this information less stringent probability cut-offs may be chosen, but for high confidence the error rate should not exceed 1 % for iProphet and 2 % for ProteinProphet analyses.

17. Do not check option "Calculate cut-off from natural N-termini" for analysis of the test dataset (*see* **Note 11**), since it does not provide sufficient data for model generation.

Acknowledgements

We thank C.M. Overall (UBC Vancouver) and S. Werner (ETH Zurich) for their support and fruitful discussions on our projects. A special thanks goes to the Functional Genomics Center Zurich (FGCZ) for excellent support in mass spectrometry. This work was supported by grants from The Swiss Initiative in Systems Biology (SystemsX.ch; BIP 2011/128), the Swiss National Science Foundation (31003A_140726), and by funds from the ETH Zurich.

References

1. Wilhelm M, Schlegl J, Hahne H et al (2014) Mass-spectrometry-based draft of the human proteome. Nature 509:582–587

2. Kim MS, Pinto SM, Getnet D et al (2014) A draft map of the human proteome. Nature 509:575–581

3. Ahrens CH, Brunner E, Qeli E et al (2010) Generating and navigating proteome maps using mass spectrometry. Nat Rev Mol Cell Biol 11:789–801

4. Quesada V, Ordóñez GR, Sánchez LM et al (2009) The Degradome database: mammalian proteases and diseases of proteolysis. Nucleic Acids Res 37:D239–D243

5. Puente XS, Sánchez LM, Overall CM et al (2003) Human and mouse proteases: a comparative genomic approach. Nat Rev Genet 4:544–558

6. Turk B, Turk DSA, Turk V (2012) Protease signalling: the cutting edge. EMBO J 31:1630–1643

7. Fortelny N, Cox JH, Kappelhoff R et al (2014) Network analyses reveal pervasive functional regulation between proteases in the human protease web. PLoS Biol 12, e1001869

8. Selkoe DJ, Wolfe MS (2007) Presenilin: running with scissors in the membrane. Cell 131:215–221

9. De Strooper B (2007) Loss-of-function presenilin mutations in Alzheimer disease. Talking point on the role of presenilin mutations in Alzheimer disease. EMBO Rep 8:141–146

10. Haass C, Kaether C, Thinakaran G et al (2012) Trafficking and proteolytic processing of APP. Cold Spring Harb Perspect Med 2:a006270

11. Wellington CL, Leavitt BR, Hayden MR (2000) Huntington disease: new insights on the role of huntingtin cleavage. J Neural Transm Suppl, 1–17

12. Kleifeld O, Doucet A, Prudova A et al (2011) Identifying and quantifying proteolytic events and the natural N terminome by terminal amine isotopic labeling of substrates. Nat Protoc 6:1578–1611

13. Kleifeld O, Doucet A, auf dem Keller U et al (2010) Isotopic labeling of terminal amines in complex samples identifies protein N-termini and protease cleavage products. Nat Biotechnol 28:281–288

14. auf dem Keller U, Prudova A, Gioia M et al (2010) A statistics-based platform for quantitative N-terminome analysis and identification of protease cleavage products. Mol Cell Proteomics 9:912–927

15. Schlage P, Egli FE, Nanni P et al (2014) Time-resolved analysis of the matrix metalloproteinase 10 substrate degradome. Mol Cell Proteomics 13:580–593

16. auf dem Keller U, Prudova A, Eckhard U et al (2013) Systems-level analysis of proteolytic events in increased vascular permeability and complement activation in skin inflammation. Sci Signal 6:rs2

17. Prudova A, auf dem Keller U, Butler GS et al (2010) Multiplex N-terminome analysis of MMP-2 and MMP-9 substrate degradomes by iTRAQ-TAILS quantitative proteomics. Mol Cell Proteomics 9:894–911

18. Tholen S, Biniossek ML, Gansz M et al (2013) Deletion of cysteine cathepsins B or L yields differential impacts on murine skin proteome and degradome. Mol Cell Proteomics 12:611–625

19. Deutsch EW, Mendoza L, Shteynberg D et al (2010) A guided tour of the Trans-Proteomic Pipeline. Proteomics 10:1150–1159

20. auf dem Keller U, Overall CM (2012) CLIPPER—an add-on to the Trans-Proteomic Pipeline for the automated analysis of TAILS N-terminomics data. Biol Chem 393:1477–1483

21. Jefferson T, auf dem Keller U, Bellac C et al (2013) The substrate degradome of meprin

metalloproteases reveals an unexpected proteo-lytic link between meprin beta and ADAM10. Cell Mol Life Sci 70:309–333

22. Shteynberg D, Deutsch EW, Lam H et al (2011) iProphet: multi-level integrative anal-ysis of shotgun proteomic data improves pep-tide and protein identification rates and error estimates. Mol Cell Proteomics 10: M111.007690

23. Nesvizhskii AI, Keller A, Kolker E et al (2003) A statistical model for identifying proteins by tandem mass spectrometry. Anal Chem 75:4646–4658

24. Keller A, Nesvizhskii AI, Kolker E et al (2002) Empirical statistical model to estimate the accuracy of peptide identifications made by

MS/MS and database search. Anal Chem 74:5383–5392

25. Lange PF, Huesgen PF, Overall CM (2012) TopFIND 2.0—linking protein termini with proteolytic processing and modifications alter-ing protein function. Nucleic Acids Res 40: D351–D361

26. Lange PF, Overall CM (2011) TopFIND, a knowledgebase linking protein termini with function. Nat Methods 8:703–704

27. Schilling O, auf dem Keller U, Overall CM (2011) Factor Xa subsite mapping by proteome-derived peptide libraries improved using WebPICS, a resource for proteomic iden-tification of cleavage sites. Biol Chem 392:1031–1037

Neuromethods (2016) 114: 209–220
DOI 10.1007/7657_2015_87
© Springer Science+Business Media New York 2015
Published online: 08 November 2015

Proteolytic Processing of Neuropeptides

Lloyd D. Fricker

Abstract

Proteolytic cleavage of neuropeptide precursors is a major post-translational modification that is essential for the production of all biologically active peptides. Differential processing of precursors can produce peptides with unique biological activities. Even differences of a single amino acid can cause large changes in biological activity. Therefore, it is important to understand the precise molecular form of the peptide that is produced in a particular cell or tissue. For this, mass spectrometry-based peptidomic approaches are ideal. Unlike older radioimmunoassay-based detection techniques, peptidomics methods can measure the precise form of each peptide and can readily distinguish between longer and shorter forms of the same peptide. In addition, peptidomic methods are not limited to known peptides and can detect hundreds of different peptides in a single experiment. Comparison between two or more groups of samples is possible with quantitative peptidomic methods. The use of quantitative methods allow for differences in levels among tissues, cell types, or between wild-type and mutant animals to be determined. This review describes a method for quantitative peptidomics using isotopic labels based on trimethylammonium butyrate, which can be synthesized in five different isotopic forms, allowing multivariate analysis of five different samples in a single liquid chromatography/mass spectrometry run.

Keywords: Prohormone convertase, Proprotein convertase, Carboxypeptidase, Peptidomics, Proteomics, Peptidase, Protease

1   Introduction

Neuropeptides are produced from precursor proteins that require processing by endo- and exopeptidases. Endopeptidases include furin, prohormone convertases (PC), and related enzymes. Furin and furin-like enzymes are located primarily in the trans Golgi network, while the PCs are present within secretory vesicles and perform the majority of the cleavages of neuropeptide precursors. There are two major PCs: PC1 (also known as PC3 and typically referred to as PC1/3) and PC2 [1, 2]. Following the endopeptidase step, most bioactive peptides require an additional step mediated by exopeptidases carboxypeptidase E (CPE) and carboxypeptidase D (CPD). These enzymes cleave C-terminal basic residues from the prohormone convertase reaction products [3–5]. Some peptides require further modifications such as C-terminal amidation [6].

Mass spectrometry can be used to determine the precise forms of peptides present in a biological sample [7, 8]. Quantitative mass spectrometry can determine the relative levels of peptides in two or more different samples. This allows for a range of applications. For example, strains of mice lacking a particular processing enzyme can be compared to wild-type littermates [9–14]. Peptides that are missing from the mutant mice are likely to be products of the enzyme, while those peptides that accumulate to high levels in the mutant mice are likely to be substrates of that enzyme. However, it is possible that peptide levels change for other reasons, and in vitro studies are needed to verify that an enzyme can perform the cleavages predicted from the analysis of mutant mice [5]. These in vitro studies can also use a similar mass spectrometry-based peptidomics approach to measure the effect of incubating purified enzymes with extracts of peptides, versus control incubations in the absence of enzyme [15–17]. The combination of animal studies and in vitro studies collectively reveal the substrates and products of an individual processing enzyme [5].

There are two basic approaches for quantitative peptidomics. One involves label-free measurements of the signal strength of each peptide [18, 19]. For this, many technical replicates are required, as well as several biological replicates. Thus, a large number of liquid chromatography/mass spectrometry (LC/MS) runs must be performed for each experiment. While large changes in peptide levels can be readily detected using a label-free approach, it is much more difficult to detect small changes in the levels of peptides. The other general approach is to label the peptides in each sample with a stable isotope so that the resulting mass of each peptide is distinct [20, 21]. The various samples are combined and a single LC/MS run is required for each set of samples. Although multiple LC/MS runs still need to be performed to obtain a suitable number of biological replicates, there is no need to perform multiple technical replicates, and the overall number of LC/MS runs is greatly reduced in the isotopic label approach, compared to the label-free approach. Under optimal conditions, changes in peptide levels as small as 10 % can be detected using the isotopic label approach [22].

A number of different isotopic labels have been described in the literature [22–29]. The protocol described below labels peptides with trimethylammonium butyrate (TMAB) [30]. This compound is a quaternary amine, meaning that it contains a permanent positively-charged amine group. It is easily synthesized from gamma-aminobutyric acid and methyl iodide, which is commercially available with normal isotope composition (i.e. >99 % hydrogen), and with 1, 2, or 3 atoms of deuterium per molecule. Because three methyl groups are incorporated into the compound, the resulting product has a mass either 3, 6, or 9 Da heavier than the product made with the non-deuterated methyl iodide [23]. In addition to these compounds, a fifth reagent can be synthesized using methyl iodide containing three deuteriums and one ^{13}C atom—the resulting mass is 12 Da heavier than the form made with regular methyl iodide. Peptides labeled with each of the five distinct isotopic forms of the TMAB reagent co-elute from HPLC. This is not usually the case for deuterated compounds due to the difference between hydrogen bonds and deuterium bonds. But in the case of the TMAB labels, the positive charge of the quaternary amine adjacent to the methyl group reduces the formation of hydrogen bonds (or deuterium bonds) and the label doesn't affect the retention time of the peptide [30]. Co-elution of all isotopic forms allows for more accurate quantification of the signals [22].

All peptidomic methods require special care in the preparation of samples. Older methods such as radioimmunoassays were less sensitive to impurities and typically started with animal tissue that had been dissected and frozen [31]. The tissue would be thawed in boiling acidic solutions and then peptides extracted and measured using an antiserum-based assay. Attempts to use these approaches for peptidomics failed for multiple reasons [20, 21]. One problem is the post-mortem changes in peptides that occur within minutes of death. Some proteins are rapidly degraded within several minutes of death, possibly due to ischemic conditions in brain. The protein degradation leads to a very high background of peptides derived from these proteins, which obscure the signals from the neuropeptides. In addition, neuropeptide levels decrease within minutes of death, possibly due to secretion of the peptide-containing granules and subsequent degradation by extracellular peptidases [32–35]. A third problem is that extraction in boiling acid leads to breakdown of amide bonds in proteins, primarily at Asp residues, with Asp-Pro sequences especially sensitive to acidic conditions [36]. Thus, the standard methods of sample preparation and peptide extraction used for decades to study neuropeptides were not applicable for mass spectrometry-based peptidomic approaches. A simple solution to these problems is to rapidly heat the brain prior to removal from the skull and dissection. This can be accomplished using microwave irradiation, either with a focused-beam device that kills mice within seconds [37], or by decapitating the mice and placing the head into a conventional microwave oven for several seconds [9]. Other approaches using rapid heating of the brain have also been successful [38]. Once the brain is heat-inactivated (80 °C), protease activity is eliminated and there is no need to extract peptides in hot acid. Instead, peptides can be extracted in hot water followed by ice-cold acid to precipitate proteins; this combination does not result in chemical breakdown of proteins [36].

The following protocol describes the basic method of sample preparation, peptide extraction, isotopic labeling, and mass spectrometry analysis.

2   Materials

High-purity water (such as Milli-Q distilled water system from Millipore).

0.1 M Hydrochloric Acid (Pierce).

0.4 M NaH_2PO_4 (Sigma).

TMAB-NHS compounds, synthesized as described [23].

1.0 M NaOH (Sigma).

Dimethylsulfoxide (Sigma).

NH_2OH HCl (Sigma).

Glycine (Sigma).

Acetonitrile, HPLC grade (Fisher Scientific).

Trifluoroacetic acid (Pierce).

Ultrasonic processor W-380 (Ultrasonic Inc., Farmingdale, NY, USA).

Low retention microcentrifuge tubes (Eppendorf).

Hydrion pH Papers, 8.0–9.5 (Micro Essential Laboratory).

Amicon Ultra 4 mL Ultracel 10,000 molecular weight cut-off Centrifugal Filter Devices (Millipore).

PepClean™ C-18 spin column (Pierce).

3   Methods

3.1 Sample Preparation

1. Sacrifice the animals and heat-inactivate the tissue. Mice can be sacrificed by cervical dislocation followed by decapitation. The heads are immediately placed in a conventional microwave oven until the internal temperature of the brain reaches 80 °C [9]. This needs to be determined for each microwave oven used, as there is considerable variability among brands and models. It is necessary to find a spot in the oven that yields consistent results. This can be done using a small beaker of water and measuring the temperature change over multiple tests. Once a consistent spot in the microwave is identified, the length of time to raise the brain temperature to 80 °C needs to be determined with animals. For this, it is best to use a digital thermometer and insert the probe into the skull immediately after removing from the microwave. After obtaining consistent results in several animals, it is not necessary to check the temperature of the mice brains to be used for peptidomics; insertion of the probe will damage the brain. An alternative to a conventional microwave oven is to use a device that employs focused microwaves to sacrifice the animals within seconds [37].

However, these devices require immobilizing the animals in tubes, causing stress that may induce changes in neuropeptide levels. A third approach is to decapitate the animals, as above, and quickly remove the brain and heat to 80 °C with a specific device designed for this purpose [38].

2. Dissection. After allowing the brain to cool, it needs to be removed from the skull and dissected. Unlike a fresh brain, after heat inactivation the brain becomes much more fragile and prone to nicking. Dissection is best done using a razor blade to cut the brain into coronal sections. Using measurement references from Paxinos and Franklin [39], prefrontal cortex is obtained by cutting at Bregma 1.94, additional coronal cuts are made at Bregma 0.00 and −3.00. The striatum (including the caudate putamen, nucleus accumbens, septum, and ventral palladium) is dissected from the 1.94–0.00 section by removing the cortex. The section of Bregma 0.00 to −3.00 is dissected into the hippocampus, thalamus, amygdala, and hypothalamus. The cortex is obtained from this section as well as the previous section (containing striatum). The cerebellum is dissected by removing the forebrain and brainstem sections.

3. Tissue storage. Tissue should be stored in low retention tubes—loss of peptides can occur if normal tubes are used. We typically use 2 mL microfuge tubes, which allows for efficient sonication (*see* **step 4**). Some tubes contain impurities that can interfere with the mass spectrometry, and we routinely pre-wash the tubes with distilled water before using. Tissue can be stored in a freezer at −70 °C for several months. In our experience, tissue stored for several years shows a decrease in the number of peptides obtained from the sample.

4. Extraction of peptides. Tissue is sonicated using a probe type device (not a bath sonicator). The precise conditions will depend on the manufacturer and model. For the sonicator we use, typical conditions are sonication for 20 s at 1 pulse/s at duty cycle 3, 50 % output, in ice-cold water (5 μL of water per μg of tissue, with a minimum of 200 μL water). The sonicator must be rinsed between tissue extractions to avoid cross-contamination of the samples.

The tubes containing the homogenates are incubated at 70 °C in a water bath for 20 min, cooled on ice for 15 min, and combined with 1/10 volume of ice-cold 0.1 M HCl to a final concentration of 10 mM HCl. It is important that the extracts are ice-cold before adding acid to prevent acid-labile peptide bonds from breaking. After addition of acid, the samples are mixed on a vortex mixer and returned to the ice bath for 15 min. Samples are then centrifuged at 13,000 × g for 40 min at 4 °C. The supernatant is transferred to a new low-retention tube that has been pre-washed with water. The pH of the extracts is adjusted to 9.5 by the addition of 0.4 M phosphate buffer (pH 9.5). The extracts can be stored at −70 °C.

3.2 Isotopic Labeling of the Peptides

With five distinct isotopic forms of the TMAB reagent, it is possible to compare five samples in a single experiment. In some of our experiments, we have compared one control with four different experimental groups. However, in most experiments we compare 2–3 biological replicates of an experimental group (i.e. a mouse with a gene knock-out) with 2–3 biological replicates of the control group (i.e. wild-type mice). This is repeated a second time, using different isotopic labels for the biological replicates, and leading to a total of five biological replicates of the experimental group and five biological replicates of the control group—a sufficient number of replicates for most experiments. It is essential to switch TMAB reagents between groups to control for potential problems with the reactivity of a particular reagent. It is not necessary to perform technical replicates; they are usually much smaller than the biological replicates.

1. Labels. The TMAB-NHS labeling reagents are dissolved in DMSO at 350 μg/μL. Typically, a total of 5 mg of TMAB-NHS reagent is used for labeling of each mouse brain region (i.e. 5 mg for each hypothalamus, striatum, etc present in the tube). If multiple mice are included in each group, scale up accordingly.

2. Labeling. The TMAB-NHS reagent is labile in water. Therefore, instead of adding the entire amount of reagent at one time, it is better to add many smaller aliquots over several hours (typically 3–4 h), adjusting the pH after each addition. Each round of labeling consists of adding one-seventh of the label volume to the sample. Samples are incubated at room temperature for 10 min before the pH is adjusted to 9.5 with 1.0 M NaOH, using pH paper to test the pH of each sample by blotting <1 μL of sample onto the paper. After adjusting the pH, incubate the samples another 10 min at room temperature before the next round of label is added. This process is repeated six times (for a total of seven rounds). After the final addition of TMAB-NHS reagent, the samples are incubated for another 10–30 min.

3. Quenching. Prior to combining the samples, it is essential to quench any unreacted TMAB-NHS reagent. For this, 10 μL of 2.5 M glycine is added per 5 mg of TMAB-NHS reagent and the mixture is incubated at room temperature for 40 min.

3.3 Peptide Purification

1. Pool and filter. Combine the samples labeled with the five different isotopic TMAB reagents. The samples are then applied to Amicon Ultra 4 mL Ultracel filters to remove proteins >10 kDa. Before use, the filters should be washed with

2 mL water to remove any glycerol or other substances present. Filtration is performed according to the manufacturer's instructions. The flow through is collected and saved—this contains the peptides and proteins smaller than 10 kDa. The material remaining behind in the filter is the proteins >10 kDa, which can be discarded or if desired, used for other analyses.

2. Removal of TMAB groups from tyrosine. The TMAB-NHS reagent labels the side chain of tyrosine as well as amines. To remove TMAB from the tyrosine residue, the filtrate is treated with hydroxylamine (NH_2OH). First, adjust the pH of the filtrate to 9.0 using 1.0 M NaOH. Then, add 2.0 M hydroxylamine in DMSO to the filtrate at a ratio of 7.5 μL of hydroxylamine solution for every 25 mg of total TMAB label in the filtrate. Keep in mind that samples have been pooled, and if 5 mg TMAB reagent was used for each of five reactions, the total TMAB is now 25 mg. The reaction is carried out in three rounds, with one third of the hydroxylamine solution added in each round. After hydroxylamine is added, the mixture is incubated for 10 min at room temperature. The pH of the reaction is adjusted back to 9.0 using 1.0 M NaOH. The addition of hydroxylamine followed by incubation and pH adjustment is performed two more times, for a total of three rounds. After this step, the solution can be stored at −70 °C.

3. Desalting. To remove salts from the sample, we use PepClean™ C-18 spin columns (Pierce). In a typical experiment, we combine resin from two C-18 columns into one column (by pouring the resin from one column into the other). The remainder of the procedure follows the manufacturers' instructions using solutions made with acetonitrile and trifluoroacetic acid. Peptides are eluted with 80 μL of 70 % acetonitrile and 0.1 % trifluoroacetic acid in water. The eluates are frozen, concentrated to 10–20 μL in a vacuum centrifuge, and stored at −70 °C until analysis.

3.4 Liquid Chromatography and Mass Spectrometry (LC/MS)

In our experience, optimal results are obtained by chromatography on a reverse phase column with direct electrospray ionization mass spectrometry on a quadrupole time-of-flight (q-TOF) instrument. We have used a variety of LC systems. The precise procedure will depend on the equipment available. A typical protocol for LC/MS analysis on a quadrupole time-of-flight mass spectrometer is described below:

1. Samples are thawed and briefly centrifuged in a microfuge to remove particulates.

2. An aliquot (typically 2–5 μL) is injected onto a Symmetry C18 trapping column (5 μm particles, 180 μm i.d. × 20 mm, Waters, USA).

3. The sample is desalted online for 15 min.

4. The trapped peptides are separated by elution with a water/acetonitrile 0.1 % formic acid gradient through a BEH 130—C18 column (1.7 μm particles, 100 μm i.d. × 100 mm, Waters, USA), at a flow rate of 600 nL/min.

5. Data are acquired in data-dependent mode and selected peptides dissociated by collisions with argon, using standard procedures. For optimal MS/MS analysis of TMAB-labeled peptides, use higher collision energy values than typically used for nonlabeled peptides.

3.5 Data Analysis: Quantification of Relative Levels

1. Using the software program for your mass spectrometer, open the data file to observe the MS spectra. Zoom in on a peak group and scroll through the spectra, making sure that all five peaks co-elute. If not, then they probably aren't a related peak set. If they do co-elute, determine which spectrum has the strongest signal strength. Several spectra can be averaged together if this improves the signal to noise ratio. Finally, measure the peak intensity of the monoisotopic peak and the peak containing one atom of ^{13}C. Intensity measurements can be made with most software programs, or can simply be performed with a ruler, measuring the increase in signal over the background level. We find it optimal to average the peak intensity of the monoisotopic peak and the peak containing one atom of ^{13}C so that the peak intensity is based on multiple points and not a single peak.

2. Log the data into a spreadsheet. In addition to peak intensity, include the mass/charge values, elute time, charge state, and number of isotopic tags. It is important to calculate the mass of the unmodified peptide in order to allow for comparison to databases. Because the tags were added to allow the quantification of the peptide, the mass of the tag needs to be subtracted in order to represent the endogenous form of the peptide. The mass of the peptide without isotopic tags or protons can be calculated from the following formula:

$$\text{mass of unmodified peptide} = (m/z \cdot z) - (c \cdot T) - (1.008 \cdot (z - T))$$

where m/z is the observed mass to charge value for the monoisotopic peak, z is the charge state, c is the mass of the TMAB tag (128.118 for D0-TMAB, 131.133 for D3-TMAB, 134.155 for D6-TMAB, 137.170 for D9-TMAB, and 140.190 for D12-TMAB), T is the number of tags incorporated, 1.008 is the mass of a proton, and $(z - T)$ is the calculation of the number of protons (i.e. the difference between the charge and the number of tags). This last part of the equation is required because the TMAB tags add a positive charge due to the quaternary amine group and the charge state is not equal to the number of protons. Typically, the mass of the unmodified peptide is taken as the average of the values determined from all of the isotopic tags.

3.6   Identification of Peptides by MS/MS Sequencing

Interpretation of MS/MS data is initially performed by computer-assisted searching of databases consisting of proteins or translated cDNA. The Mascot program allows for the identification of peptides labeled with the TMAB reagents. This program only has four of the five TMAB labels included as options (the D12-TMAB is missing from the list of modifications to search for). The TMAB labels are called "GIST" in the Mascot program, and there are separate modifications for N-terminal TMAB and Lys TMAB of each isotopic form, resulting in a total of eight choices. The Mascot program considers the neutral loss of TMA from the peptides during collision-induced dissociation; this causes the loss of 59 Da from peptides labeled with one D0-TMAB tag, 62 Da from peptides labeled with one D3-TMAB tag, 65 Da from peptides labeled with one D6-TMAB tag, and 68 Da from peptides labeled with one D9-TMAB. Following Mascot searches, it is essential to perform manual interpretation to eliminate false positives. Important criteria to confirm the peptides identified by Mascot include:

1. The isotopic form of TMAB matched by Mascot is correct, based on analysis of the peak set. Although this may seem obvious, Mascot does not consider the peak set and know which of the individual peaks correspond to each of the isotopic forms (i.e. D0, D3, etc.). When using the five isotopic forms of TMAB, there is a 1 in 5 chance that a false positive labeled with one tag is correct, and 4/5 of the false positives will fail this test. If a peptide is labeled with two tags (i.e. there is a lysine residue as well as the free primary amine), there is a 1 in 25 chance that a false positive has the correct tags and 24/25 false positives will fail this test. If a peptide contains two lysine residues and a free N-terminal amine and is labeled with three tags, there is a 1 in 125 chance that a false positive will have the correct number of tags, and 124/125 false positives will fail this test. Thus, confirming that the isotopic TMAB form in the observed peak set corresponds to the predicted Mascot match is a simple and necessary step that will eliminate the majority of false positives.

2. The number of tags incorporated into the peptide matches the number of free amines (N-terminus and side chains of Lys).

3. If multiple tags, all should be the same isotopic form on a particular peptide (i.e. all D0-TMAB, or D9-TMAB, and not one D0-TMAB and one D9-TMAB on the same form of a peptide). As with criteria #1, Mascot does not consider that

the labeling was performed in separate tubes and therefore peptides must be labeled with only a single isotopic form.

4. The Mascot score is either the top score of all potential peptides, or the other peptides with comparable scores can be excluded by the other criteria listed here, leaving only one peptide that matches all criteria.

5. The majority (>80 %) of the major MS/MS fragment ions match predicted a, b, or y ions, or precursor ions with loss of trimethylamine.

6. The mass accuracy of the fragment ions is within the accepted specification for the q-TOF instrument used for the analysis.

7. A minimum of five fragment ions match b or y ions. For small peptides, this can be a problem.

8. The charge state of the observed ion matches the expected charge state of the peptide sequence. This is not always precise, and some peptides with a large number of positive charges will appear with fewer protons. Conversely, some peptides that should have only a single positive charge will appear on occasion with two positive charges. However, if the charge state of the observed ion is different than the expected charge state based on peptide sequence, the other criteria should be very solid in order to consider the identification valid.

4 Notes

The quantitative peptidomics technique is able to identify hundreds of peptides in mouse brain regions and other tissues. However, the method is not able to quantify every peptide in a sample. One problem is peptides lacking a free amine such as peptides with an N-terminal acetyl or pyroglutamyl group. If these peptides also lack an internal lysine residue, they will not be labeled by the TMAB reagent and will appear as a single peak. While it is possible to detect and identify these peptides from the MS/MS data, the absence of TMAB tags means that their relative levels cannot be quantified using this technique. For other peptides, intrinsic factors can result in low ionization efficiency and failure to detect the peptide. A particular problem is that very small or very large peptides are difficult to detect, especially if their mass/charge ratios are outside the range of the instrument. The dynamic range of peptide levels in biological samples varies several orders of magnitude, and low abundance peptides are difficult to detect above the background.

Small molecule and polymeric contaminants in the sample can greatly interfere with the MS analysis. In some cases, the contaminants completely overwhelm the signal from tissue-derived peptides. It is important to use high quality ultrapure deionized water.

Some brands of microfuge tubes and filtration devices contain polymeric contaminants that appear as polyethylene glycol-related compounds when analyzed on MS. It is important to use low-retention microfuge tubes and pipette tips in order to avoid loss of sample during the various procedures. The tubes should be washed with ultrapure deionized water and then dried before use.

All solutions should be freshly prepared with ultrapure deionized water to avoid contamination from small organic molecules that can interfere with the mass spectrometry. High quality reagents should be purchased. It is important to clean the Centricon filters before filtering peptides. The filters usually contain glycerol and other substances, which need to be washed off before using. Washing with ultrapure deionized water usually solves this problem.

The most time-consuming part of the procedure is the data analysis. Unfortunately, automated programs for the quantitation of TMAB-labeled peptides have not been successful, and the data need to be manually analyzed. While computers can help with the identification of peptides from the MS/MS spectra, these data need to be manually interpreted to avoid false positives.

References

1. Zhou A, Webb G, Zhu X et al (1999) Proteolytic processing in the secretory pathway. J Biol Chem 274:20745–20748

2. Seidah NG, Chretien M (1994) Pro-protein convertases of subtilisin/kexin family. Meth Enzymol 244:175–188

3. Song L, Fricker LD (1995) Purification and characterization of carboxypeptidase D, a novel carboxypeptidase E-like enzyme, from bovine pituitary. J Biol Chem 270:25007–25013

4. Fricker LD, Snyder SH (1982) Enkephalin convertase: purification and characterization of a specific enkephalin-synthesizing carboxypeptidase localized to adrenal chromaffin granules. Proc Natl Acad Sci U S A 79:3886–3890

5. Sapio MR, Fricker LD (2014) Carboxypeptidases in disease: insights from peptidomic studies. Proteomics Clin Appl 8:327–337

6. Prigge ST, Mains RE, Eipper BA et al (2000) New insights into copper monooxygenases and peptide amidation: structure, mechanism and function. Cell Mol Life Sci 57:1236–1259

7. Fricker LD, Lim J, Pan H et al (2006) Peptidomics: identification and quantification of endogenous peptides in neuroendocrine tissues. Mass Spectrom Rev 25:327–344

8. Hummon AB, Amare A, Sweedler JV (2006) Discovering new invertebrate neuropeptides using mass spectrometry. Mass Spectrom Rev 25:77–98

9. Che FY, Lim J, Biswas R et al (2005) Quantitative neuropeptidomics of microwave-irradiated mouse brain and pituitary. Mol Cell Proteomics 4:1391–1405

10. Pan H, Nanno D, Che FY et al (2005) Neuropeptide processing profile in mice lacking prohormone convertase-1. Biochemistry 44:4939–4948

11. Pan H, Che FY, Peng B et al (2006) The role of prohormone convertase-2 in hypothalamic neuropeptide processing: a quantitative neuropeptidomic study. J Neurochem 98:1763–1777

12. Zhang X, Che FY, Berezniuk I et al (2008) Peptidomics of Cpe(fat/fat) mouse brain regions: implications for neuropeptide processing. J Neurochem 107:1596–1613

13. Zhang X, Pan H, Peng B et al (2010) Neuropeptidomic analysis establishes a major role for prohormone convertase-2 in neuropeptide biosynthesis. J Neurochem 112:1168–1179

14. Wardman JH, Zhang X, Gagnon S et al (2010) Analysis of peptides in prohormone convertase 1/3 null mouse brain using quantitative peptidomics. J Neurochem 114:215–225

15. Lyons PJ, Fricker LD (2011) Peptidomic approaches to study proteolytic activity. Curr Protoc Prot Sci Chapter 18:Unit 18, 13

16. Lyons PJ, Fricker LD (2010) Substrate specificity of human carboxypeptidase A6. J Biol Chem 285:38234–38242

17. Tanco S, Zhang X, Morano C et al (2010) Human carboxypeptidase A4: characterization of the substrate specificity and implications for a role in extracellular peptide processing. J Biol Chem 285:18385–18396

18. Bantscheff M, Schirle M, Sweetman G et al (2007) Quantitative mass spectrometry in proteomics: a critical review. Anal Bioanal Chem 389:1017–1031

19. Old WM, Meyer-Arendt K, Aveline-Wolf L et al (2005) Comparison of label-free methods for quantifying human proteins by shotgun proteomics. Mol Cell Proteomics 4:1487–1502

20. Gelman JS, Wardman JH, Bhat VB, Gozzo FC, Fricker LD (2012) Quantitative peptidomics to measure neuropeptide levels in animal models relevant to psychiatric disorders. Methods Mol Biol 829:487–503

21. Wardman J, Fricker LD (2011) Quantitative peptidomics of mice lacking peptide-processing enzymes. Methods Mol Biol 768:307–323

22. Che FY, Fricker LD (2005) Quantitative peptidomics of mouse pituitary: comparison of different stable isotopic tags. J Mass Spectrom 40:238–249

23. Morano C, Zhang X, Fricker LD (2008) Multiple isotopic labels for quantitative mass spectrometry. Anal Chem 80:9298–9309

24. Simons BL, Wang G, Shen RF et al (2006) In vacuo isotope coded alkylation technique (IVICAT); an N-terminal stable isotopic label for quantitative liquid chromatography/mass spectrometry proteomics. Rapid Commun Mass Spectrom 20:2463–2477

25. Brancia FL, Montgomery H, Tanaka K et al (2004) Guanidino labeling derivatization strategy for global characterization of peptide mixtures by liquid chromatography matrix-assisted laser desorption/ionization mass spectrometry. Anal Chem 76:2748–2755

26. Hsu JL, Huang SY, Chow NH et al (2003) Stable-isotope dimethyl labeling for quantitative proteomics. Anal Chem 75:6843–6852

27. Ong SE, Blagoev B, Kratchmarova I et al (2002) Stable isotope labeling by amino acids in cell culture, SILAC, as a simple and accurate approach to expression proteomics. Mol Cell Proteomics 1:376–386

28. Zhang R, Regnier FE (2002) Minimizing resolution of isotopically coded peptides in comparative proteomics. J Proteome Res 1:139–147

29. Zhu H, Pan S, Gu S et al (2002) Amino acid residue specific stable isotope labeling for quantitative proteomics. Rapid Commun Mass Spectrom 16:2115–2123

30. Zhang R, Sioma CS, Thompson RA et al (2002) Controlling deuterium isotope effects in comparative proteomics. Anal Chem 74:3662–3669

31. Chard T (1987) An introduction to radioimmunoassay and related techniques; Burdon RH, Van Knippenberg PH, editors. Elsevier, Amsterdam, pp 1–255

32. Nylander I, Stenfors C, Tan-No K et al (1997) A comparison between microwave irradiation and decapitation: basal levels of dynorphin and enkephalin and the effect of chronic morphine treatment on dynorphin peptides. Neuropeptides 31:357–365

33. Mathe AA, Stenfors C, Brodin E et al (1990) Neuropeptides in brain: effects of microwave irradiation and decapitation. Life Sci 46:287–293

34. Theodorsson E, Stenfors C, Mathe AA (1990) Microwave irradiation increases recovery of neuropeptides from brain tissues. Peptides 11:1191–1197

35. Galli C, Racagni G (1982) Use of microwave techniques to inactivate brain enzymes rapidly. Methods Enzymol 86:635–642

36. Che FY, Zhang X, Berezniuk I et al (2007) Optimization of neuropeptide extraction from the mouse hypothalamus. J Proteome Res 6:4667–4676

37. Svensson M, Skold K, Svenningsson P et al (2003) Peptidomics-based discovery of novel neuropeptides. J Proteome Res 2:213–219

38. Scholz B, Skold K, Kultima K et al (2011) Impact of temperature dependent sampling procedures in proteomics and peptidomics – a characterization of the liver and pancreas post mortem degradome. Mol Cell Proteomics 10: M900229MCP900200

39. Paxinos G, Franklin KBJ (2001) The mouse brain in stereotaxic coordinates. Academic, San Diego, CA

Neuromethods (2016) 114: 221–242
DOI 10.1007/7657_2015_93
© Springer Science+Business Media New York 2015
Published online: 08 November 2015

Isotope Dilution Analysis of Myelin Basic Protein Degradation After Brain Injury

Andrew K. Ottens

Abstract

Functional diversity within the proteome has expanded up the phylogenic tree with the introduction of multiple isoforms expressed varyingly across the lifespan. The neuroproteome is no exception with selective isoforms differentiating neurodevelopment from maturation and, as we will discuss, in response to brain injury. However, isoform characterization requires selective analysis of alternatively spliced sequences missed by non-targeted proteomic methods. Isotope dilution analysis provides an answer whereby isoform-selective peptides are specifically targeted for absolute quantification. Synthesized isotopically labeled peptide standards provide an internal reference when monitored simultaneously with the endogenous targets via multiple reaction monitoring on a tandem mass spectrometer. Provided is a detailed description of isotope dilution methodology exemplified in monitoring the proteolytic degradation of myelin basic protein isoforms in response to brain injury.

Keywords: Isotope dilution analysis, Mass spectrometry, TBI, Myelin, Isoforms, MBP, Neuroproteomics, Proteolysis

1 Introduction

Proteomic research, in developing biomarkers for example, necessitates the use of sensitive and specific quantitative assays. High-throughput technologies that assess multiple markers are of particular interest. Targeted mass spectrometry is one such technique, recognized as the 2013 method of the year by Nature Methods for enabling precise proteomic quantification [1]. Isotope dilution analysis (IDA), or stabile isotope dilution analysis, takes targeted mass spectrometry a step further by providing absolution quantification [2, 3]. IDA combines the use of isotopically labeled internal standards with multiple reaction monitoring mass spectrometry (MRM-MS) [4]. Peptides are chosen to each protein of interest, which are selectively profiled by isolating their precursor mass-to-charge (m/z) and detecting a molecule-specific series of corresponding product ions (ion transitions). Thereby, MRM-MS offers unprecedented target specificity while maintaining assay sensitivity due to the noise reduction inherent to tandem mass spectrometry (MSMS). Further, MRM-MS methods are readily customized to just about any protein target with effective quantification of multiple targets. Case-in-point, Abbotiello et al. demonstrated <20 % variability across 11 different laboratories in analyzing 125 peptide targets within plasma, substantiating the clinical potential for IDA biomarker assays [4].

Our own interest in IDA began with the need to characterize myelin basic protein (MBP) proteolytic degradation following traumatic brain injury (TBI) [5]. MBP has been regarded as a brain injury biomarker for over 35 years [6], yet it has not been as widely studied in neurotrauma due to issues with assay specificity and sensitivity [7, 8]. Part of the complication arises from MBP's multiple isoforms, which vary in their expression and function throughout myelin formation, compaction, and stabilization [9, 10]. Thus, understanding MBP dynamics in neurodevelopment [11], neurodegenerative disease [11, 12], and neurotrauma [13, 14] requires an account of the ratio across these isoforms.

Differential splicing of exons 2, 5, and 6 of the MBP gene results in [11, 15] the eight possible isoforms depicted in Fig. 1 [16]. Thus far, mRNA for all eight has been sequenced [17], while only five (21.6, 18.6, 17.3, 17.2, and 14.0 kDa) have been identified as protein [18–20]. MBP isoform expression varies across the lifespan with respect to their function, as can be classified into three groups [11]. Isoforms 2, 4, 6, and 8 lacking exon 5 (Fig. 1) are linked with embryonic myelin formation, while those without exon 5 are most common in adulthood [17]. Isoforms 1, 2, 5, and 6 containing exon 2 are associated with postnatal myelin development as well as remyelination in aging [10, 19, 20] and disease [21]. Isoforms 3, 4, 7, and 8 lacking exon 2 appear with myelin maturation in adolescence. The functional significance of exon-6, however, is not well understood, other than that those containing it are less basic [10, 20]. Importantly, all eight isoforms share a common calpain cleavage motif between Phe114 and Lycl15 that is targeted with neurodegeneration, such as with multiple sclerosis and TBI [12, 14]. Thus, MBP proteolytic products may be particularly informative as biomarkers of myelin degeneration if readily detected within biofluids. To this end, our IDA studies demonstrated that MBP breakdown products are detected within cerebrospinal fluid following TBI [5]. More recently, others have established MBP blood assays to detect fragments shed following TBI, which hold promise as specific assessments of myelin degeneration [8, 22].

Herein we detail the optimization and implementation of IDA methodology as used in profiling the ratio of MBP isoforms and their calpain degradation following brain injury. The resultant assay provided picomolar detection from brain tissue and has been effectively instrumented in assessing biofluids [5]. The described IDA

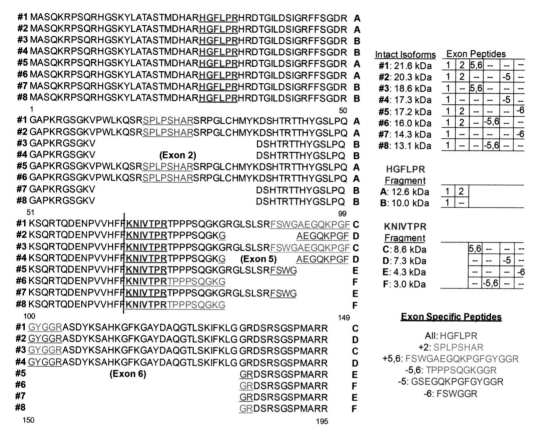

Fig. 1 Rat MBP and marker peptides. Indicated are the sequences for each isoform (#1–8), the calpain cleavage site (*vertical red line*) with associated proteolytic fragments (A–F), and the select marker peptides (*underlined* and *colored*) of myelin basic protein. The molecular mass and associated variable transcription regions of the eight known isoforms and six predicted calpain fragments of MBP are tabulated. Two peptides used for pan-isoform MBP IDA quantification analysis, HGFLPR (*magenta*) and KNIVTPR (*red*), are shown in *bold*. Listed are six exon-specific peptides for differentiating the MBP isoforms. Reproduced from Ottens et al. [5] with modification and permission from Wiley & Sons

approach is readily adapted to other proteins with the synthesis of new isotopically labeled standards—a considerable benefit in studying the complex neuroproteomic response to brain injury and disease [23, 24]. MRM-MS ion transition lists are then simply updated and can be engineered for multiple proteins employing commercial or open access software [25]. IDA is highly amenable to large-cohort biomarker studies, providing greater target specificity than immunological-based assays. IDA is particularly advantageous in that it can assess multiple markers simultaneously without compromising sensitivity.

2 Materials

2.1 Synthesis of Isotopically Labeled Peptides

1. Isotopic amino acids (Cambridge Isotope Laboratories, Andover, MA) (*see* **Note 1**)

2. Purified or enriched standard of target protein (e.g., MBP Bovine, p/n M1891, Sigma-Aldrich, St. Louis, MO)

3. Sequences for target protein isoforms (see http://www.uniprot.org/)

2.2 Brain Lysis and Protein Extraction

1. Fresh frozen brain tissue (*see* **Note 2**)

2. Cryostat or a −20 °C chest-style freezer

3. Lysis Buffer: 1 % Triton X-100 (v/v), 20 mM Tris–HCl (pH 7.5), 150 mM NaCl, 50 mM EDTA, 50 mM EGTA, 1 mM dithiothreitol, Complete Mini Protease Inhibitor Cocktail tablet (Roche, Indianapolis, IN), 1 mM sodium vanadate (*see* **Note 3**)

4. 1.5 mL LoBind microcentrifuge tubes (Eppendorf, Hauppauge, NY)

5. Disposable Pellet Pestles (Fisher Scientific, Pittsburg, PA)

6. Refrigerated microcentrifuge (e.g., ThermoScientific Legend RT, Marietta, OH)

7. Refrigerated tube rotator (e.g., VWR Tube Rotator Unit placed within a laboratory refrigerator, Radnor, PA)

8. Ice bath

9. 0.1 μm pore-size Ultrafree filter units (EMDMillipore, Billerica, MA)

10. Detergent-compatible protein assay kit (e.g., DC Protein Assay, Bio-Rad, Hercules, CA)

2.3 MBP Gel Electrophoresis

1. Denaturing sample buffer (e.g., NuPAGE LDS sample buffer, Life Technologies, Carlsbad, CA)

2. Reducing agent (e.g., NuPAGE Sample Reducing Agent, Life Technologies)

3. Pre-cast gel for resolution of low-mass proteins (e.g., 10 % NuPAGE Bis-Tris gel used with MES Running Buffer, Life Technologies)

4. Antioxidant for adding into running buffer (e.g., NuPAGE Antioxidant, Life Technologies)

5. Prestained protein standard for low molecular mass correlation (e.g., Dual Xtra, Bio-Rad)

6. Purified MBP, Bovine as positive control and marker (Sigma-Aldrich)

7. Pre-cast gel running tank and compatible power supply (e.g., XCell SureLock and Novax Power Supply, Life Technologies)

8. Coomassie Blue R250 (Bio-Rad)

9. Destain solution: 40 % ethanol/10 % acetic acid in nanopure water.

10. Digital gel imager with software (e.g., In-Vivo F Pro with MI software, Bruker, Billerica, MA).

2.4 In-Gel Digestion and Sample Preparation

1. LC/MS grade water and acetonitrile (Optima solvents, Fisher Scientific)

2. A clean surface for gel cutting (e.g., one-half of the disposable plastic gel cassette)

3. Surgical steel scalpel (#15)

4. White light box

5. 1.5 mL LoBind microcentrifuge tubes (Eppendorf)

6. 50 mM ammonium bicarbonate

7. 50:50 Solution: 1:1 (v/v) mixture of 50 mM ammonium bicarbonate and acetonitrile (LC/MS grade)

8. Vortex with microcentrifuge tube adaptor

9. Gel loader tips (non-filtered)

10. Speed vacuum concentrator (e.g., SPD-1010, ThermoScientific)

11. Clostripain Digest Solution: 25 ng/μL clostripain (a.k.a. Arg-C, Worthington Biochemical, Lakewood, NJ), 50 mM ammonium bicarbonate, 20 mM calcium chloride, 2.5 mM dithiothreitol in LC/MS grade water. For digestion of unknown samples, add isotopically labeled KNIV*TPR peptide (to 10 mM) as internal standards for MBP calpain proteolysis and exon-selective peptides for IDA of MBP isoforms: HGFL*PR (all isoforms), SPL*PSHAR (with exon 2), FSW*GGR (with Exon 5 and not 6), GSEGQK*PGFGYGGR (with exon 6 and not 5), FSW*GAEGQKPGFGYGGR (with exons 5 and 6), TPPPSQ*GKGGR (lacking exons 5 and 6) (Fig. 1) (*see* **Note 4**).

12. 50:45:5 Extraction Solution: 50 % acetonitrile, 45 % (v/v) 50 mM ammonium bicarbonate, 5 % acetic acid

13. Autosampler vials for LC/MSMS systems (LC/MSMS clean with low peptide binding characteristics)

2.5 Reversed-Phase Liquid Chromatography (RPLC)

1. Mobile Phase A: LC/MS grade water with 0.1 % formic acid (v/v)

2. Mobile Phase B: LC/MS grade acetonitrile with 0.1 % formic acid (v/v)

3. Reversed-phase trap and capillary analytical columns (can be commercial or self-packed). We presently use Waters (Milford, MA) Symmetry C18 trap (2 cm × 180 μm i.d.) and Waters HSS T3 NanoAcquity analytical (15 cm × 75 μm i.d.) columns.

4. Ultra or high performance liquid chromatography system with autosampler, binary high-pressure gradient programmable pump, and trapping valve. The autosampler should provide precise (<10 % error) sample injection at 1 μL volumes. The gradient pump should provide precise mobile phase mixing at low flow rates around 250 μL/min. A column oven (kept at 55 °C) may improve separation performance and reproducibility. We presently use Waters NanoAcquity UPLC systems.

2.6 Multiple Reaction Monitoring Mass Spectrometry (MRM-MS)

1. Tandem mass spectrometer with DDA and MRM analysis modes (triple quadruple instruments are used commonly, such as a Waters Xevo TQMS, though ion trap or qTOF-based instruments with a linear response over four orders in magnitude may also be used).

2. Quality control protein digest standard (e.g., HeLa Protein Digest, Life Technologies)

3. Reconstitution/dilution buffer: 0.1 % formic acid (v/v) in LC/MS grade water

4. Isotopically labeled peptide standards

2.7 Quantitative Analysis of MBP Calpain Degradation

1. Skyline (MacCoss Lab, Seattle, WA, https://skyline.gs.washington.edu)

2. Excel (Microsoft, Redmond, WA, http://office.microsoft.com)

3 Methods

3.1 Synthesis of Isotopically Labeled Peptides

1. Determine the amino acid sequences for each target protein isoform. Online resources such as www.uniprot.org provide convenient access to verified as well as hypothetical protein sequences for model organisms and humans.

2. Align isoform sequences (www.uniprot.org/align), as demonstrated with the eight possible MBP isoforms (Fig. 1). Note the alternative splicing regions (e.g., exons 2, 5, and 6 in Fig. 1).

3. Identify candidate peptides that selectively cover the alternative sequence regions (see color-coded peptides in Fig. 1) (*see* **Note 5**). Also, select peptide sequences covering known posttranslational modifications of interest. For example, we sought to quantify the calpain cleavage of MBP via the product peptide KNIVTPR (shown in red within Fig. 1).

4. When possible, it is recommended to test the digestion of a purified or enriched standard of the target protein. Nontargeted data-dependent or data-independent methods along with protein sequence identification will discern which isoform-selective peptides are readily produced with the chosen

protease and detected on the LC/MSMS platform to be used. Results may then guide peptide selection between two or more candidate peptides covering a particular splice region.

5. Submit a list of isoform-selective peptide sequences along with the desired isotopically labeled amino acids to an institutional or commercial peptide synthesis facility (synthesis methods are beyond the scope of this chapter). For example, we incorporated $^{13}C5$ labeled L-valine into our labeled KNIV*TPR standard. Synthesis of the unlabeled peptides may also be useful when examining for interfering ions during LC/MSMS method development.

3.2 Brain Lysis and Protein Extraction

1. At −20 °C (e.g., within a cryostat), block-cut each frozen brain with surgical steel blades as necessary to isolate the anatomical region of interest (see **Note 6**).

2. Tissue should be placed into pre-chilled LoBind microcentrifuge tubes, and kept frozen at −20 °C until ready to add ice-cold lysis buffer.

3. Add 100 µL of Lysis Buffer per 1 mm³ of frozen tissue and homogenize (press and rotate 30 times) using disposable Pellet Pestles (see **Note 7**).

4. Seal tubes and place within a refrigerated tube rotator for 90 min such that the lysate is able to move end-over-end within the tube about once every 5 s.

5. Centrifuge the lysate at 14,000 × g and 4 °C for 10 min.

6. Remove supernatant and pass through an Ultrafree filter unit.

7. Measure the protein concentration using a detergent-compatible protein assay kit. Produce a concentration-balanced set of samples by dilution with lysis buffer. Keep the concentration as high as possible to facilitate loading sufficient material onto a pre-cast gel (e.g., 3 µg/µL).

3.3 MBP Gel Electrophoresis

1. Combine between 20 and 50 µg of brain lysate with sample buffer and reducing agent to achieve a final volume commiserate with the maximal loading volume for the selected pre-cast gel (e.g., 25 µL for a 10 well 1.0 mm thick NuPAGE gel) (see **Note 8**).

2. Heat samples at 70 °C for 10 min to facilitate protein denaturation. Allow to cool to room temperature prior to loading.

3. Prepare and set gel within the running tank per manufacturer instructions and fill inner and outer chambers with running buffer pre-mixed with antioxidant.

4. Load the gel with replicate TBI and control brain tissue lysate samples in an interspersed order with 10 µL of prestained

molecular mass marker spiked with 1 µg of purified bovine MBP set between replicates and in the outside lanes.

5. Ensure that the running buffer has not leaked out of the inner chamber (the wells remain covered with buffer). Connect the running tank to a power supply programmed with the optimal settings for the selected pre-cast gel (e.g., 35 min at a constant 200 V for NuPAGE MES running buffer).

6. Remove the gel and place in well-cleaned staining trays with 15 mL of Destain Solution. Place on a shaker for 30 min.

7. Remove Destain Solution and cover the gel with Coomassie R250 stain. Shake for 20 min.

8. Remove Coomassie stain and add 15 mL of Destain Solution and shake for 2 h.

9. Capture a digital image of the destained gel, being careful not to contaminate its surface. Process the gel image with appropriate 1D gel software to assign apparent molecular masses (M_a) to each 1 mm of length as calibrated against the Pre-stained Molecular Mass Marker. The Coomassie stained MBP standard should also be apparent as a band below the 25 kDa marker band (*see* **Note 9**).

3.4 In-Gel Digestion and Sample Preparation

1. Rinse the gel for 5 min in LC/MS grade water.

2. The processed image should provide an estimate as to where above the 25 kDa marker band to begin cutting. This should be at an M_a of approximately 28. Proceed to cut 1 mm bands from each lane until reaching an M_a of 3. You will end up with approximately 15 gel bands per lane (Fig. 2) (*see* **Note 10**).

3. Excise the MBP bovine standard gel bands (from the molecular mass marker lanes) for method development. These gel slices should be digested (**step 8**) without the addition of isotopically labeled peptide standards.

4. Dice each gel band into 1-mm^3 cubes and transfer them into one pre-labeled (for lane and gel band position) LoBind 1.5 mL microcentrifuge tube per band.

5. Add 150 µL of 50/50 Solution to each tube and vortex gentle for 15 min. Discard the liquid using gel loader tips (aspirate from under the gel cubes). Repeat this step until gel cubes are clear of Coomassie stain.

6. Add 30 µL of LC/MS grade acetonitrile to the gel cubes and vortex gently for 5 min.

7. Discard the acetonitrile and speed vacuum the gel cubes dry (around 15 min).

8. Rehydrate the dried gel cubes with 25 µL of Clostripain Digestion Solution (containing the isotopically labeled peptides). Close tubes tightly and incubate overnight at 37 °C.

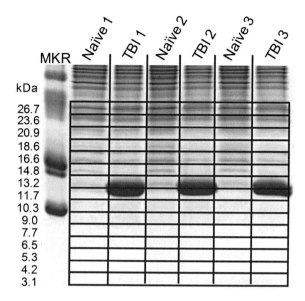

Fig. 2 Gel electrophoresis separation of naïve and TBI brain lysates. Prior to mass spectrometry analysis, brain lysates were separated by gel electrophoresis, resolving MBP isoforms and their proteolytic fragments. An adjacent Prestained Molecular Mass Marker (MKR) allowed the calculation of the apparent molecular mass (M_a) for each gel band. Slices were excised along each gel lane as indicated, between 28 and 3 M_a. TBI-induced blood brain barrier disruption produced excess hemoglobin at 12.5 M_a. Reproduced from Ottens et al. [5] with permission from Wiley & Sons

9. Add 25 μL of 50 mM ammonium bicarbonate and vortex gently for 5 min. Collect the digest into a labeled autosampler vial.

10. Add 50:45:5 Extraction Solution to the gel cubes and vortex gently for 5 min. Collect the solution into the same autosampler vial.

11. Speed vacuum the autosampler vials to dryness. Vials can be stored at −80 °C until ready for LC/MSMS analysis.

3.5 Reversed-Phase Liquid Chromatography (RPLC)

3.5.1 RPLC Method Optimization

1. Create a 25 μL mixture of the isotopically labeled MBP peptides each at 100 nM in 0.1 % (v/v) modified LC/MS grade water.

2. Configure a tandem mass spectrometer for data-dependent analysis (DDA) (*see* Section 3.6.1).

3. Begin optimizing the RPLC elution profile with a wide gradient; e.g., with 5 % mobile phase B and ending at 65 % B over 60 min (rate of change = 1 %/min).

4. The gradient range should be narrowed as possible while maintaining retention and elution of all MBP target peptides.

5. If the target peptides are fully resolved, try increasing the rate of change (e.g., to 1.5 %/min).

6. A step gradient may also be employed should certain areas of the gradient profile need to be sped up or slowed down to fully resolve the target peptides while minimizing the total gradient time.

7. Lastly, evaluate the optimized gradient profile at several flow rates (e.g., 300, 400, 500, 600 nL/min). Peak capacity, assessed as the average peak width (at 50 % height) across MBP target peptides, degrades when mobile phase velocity is either too slow or two fast for a given column.

3.5.2 RPLC Separation for Isotope Dilution Analysis of MBP

1. Operational performance of an LC/MSMS system should be affirmed throughout the experiment using a protein digest quality control standard. Results should be assessed for chromatographic peak shape, elution profile, excessive background, signal intensity, and fragmentation efficiency relative to previous quality control data.

2. Reconstitute sample peptides in 10 μL of formic acid (0.1 % v/v) modified LC/MS grade water (*see* **Note 11**). Vortex autosampler vials for 1 min at a sufficient rotational speed that just starts to move the water up the side of the vial (*see* **Note 12**).

3. Allow autosampler vials to equilibrate to the temperature of the autosampler system (i.e., if the autosampler is cooled, then the samples must be equilibrated to the desired temperature before injection).

4. Inject the sample peptides onto a RPLC trap column at a low % A (e.g., 2 % A), allowing three column volumes of mobile phase to pass through the trap to remove excess salt/buffer (*see* **Note 13**).

5. Place the trap column in-line with the capillary analytical column and begin the previously optimized gradient profile program (*see* Section 3.5.1).

6. Following each gradient separation, the analytical column should be washed with a high %B (e.g., 85%B) for three to five column volumes and then equilibrated back to the starting %B over five column volumes.

3.6 Multiple Reaction Monitoring Mass Spectrometry (MRM-MS)

1. Reconstitute a digest of one of the MBP bovine standard gel bands with 0.1 % formic acid modified water containing each isotopically labeled peptide at 20 nM (assuming a 1 μL injection volume).

3.6.1 MRM-MS Method Development

2. Analyze the MBP standard using a DDA method with a target list containing the precursor *m/z* for each endogenous and isotopically labeled MBP peptide per the table below (*see* **Note 14**).

Native peptide	Precur. *m/z*	Top *z*	Isotopically labeled peptide	AA label	Precur. *m/z*
HGFLPR	363.71	2	HGFL*PR	L(13C6,15N)	367.24
SPLPSHAR	432.74	2	SPL*PSHAR	L(13C6,15N)	436.27
FSWGAEGQKPGFGYGGR	600.95	3	FSW*GAEGQKPGFGYGGR	W(13C11,15N2)	365.57
TPPPSQGKGGR	361.20	2	TPPPSQ*GKGGR	Q(13C5,15N)	364.23
GSEGQKPFGFYGGR	743.86	2	GSEGQ*KPFGFYGGR	Q(13C5,15N)	746.89
FSWGGR	326.16	2	FSW*GGR	W(13C11,15N2)	332.72
KNIVTPR	414.26	2	KNIV*TPR	V(13C5)	416.78

3. Resulting tandem mass spectra (see examples for HGFLPR and KNIVTPR in Fig. 3) should be validated against predicted fragment ions for the target peptides. Repeat the DDA analysis over a range of collision energies (instrument dependent). The *m/z* values for the three most intense singly charged product ions at their optimal collision energy should be tabulated for each native and isotopically labeled peptide. Using HGFLPR (parent *m/z* of 363.71) as an example, the most intense product ions observed were: y1-5 at 589.35; b1-5 at 532.37; b1-4 at 455.26. In addition, tabulate a retention time window for each peptide (generally a 1 min window centered at the peptide's elution time is sufficient) (*see* **Note 15**).

4. Develop an MRM method from the above tabulated precursor/product ion *m/z* values, specifying MRM transitions for the three most-intense product ions per peptide (*see* **Note 16**). Program successive retention time windows with the respective MRM transitions for each native and isotopically labeled peptide pair (known as a scheduled MRM method). With only six MRM transitions to monitor at any given time, longer dwell times (e.g., 25–50 ms) may be used to enhance assay sensitivity, while still maintaining adequate chromatographic peak sampling (>2 Hz) (*see* **Note 17**).

5. Next, assess the developed MRM assay for quantitative performance. Reconstitute another MBP bovine standard digest as in **step 1**. From this digest, prepare an eight-step dilution series across five orders in magnitude. Analyze the dilution series (lowest concentration first) in triplicate. Tabulate the peak area for each peptide (summed across all ion transitions to the peptide) and calculate the native:labeled ratio for each peptide. Multiply each ratio by 20 nM to provide the concentration for the respective native peptide. Average the triplicate results and produce a calibration curve as in Fig. 4. The assay should

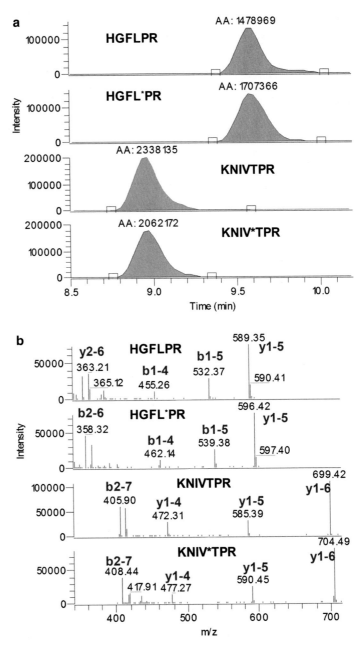

Fig. 3 RPLC-MSMS results for native and isotopically labeled MBP peptide pairs. (**a**) Chromatographic peak area results for HGFLPR and KNIVTPR MBP peptides and their corresponding isotopically labeled (denoted by *) partners are shown based on their selective precursor and product ion data. A seven-point Gaussian smoothing algorithm was applied during peak area integration. (**b**) Collected tandem mass spectra were used to confirm peptide identity. Product ions are labeled with their respective b- and y-ion assignments (in *red*). Results were acquired for a mixture of the four peptides each at 1 nM, with 2 μL injected on column. Reproduced from Ottens et al. [5] with modification and permission from Wiley & Sons

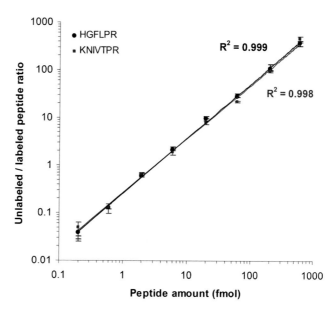

Fig. 4 Linear dynamic range for MRM-MS MBP quantification. From 0.2 to 600 fmol of HGFLPR and KNIVTPR was quantified relative to 20 fmol of their respective isotopically labeled internal standards (loaded on column). The chromatographic peak ratio remained linear across four orders of magnitude, with attomole detection. Values are reported as the mean ± SD ($n = 3$). Reproduced from Ottens et al. [5] with modification and permission from Wiley & Sons

demonstrate a linear response across much of the calibration plot, with sub fmol/μL detection expected (*see* **Note 18**).

6. Finally, it is suggested that the developed RPLC/MRM-MS methodology be evaluated against gel bands from a lane of naïve and TBI brain lysate to adjust for interferent endogenous molecules (*see* **Note 18**).

3.6.2 MRM-MS Acquisition for Isotope Dilution Analysis of MBP

1. Ensure that the instrument is properly tuned and calibrated ahead and strategically throughout the course of the study (if data will be acquired over multiple days).

2. The reconstituted gel band digests should be injected onto the RPLC/MSMS system in a treatment-interspersed order across biological replicates, so as to avoid introducing bias between the groups. In subsequent technical replicates, use a different treatment-interspersed order across the biological replicates.

3. Acquire all biological and technical replicates at a given M_a before moving to the next row of gel bands, so as to minimize variance when measuring a given MBP isoform.

4. When acquiring over multiple days, perform a quality control analysis following tuning and calibration to affirm system performance.

3.7 Quantitative Analysis of MBP Calpain Degradation

MRM-MS results are generated as the summed peak area intensity across all ion transitions for a given peptide. Open-source and commercial software are available to simplify this process and produce tabulated results that can be interrogated using statistical software. Here we will review IDA data processing using the open-source software Skyline (https://skyline.gs.washington.edu), the most compatible, freely available software for MRM-MS analysis at this time [26]. However, commercial software packages generally follow a similar MRM-MS processing workflow.

1. Beginning with a blank worksheet in Skyline, import or manually set up an ion transition list within the "Targets" pane. Most mass spectrometer control software packages are able to export the ion transition list from the used MRM-MS method (*see* **Note 19**). If building the list manually, simply type in the peptide sequences and Skyline will automatically generate the precursor and product ion masses. It may be necessary to manually select which product ion transitions were used from the pull-down next to each precursor while turning off "Auto Selected Filter Transitions". It is only necessary to enter the native peptides information as the isotopically labeled ion transitions will be automatically added in the next step.

2. Select "Settings" and "Peptide Settings": under "Digestion", specify Arg-C (a.k.a. clostripain); under "Filter", reduce the Min length to 5; under "Modifications", specify any fixed modifications (e.g., carbamidomethyl C if iodoacetamide was used during in-gel digestion), for the isotope label type, edit the list to include the appropriate isotopically labeled amino acids (e.g., select "Label:13C(6)15N(1)(L)" for peptide HGFLPR), and select heavy under Internal standard type.

3. Proceed to "Settings" and "Transition Settings": under "Instrument", adjust the m/z min, max, and tolerance as appropriate for your instrument (m/z tolerance is a critical settings); under "Full-Scan", select MS/MS filtering as a Targeted Acquisition method and specify the Product mass analyzer, isolation scheme, and an appropriate mass resolution; also under "Full-Scan", select Include all matching scans under Retention time filtering (*see* **Note 20**).

4. Proceed to "File" and "Import Results". Specify "Add single-injection replicates in files" and click "OK". Navigate to and select all replicate samples for the study (*see* **Note 21**).

5. Imported data should automatically show chromatograms for all identified native (light) and isotopically labeled (heavy) peptides across all study replicates (Fig. 5). You can select a given peptide in the "Targets" pane as well as a specific ion transition. Further, the software will automatically integrate

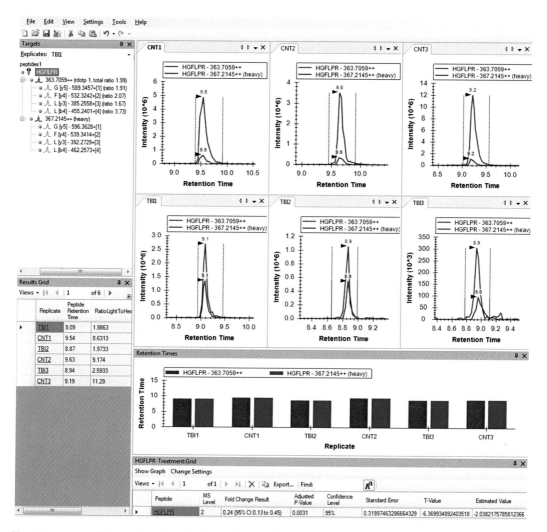

Fig. 5 Processing IDA results using Skyline software. Native and isotopically labeled MBP peptide sequences are entered into the Skyline Targets list in the upper left, with precursor and product ion m/z values generated automatically. Raw data are then added to the main window, which displays the summed chromatographic peak intensity for all product ions to each peptide. The native:labeled peptide ratio is tabulated and reported in the *lower left*, while a quality control evaluation of the retention time distribution across replicates and statistical analysis results are displayed on the *lower right*

and sum the peak areas for all product ions and tabulate the light-to-heavy peptide ratios.

6. Under "View" select: "Results Grid" to tabulate the light-to-heavy peptide ratios; "Auto Zoom, Best Fit" to focus each replicate's chromatogram onto the target peptide peak; "Retention Times, Replicate Comparisons" to evaluate the retention time consistency for the auto-picked peptide peaks, which is useful in identifying peak-picking errors.

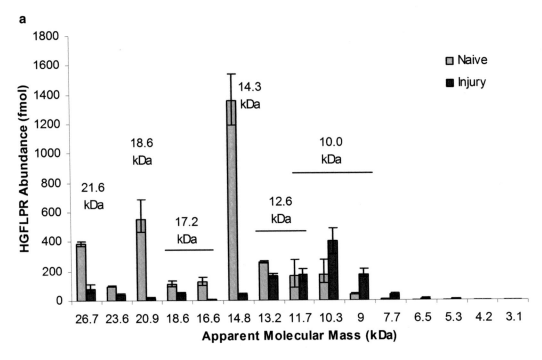

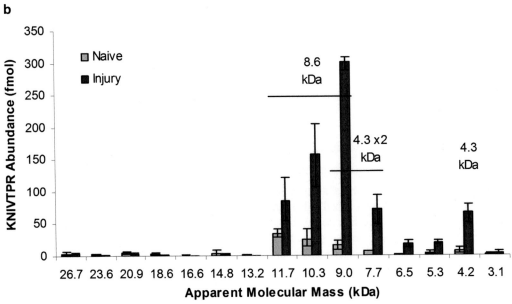

Fig. 6 Quantitative profiling of MBP degradation following TBI. (**a**) HGFLPR and (**b**) KNIVTPR peptide abundance is plotted against gel band M_a value for naïve and TBI rat cortex lysate. Also shown are (**a**) the corresponding MBP isoforms as indicated by their molecular mass, as well as their respective (**b**) calpain breakdown products. Abundance is given in fmol relative to 5 μg of brain protein digested loaded on column. Values are reported as the mean ± SD ($n = 3$). Reproduced from Ottens et al. [5] with modification and permission from Wiley & Sons

7. Proceed to "Settings" and "Document Settings": under "Annotations", "Add" three variables to define technical replicates (TechRep), biological replicates (BioRep), and treatment conditions (Treatment) as type text and value list (e.g., CNT and TBI), respectively. Apply both to "Replicates"; under "Group Comparison", "Add" a Treatment comparison by specifying Treatment under Control group annotation, CNT under Control group value, TBI under Value to compare against, BioRep under Identity annotation (for technical replicates), and ratio to Heavy under Normalization method.

8. Proceed to "View" and new "Group Comparison" to see the differential statistic results for each target peptide. Pop-up panes may be placed back onto the main Skyline window (as in Fig. 5). Result tables can be customized via the Views pull-down in the upper left corner, allowing you to add or remove column data.

9. Finally, use the Export function on a results table to transfer the data into Excel. Multiply the tabulated light-to-heavy ratios by 20 nM to calculate the absolute native peptide concentrations at each gel band M_a. Average the fmol amounts (on column) across technical replicates and then calculate the target mean and standard deviation across biological replicates for each peptide. Figure 6 shows IDA results for the peptides HGFLPR, a marker common to all MBP isoforms, and KNIVTPR, the calpain proteolysis-selective marker indicative of post-TBI degradation of MBP.

4 Notes

1. Cambridge Isotope Laboratories offers amino acids enriched with one or more heavy isotope (e.g., D, ^{13}C and ^{15}N), providing the flexibility to label any peptide and adjust the mass difference from the native form. By incorporating an amino acid with multiple enriched isotopes, for example L-leucine ($^{13}C6$, ^{15}N) with a delta mass of 7 Da, the monoisotopic peak of the labeled peptide will be well resolved from the naturally occurring peptide. Amino acids with ^{13}C and ^{15}N isotopes is preferred over deuterated amino acids when using reversed-phase chromatography, in order to avoid an elution shift between the labeled and native peptide forms. Amino acids are available with Fmoc and BOC protecting groups for ease of synthesis.

2. Brain tissues should be fresh-frozen either over liquid nitrogen or dry ice-chilled isopentane for proteomics. Time between sacrifice and freezing of the brain should be minimized and kept consistent (<5 min) to reduce postmortem artifacts within

the neuroproteome. Avoid using frozen tissue-embedding media, such as HistoPrep or OCT, as these polymeric compounds will produce unwanted chemical noise within the mass spectra.

3. When formulating your brain lysis buffer, carefully consider: (1) The buffering agent to avoid interference with downstream processes; e.g., phosphate buffers should be avoided with clostripain digestion as the required calcium will precipitate out as calcium phosphate; Tris buffers interfere with amine-reactive agents; HEPES can promote radicals and, along with MOPS, may offer varied buffering capacity with membranous preparations. (2) The additives, such as NaCl to maintain ionic strength and EDTA/EGTA to chelate metals, which may interfere with downstream separation procedures. (3) The reduction agent, as those containing thiols should be prepared fresh to maintain reducing capacity. The reducing agent TCEP, which is stable at room temperature, has become popular as a replacement for thiol-related agents. (4) The surfactant, such as Triton X-100, which is an essential component membrane protein extraction. Yet while ionic surfactants such as SDS are excellent for protein extraction, they tend to interfere with downstream separations and mass spectrometric analysis. Triton X-114 has recently been shown to provide better performance than X-110 in solubilizing membranous proteins from brain and affording a higher number of identified proteins by LC/MSMS analysis [27]. (5) Protease (e.g., Roche Complete Mini) and Phosphatase inhibitors (sodium vanadate), which limit postmortem processing of the neuroproteome. Of late, our group has switched to using the Pierce Halt Protease and Phosphatase Inhibitor Cocktail (Life Technologies, Rockford, IL), which provides broad-based inhibition as an easy-to-use single source product.

4. The Clostripain Digestion Buffer should be activated for 1 h at room temperature prior to usage. The calcium chloride is first prepared as a 1 M stock that is then added dropwise to make the digestion buffer.

5. While trypsin remains the most widely utilized protease for bottom-up proteomics, other mass spectrometry-grade proteases may be preferential in generating isoform-specific peptides. Thermo Pierce (www.piercenet.com/guide/protease-selection-chart) offers a variety of mass spectrometry-grade proteases that may provide more advantageous cleavage for a given sequence (e.g., for sequences containing too many or too few arginine residues, it may be preferred to use Asp-N for cleaving the amino side of aspartate or Glu-C for cleaving the carboxyl side of glutamate and aspartate).

6. In brain injury studies, neuroanatomy and contusional burden are both important to consider when dissecting tissues. For example, one would preferentially examine white matter, such as the corpus callosum, for myelin degeneration after TBI, both from regions proximal and distal from the contusion center.

7. Ensure that roughly the same volume of brain tissue is processed for each sample in order to minimize protein extraction variability. Use of the Pellet Pestles should be kept consistent across samples.

8. Wider pre-cast gels, such as the NuPAGE midi format, are preferred so that all biological replicates can be separated on the same gel (or two). Load as much protein as possible per lane to maximize LC/MSMS detection; e.g., the 20 well NuPAGE midi gel can hold up to 25 μL per lane (60 % of the actual volume so as to avoid spill-over between wells).

9. MBP isoforms are known to migrate slower than expected relative to Prestained Molecular Mass Markers. Thus, the 18.6 kDa isoform of MBP enriched in the purified bovine standard will appear closer to 21 kDa on the calibrated molecular mass curve.

10. Cutting 1 mm gel bands between an M_a of 28 and 3 should resolve all predicted isoforms of MBP as well as potential calpain cleavage products following TBI. Depending on one's objective, fewer bands may be required. Gel bands should be cut using a clean scalpel on a clean surface ideally placed atop a white light box.

11. The reconstitution volume should allow for replicate sample injections and sample loss between injections. An individual injection volume for a capillary chromatography column would typically range between 1 and 5 μL, depending on the precision of an autosampler system and the internal diameter of the column. Multiply the injection volume by five for a suggested reconstitution volume.

12. The rotational speed of a vortex should always be optimized for the sample vessel and volume, such that the liquid swirls within the vessel without lifting too high up the side. Peptides dried by speed vacuum will largely concentrate at the bottom of the vessel. Thus, adjust the vortex speed to allow the liquid to swirl towards the bottom of the vessel to facilitate maximal peptide reconstitution.

13. An RPLC trap column should contain a reversed-phase media with slightly less retentive properties than the media of the analytical column in order to allow peptide refocusing at the head of the analytical column during gradient elution.

14. Operational parameters for DDA acquisition should be provided by the instrument vendor. DDA methods generally allow the user to specify a precursor ion m/z target list. MSMS spectra are typically acquired between m/z 150 and 2000. Adjust the parameters to allow for multiple MSMS events (e.g., top six events) such that any co-eluting target peptides and their isotopically labeled partners will be fragmented. Given the use of a target list, it is advisable that dynamic exclusion be turned off.

15. RPLC method optimization (*see* Section 3.5.1) ideally would resolve target peptides from one another so that 100 % of the MS duty cycle is devoted to each peptide pair. However, most modern instruments are sufficiently sensitive to simultaneously profile several peptides in MRM-MS mode for a more rapid assay.

16. While a given peptide could be quantified with a single MRM transition, a minimum of three ion transitions are suggested to provide sufficient selectivity. Most instruments employ a quadruple mass filter for precursor m/z selection, which require an isolation window between 0.5 and 3 Da to avoid a loss in sensitivity. As such, the possibility remains that multiple precursors may be fragmented together and that have common product ions. In practice, this issue was observed often enough that we always employ two to three transitions per peptide, allowing us to validate the calculated native:labeled peptide ratio across the ion transitions. In addition, summing the chromatography peak area for three product ions increases the sensitivity of the assay.

17. Vendor-specific (e.g., PinPoint, ThermoScientific) and open-source (e.g., Skyline: https://skyline.gs.washington.edu) software can be used to automate MRM method generation [25]. DDA data is searched against an MBP isoform database, then MRM generation software is used to select the target peptides and their ion transitions.

18. The calibration plot should remain linear across at least three orders in magnitude. If this is not observed, the native:labeled ratio for each of the three product ions should be compared at each concentration to assess whether one or more transitions is perturbed by another endogenous signal. The RPLC and/or MRM methods should be adjusted to exclude any interferent signal.

19. As mentioned in **Note 17**, Skyline may also be used in generating the MRM-MS method. In that case, the saved "Skyline Document" already contains the required MRM transition list for your study, and results can be added without further modification of the "Targets" pane.

20. Refer to the Skyline tutorials for further information if you wish to import search results and retention time information for targeted peptides.

21. Skyline, in collaboration with ProteoWizard, is able to read raw data formats from all major mass spectrometer vendors as well as mzML, mzXML, and MZ5 formats.

References

1. Marx V (2013) Targeted proteomics. Nat Methods 10:19–22

2. Barr JR, Maggio VL, Patterson DG Jr et al (1996) Isotope dilution-mass spectrometric quantification of specific proteins: model application with apolipoprotein A-I. Clin Chem 42:1676–1682

3. Gerber SA, Rush J, Stemman O, Kirschner MW, Gygi SP (2003) Absolute quantification of proteins and phosphoproteins from cell lysates by tandem MS. Proc Natl Acad Sci U S A 100:6940–6945

4. Abbotiello SE, Schilling B, Mani DR et al. (2015) Large-scale inter-laboratory study to develop, analytically validate and apply highly multiplexed, quantitative peptide assays to measure cancer-relevant proteins in plasma. Mol Cell Proteomics pii: mcp.M114.047050

5. Ottens AK, Golden EC, Bustamante L et al (2008) Proteolysis of multiple myelin basic protein isoforms after neurotrauma: characterization by mass spectrometry. J Neurochem 104:1404–1414

6. Thomas DG, Palfreyman JW, Ratcliffe JG (1978) Serum-myelin-basic-protein assay in diagnosis and prognosis of patients with head injury. Lancet 1:113–115

7. Berger RP, Adelson PD, Richichi R, Kochanek PM (2006) Serum biomarkers after traumatic and hypoxemic brain injuries: insight into the biochemical response of the pediatric brain to inflict brain injury. Dev Neurosci 28:327–335

8. Berger RP, Dulani T, Adelson PD et al (2006) Identification of inflicted traumatic brain injury in well-appearing infants using serum and cerebrospinal markers: a possible screening tool. Pediatrics 117:325–332

9. Voskuhl RR, Robinson ED, Segal BM et al (1994) HLA restriction and TCR usage of T lymphocytes specific for a novel candidate autoantigen, X2 MBP, in multiple sclerosis. J Immunol 153:4834–4844

10. Kruger GM, Diemel LT, Copelman CA, Cuzner ML (1999) Myelin basic protein isoforms in myelinating and remyelinating rat brain aggregate cultures. J Neurosci Res 56:241–247

11. Boggs JM (2006) Myelin basic protein: a multifunctional protein. Cell Mol Life Sci 63:1945–1961

12. Schaecher KE, Shields DC, Banik NL (2001) Mechanism of myelin breakdown in experimental demyelination: a putative role for calpain. Neurochem Res 26:731–737

13. James T, Matzell D, Bartus R et al (1998) New inhibitors of calpain prevent degradation of cytoskeletal and myelin proteins in spinal cord in vitro. J Neurosci Res 51:218–222

14. Liu MC, Akle V, Zheng W et al (2006) Extensive degradation of myelin basic protein isoforms by calpain following traumatic brain injury. J Neurochem 98:700–712

15. Harauz G, Ishiyama N, Hill CM et al (2004) Myelin basic protein – diverse conformational states of an intrinsically unstructured protein and its roles in myelin assembly and multiple sclerosis. Micron 35:503–542

16. Kimura M, Sato M, Akatsuka A et al (1998) Overexpression of a minor component of myelin basic protein isoform (17.2 kDa) can restore myelinogenesis in transgenic shiverer mice. Brain Res 785:245–252

17. Nakajima K, Ikenaka K, Kagawa T et al (1993) Novel isoforms of mouse myelin basic protein predominantly expressed in embryonic stage. J Neurochem 60:1554–1563

18. Givogri MI, Bongarzone ER, Campagnoni AT (2000) New insights on the biology of myelin basic protein gene: the neural-immune connection. J Neurosci Res 59:153–159

19. Boggs JM, Rangaraj G, Koshy KM, Mueller JP (2000) Adhesion of acidic lipid vesicles by 21.5 kDa (recombinant) and 18.5 kDa isoforms of myelin basic protein. Biochim Biophys Acta 1463:81–87

20. Akiyama K, Ichinose S, Omori A, Sakurai Y, Asou H (2002) Study of expression of myelin basic proteins (MBPs) in developing rat brain using a novel antibody reacting with four major isoforms of MBP. J Neurosci Res 68:19–28

21. Capello E, Voskuhl RR, Mctarland HF, Rane CS (1997) Multiple sclerosis: re-expression of developmental gene in chromic lesions

correlates with remyelination. Ann Neurol 41:797–805

22. Berger RP, Bazaco MC, Wagner AK, Kochanek PM, Fabio A (2010) Trajectory analysis of serum biomarker concentrations facilitates outcome prediction after pediatric traumatic and hypoxemic brain injury. Dev Neurosci 32:396–405

23. Cortes DF, Landis MK, Ottens AK (2012) High-capacity peptide-centric platform to decode the proteomic response to brain injury. Electrophoresis 33:3712–3719

24. Lizhnyak PN, Ottens AK (2015) Proteomics: in pursuit of effective traumatic brain injury therapeutics. Expert Rev Proteomics 12:75–82

25. Colangelo CM, Chung L, Bruce C, Cheung KH (2013) Review of software tools for design and analysis of large scale MRM proteomic datasets. Methods 61:287–298

26. MacLean B, Tomazela DM, Shulman N et al (2010) Skyline: an open source document editor for creating and analyzing targeted proteomics experiments. Bioinformatics 26:966–968

27. Shevchenko G, Musunuri S, Wetterhall M, Bergquist J (2012) Comparison of extraction methods for the comprehensive analysis of mouse brain proteome using shotgun-based mass spectrometry. J Proteome Res 11:2441–2451

Neuromethods (2016) 114: 243–262
DOI 10.1007/7657_2015_91
© Springer Science+Business Media New York 2015
Published online: 08 November 2015

Identification and Characterization of Protein Posttranslational Modifications by Differential Fluorescent Labeling

Eric D. Hamlett, Cristina Osorio, and Oscar Alzate

Abstract

Analysis of posttranslational modifications (PTMs) of proteins is of major interest in current biomedical research due to the multiple roles played by PTMs in protein regulation and function. In general this is a complex problem because first and foremost there are large numbers of potential PTMs in every protein that result in myriads of possible combinations of functional enzymes. In this chapter we present a general method for PTM analysis based on specifically labeling modified and unmodified samples with distinguishable fluorescent dyes followed by protein fractionation. The method is open to further refinements; in particular there are no computer programs tailored for this type of analysis, and in multiple cases little or nothing is known about a specific PTM, about how to alter such modification, or about successful labeling of the target amino acid. We present this chapter with two goals in mind, first to share with other scientists some of our experiences in this field, and second to invite those interested in the subject to bring their own contributions to an area that should be further explored and enhanced in order to create a large tool kit for PTM analysis. Although we present just four examples of this technology, in principle, any PTM can be targeted for analysis using the general principles delineated here.

Keywords: Differential in gel electrophoresis (DIGE), Oxidation, Palmitoylation, Ubiquitination, Phosphorylation, Protein isoforms, DeCyder, Quantitative intact proteomics (QIP)

1 Introduction

1.1 PTMs: A Cell's Modify and Conquer

Selection processes have resulted in a human genome having a modest set of protein-coding genes. The 2004 consortium of the human genome reported less than 25,000 protein-coding genes comprising only 1.2 % of the total euchromatic genome landscape [1]. However, human cells have found compensatory mechanisms to expand upon the proteome. At the gene level, mutation and splicing expand protein-coding outcomes. Scattered over individual genomes are over 10 million common single nucleotide polymorphisms (SNPs) [2]: some leading to coding changes in proteins while others are either synonymous or silent, and lead to no appreciable differences. Random SNP mutations in protein-coding genes are highly affected by intron/exon density. Gene SNPs within exons may not always lead to protein changes. From studies in yeast, the effect of a mutation is limited, due to the degenerate nature of the amino acid code, so that only 12 % of protein-coding SNPs are actually deleterious [3]. This ratio might not apply to the human genome landscape since SNP sensitivity is far more complex. Gene SNPs within introns are likely silent unless they affect regulating or splicing mechanisms. Alternative splicing of gene transcription products allows an individual gene to produce multiple protein isoforms from a single coding space. Splicing mechanisms frequently influence which gene promoters and transcription start sites are selected for the protein code or polycistronic protein codes [4]. Recent evidence has revealed that over 73 % of human genes are alternatively spliced with half of these altering the reading frame of the protein code [5]. All gene-level modifications encompass pre-translational events that significantly expand the possibilities for total protein-coding outcomes and create a broader protein base for the cell's most diverse biochemical tool, **Posttranslational Modification** (PTM) [6].

In 2011, a survey of the non-redundant Swiss-Prot database (http://www.uniprot.org/) revealed at least 431 different types of PTMs with 87,308 experimentally observed and 234,938 putative modifications on 530,264 proteins [7]. The incidence of PTMs may be far greater as over 2 million putative PTMs are predicted by statistically driven algorithms [8]. Frequency analysis of the most commonly observed modifications revealed that phosphorylation, glycosylation, and acetylation dominate the PTM landscape. The top 15 most seen PTMs included 88 % of all the observations. However, a high incidence of PTM does not necessarily correlate with a stronger functional role. Even the rarest PTMs, often called orphans, are absolutely critical in catalytic reactions and enzyme regulation [9]. The plethora of PTMs greatly expands the total protein-coding outcomes that comprise the proteome.

Many techniques have been either implemented or modified to study and characterize PTMs. In this chapter we present our approaches using a modified version of the differential in gel electrophoresis (DIGE) method. In DIGE-based quantitative intact proteomics (QIP) a sample of interest is labeled with a fluorescent probe and compared in the same gel with a control sample, which is also labeled with a fluorescent probe of a different wavelength. In addition, equimolar amounts of both samples are pooled and then labeled with a third fluorophore. All samples are fractionated by 2D gel electrophoresis and the images are individually resolved via the corresponding wavelength of each fluorescent label [10]. An important feature of the DIGE approach is that the fluorescent labels do not induce significant changes to the protein's molecular weight and isoelectric point.

Our approaches are based on the following principle: target any PTM for analysis, find a method to label such PTM or a treatment that will make the modification suitable for fluorescent labeling, and fractionate the samples as in the DIGE approach. This methodology allows many variations in which it is possible to compare multiple samples, to determine the presence of specific PTMs and the changes in protein expression. The experiments presented next have been performed in our laboratory. In most cases data analyses have been performed with the GE DeCyder software [11]; although this computer program is specifically tailored for DIGE-based QIP, there are multiple ways in which the variables can be assigned to obtain significant results.

2   Theory and Methods

2.1   General Methodology

The general methodology is the following (Fig. 1a, b):

1. Select the posttranslational modification (PTM) to be analyzed;

2. Determine whether:

   (a) This PTM can be fluorescently labeled; or

   (b) This PTM can be altered to generate a reactive group that can be fluorescently labeled;

3. Purify protein samples;

4. Split samples in two identical aliquots for **PTM analysis** (Fig. 1a); or three identical aliquots for **PTM and Quantitative analysis** (Fig. 1b);

5. Treat one of the aliquots to remove the corresponding modification (for instance, to dephosphorylate proteins, treat the sample with phosphatases)—Altered sample, **A**;

6. Treat the other aliquot under identical conditions without removing the modification (i.e., for phosphorylation analysis treat the samples as in 5, without adding protein phosphatases)—Unaltered sample, **U**;

7. Label equal amounts of sample **U** with Cy3 and sample **A** with Cy5;

   See **Note 1**.

8. Remove unreacted labels and determine protein concentration;

9. Run a 2D fluorescent gel;

10. Analyze images with DeCyder 2D;

11. After gel analysis, protein spots of interest can be cut from a pick gel and analyzed by mass spectrometry (MS) for confirmation of the PTM or characterization of the modification sites.

This methodology can be applied to study any PTM provided that specific methods for altering the PTM are identified, and that the alteration is sufficient to generate detectable changes in the

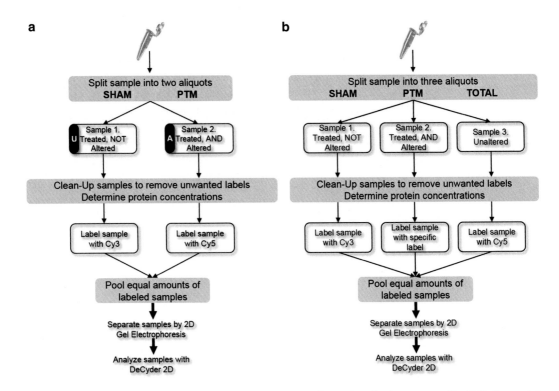

Fig. 1 General methodology. (**a**) To determine if a group of proteins displays a specific PTM the protein samples should be prepared via protocol, and the target PTM should be modified with the appropriate method. For instance, to analyze palmitoylated proteins the samples should be treated with hydroxylamine hydrochloride [12]; or to analyze phosphorylated proteins the samples should be treated with phosphatases. A sham sample should be prepared along with the target sample under exactly the same experimental conditions. After treatments, the proteins are usually cleaned up again, to remove unreacted labels, and the concentration is determined. Equal amounts of treated and untreated samples are individually labeled with Cy dyes, i.e., treated samples are labeled with Cy5 (sample A), and untreated samples with Cy3 (sample U). It is important to identify a different type of fluorophore when the modification itself can be specifically labeled, such as maleimide for disulfide bridges, or Alexa Fluor for oxidized proteins [13]. Fluorescently labeled proteins are separated by 2D gel electrophoresis, and the resulting gels are scanned with a Typhoon TRIO+ scanner, followed by image analysis with DeCyder 2D. (**b**) For quantitative analysis as well as analysis of the PTM of interest, three samples should be prepared: (a) a treated sample that will show the presence of the target PTM, (b) a sham sample that will be used to compare the presence of the PTM, and (c) an untreated sample that will be used to determine changes in protein expression and posttranslational modifications. The latter approach is more complex as it requires a larger number of gels in order to obtain statistical significant differences [14]

protein's molecular weight, or isoelectric point, or both [11]. The following pages explain four examples in which these methods have been successfully applied.

2.2   Analysis of Protein Palmitoylation

2.2.1   Materials

1. Lysis buffer (8 M Urea, 4 % CHAPS, 30 mM Tris, pH 8.5).
2. Protease and phosphatase inhibitors (Sigma-Aldrich).
3. Freshly prepared 100 mM NEM (N-ethylmaleimide).
4. Hydroxylamine chloride (Sigma).

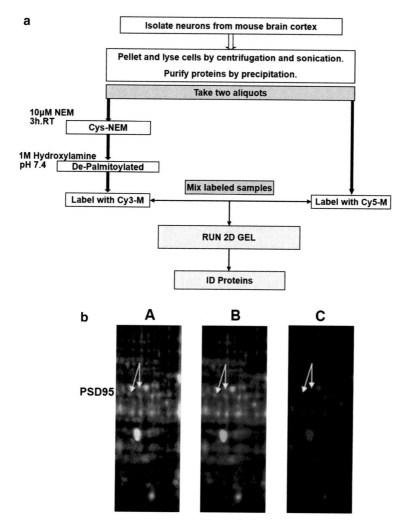

Fig. 2 Analysis of protein palmitoylation via modified DIGE methodology. (**a**) Methodology following the general guidelines described in Fig. 1 with specific details for targeting protein palmitoylation. (**b**) Overlay image displaying proteins labeled with Cy3-M and Cy5-M (*panel a*). Several protein spots are seen in *green*, which correspond to proteins that were depalmitoylated with hydroxylamine (not visible in *panel c*). The *arrows* indicate proteins identified by mass spectrometry

5. Cy3 and Cy5 maleimide (Cy3-M, Cy5-M, GE Healthcare).

6. Cy3 and Cy5 (GE Healthcare).

7. GE Clean-up Kit (GE Healthcare).

2.2.2 Methods

1. Prepare samples of interest according to the established protocols. Resuspend cell pellets in lysis buffer containing protease and phosphatase inhibitors.

2. Extract proteins and determine protein concentration.

3. Prepare fresh 100 mM NEM.

4. Add 1 μL of 100 mM NEM to 100 μL of protein sample (concentration ~ 1 μg/μL), reach a final concentration of 10 μM NEM.

5. Incubate the sample in the dark at RT for 3 h or overnight at 4 °C with gentle rocking.

6. Clean up the sample using the 2D clean-up kit (GE Healthcare) or the cold chloroform/methanol method (Section "Sample Cleaning") to remove unbound NEM.

7. Resuspend the pellet in lysis buffer (50–100 μL) and split the samples in two identical fractions. These two samples are identified as U (Unaltered) and A (Altered) samples.

8. To sample A add an equal volume of 1 M freshly prepared hydroxylamine hydrochloride. Treatment with hydroxylamine hydrochloride should result in depalmitoylated proteins [12].

9. To the U sample add an equal volume of lysis buffer.

10. Incubate both samples at room temperature for 1 h with rocking.

11. Clean up each sample via the clean-up method.

12. Resuspend each final pellet in 25–30 μL lysis buffer.

13. Add 1 μL of Cy5-maleimide dye to the A sample and vortex briefly.

14. Always protecting the sample from light, flush the tube containing the A sample with nitrogen gas, and incubate at RT for 2 h, and then overnight at 4 °C with gentle rocking.

15. The next day, clean up sample A to remove unbound Cy5-maleimide and label the U sample (control; i.e., palmitoylated) with Cy3-maleimide.

16. Pool both samples, shake them gently and add a volume of 2×-buffer (Section "Modified DIGE") equivalent to the total volume of the pooled samples. Run a 2D gel as explained below (Section 2.6.2).

17. Scan the gel as explained (Section "Gel Imaging"), and use the images for differential analysis as explained in Section "DeCyder-Based Differential Protein Modification Analysis." In principle, all spots that have a red labeling should have been palmitoylated in the original sample. The analysis has to be performed with great care because the Cy5-maleimide introduces a pI shift that depends on the number of palmitoylated amino acids.

    See **Note 2**.

2.3 Analysis of Protein Ubiquitination in Postsynaptic Densities

2.3.1 Materials

1. Dissection media: 80 mL Hanks Balanced Salt Solution, 20 mL FBS. FBS stands for Fetal Bovine Serum.

2. Digestion solution: 50 mg trypsin, 6 mg DNAse (Sigma-Aldrich, Cat# D5025) dissolved in pre-warmed (37 °C) Dissection media.

3. Trypsin (Life Technologies, Cat# 27250-018). For High and Low Inhibitor.

4. Trypsin inhibitor (specific source or grade?).

5. AraC (cytosine-1-β-D-arabinofuranoside—Sigma C1768).

6. Lysis buffer: 8 M Urea, 2 M thiourea, 4 % CHAPS, 20 mM Tris, pH 7.5, 1 % DTT, 0.5 % IPG buffer.

7. 1 M Tris–HCl buffer, pH 7.4.

8. 0.5 M $MgCl_2$.

9. 588 µM ubiquitin aldehyde (Santa Cruz Biotech, Santa Cruz, CA, USA; prepare fresh).

10. 188.7 mM AMP-PNP (Adenylyl-imidodiphosphate, tetra-lithium salt; Santa Cruz).

11. 10 mM MG132 stock (Boston Biochem, Boston, MA, USA). Store at −20 °C.

12. 0.5 M ATP stock (Sigma-Aldrich, St. Louis, MO). Store at −20 °C.

13. 0.5 M Creatine phosphate stock (EMD Biochemicals, Gibbstown, NJ, USA). Store at −20 °C.

14. 2× Ubiquitin conjugation buffer (100 mM Tris–HCl, pH 7.8, 10 mM $MgCl_2$, 1 mM DTT, 150 µM MG132, 4 µM ubiquitin aldehyde, 5 mM AMP-PNP, 20 mM ATP, 20 mM creatine phosphate, and 0.5 mg/mL creatine phosphokinase). Prepare fresh as required.

15. 1 M HEPES buffer.

16. Isopeptidase-T (Enzo Life Sciences, Plymouth Meeting, PA, USA).

17. UCH-L3 (R&D Systems, Minneapolis, MN, USA).

18. De-ubiquitinating buffer (25 mM HEPES, pH 7.4, 10 mM DTT, isopeptidase-T and UCH-L3). Prepare fresh as required.

19. Phosphatase inhibitors (Sigma-Aldrich, Cat# P5726).

2.3.2 Methods

Sample Preparation

The following protocol was developed using cultured cortical neurons from 18-days-old rat embryos (E18).

1. Dissect cortical tissues in Dissection media, and transfer five cortexes to 15 mL conical tubes with Hanks solution.

2. Spin down at 106.2 × *g* for 20 s.

3. Prepare and filter-sterilize the digestion solution.

4. Incubate culture at 37 °C for 5 min. Shake tubes gently two times during incubation.

5. Centrifuge at $106.2 \times g$ for 20 s.

6. Prepare High inhibitor: dissolve 26 mg of trypsin inhibitor in 5.0 mL pre-warmed (37 °C) Dissection media.

7. Prepare Low inhibitor: mix 0.5 mL of High inhibitor in 4.5 mL Dissection media.

8. Wash pelleted cortexes with filter sterilized Low inhibitor.

9. Centrifuge as in 5.

10. Replace the Low inhibitor with filter sterilized High inhibitor. Repeat this step.

11. Pellet cells by mild centrifugation as above and remove High inhibitor.

12. Wash cells twice with 10 mL ice-cold Dissection media.

13. Centrifuge as above.

14. Dissolve 6 mg of DNAse in 5.0 mL ice-cold Dissection media, and use it to titrate the mixture 10–12 times.

15. Pellet cells as above, and resuspend in 5.0 mL Hanks media.

16. Count cells and plate appropriate amount on coated 60 mm tissue culture dishes in 3.0 mL pre-equilibrated plating media with 1:1000 dilution of AraC (10 mM).

17. After 48 h replace 2 mL of the plating media with feeding media with 1:1000 dilution of FUDR (5-Fluoro-2′-deoxyuridine; Sigma-Aldrich, Cat# F0503). Repeat this **step 4** days later.

18. Allow cells to grow for 1 week, then remove from dishes with scraper, and resuspend in 1 mL lysis buffer containing protease and phosphatase inhibitors.

19. Grind cells in cold lysis buffer, followed by gently sonication for 30 s.

20. Spin down ($17709.1 \times g$) resulting solution in ultracentrifuge to remove cell debris.

21. Resuspend pellets in an appropriate amount of lysis buffer with protease and phosphatase inhibitors and determine protein concentration.

De-ubiquitination and Labeling

1. Take a portion of the sample prepared above and divide it into two aliquots (the amount of protein should be around 120 μg in 20 μL).

2. Clean each aliquot as described (Section "Sample Cleaning"), resuspend the resulting pellets as indicated below.

3. To one aliquot add 20 μL of 2× ubiquitin conjugation buffer and 20 μL of water and mix. Incubate the mixture at 37 °C with shaking for 1 h. Sample U.

4. To the other aliquot add 40 μL of de-ubiquitination buffer and mix. Incubate this solution at room temperature for 1 h. Sample A.

5. To each sample add an equal volume of lysis buffer containing protease and phosphatase inhibitors and incubate at 4 °C overnight.

6. Label sample U with Cy3 and sample A with Cy5 dyes as described below (Section "Modified DIGE").

7. After labeling mix the samples together.

8. Run a 2D gel as described in Section 2.6.2.

9. Scan the gel as described in Section "Gel Imaging" to determine changes in protein migration as result of the de-ubiquitination treatment.
   See **Note 3**.

2.4   Analysis of Protein S-Nitrosylation (S-NO)

1. S-NO labeling buffer: 8 M Urea, 4 % CHAPS, 20 mM Tris pH 7.3.

2. MMTS.

2.4.1   Materials

3. DMF.

4. Ascorbic acid.

5. Lysis buffer: 8 M Urea, 4 % CHAPS, 30 mM Tris pH 8.5.

2.4.2   Methods

1. Extract proteins from cells of interest using standard protocols, and resuspend the final pellets in S-NO labeling buffer.

2. Determine protein concentration. The sample should contain at least 1 μg/μL.

3. Take 100 μL and perform a quick precipitation with cold methanol/chloroform (Section "Sample Cleaning").

4. Split the sample into two identical fractions: U (U̲nmodified, to be used as control), and A (A̲ltered, to be used as the target sample for S-NO analysis).

5. Prepare fresh 2 M MMTS by dissolving 200 mg in 792.4 μL DMF. This solution can be stored at 4 °C. Avoid light exposure.

6. Add MMTS from stock to sample A to reach a 20 mM final concentration, and incubate the sample at RT for 1 h always protected from light.

7. Clean up each sample via clean-up method (Section "Sample Cleaning").

Fig. 3 Analysis of protein *S*-nitrosylation. Proteins that are denitrosylated and labeled with Cy5-maleimide are shown in *red*; proteins that are not treated appear in *green*

8. Resuspend the final pellets in *S*-NO labeling buffer.

9. Prepare 2.5 M ascorbic acid by dissolving 0.495 g in 1.0 mL water.

10. Add ascorbic acid from stock to sample A. Final concentration should be 3 mM.

11. Clean up each sample (Section "Sample Cleaning").

12. Resuspend both samples in *S*-NO labeling buffer.

13. Label sample A with 0.5–1.0 µL Cy5-maleimide. At this point replace air on tube with nitrogen by flushing air out with a stream of nitrogen and quickly closing the tube, then covering the tube cap with parafilm.

14. Incubate labeled sample at RT for 2 h protected from light, then continue the reaction overnight at 4 °C with gentle rocking.

15. The next day, clean up both samples using 2D clean-up kit, and resuspend the final pellets in lysis buffer.

16. Label the U sample with Cy3 via protocol. Stop the reaction with Lysine.

17. Pool both samples and run a 2D gel as explained in Section "Modified DIGE," *S*-Nitrosylated proteins should appear red.

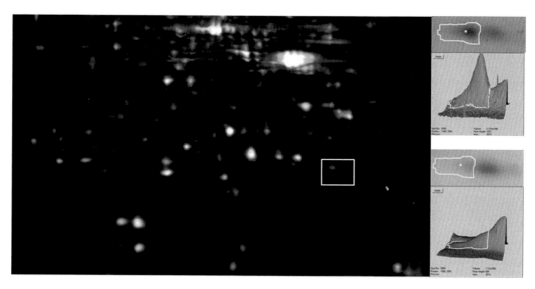

Fig. 4 Analysis of protein phosphorylation. Phosphatase-treated proteins are labeled with Cy5 (shown in *red*). The *insets* display the two spots shown in the *white rectangle*. The *top panel* shows the untreated spot in which it is clear that the isoelectric migration of the protein is changed compared to the *lower panel* in which the treated sample is shown

2.5 Analysis of Protein Phosphorylation

2.5.1 Materials

1. Lysis buffer: 8 M Urea, 4 % CHAPS, 30 mM Tris, pH 8.5.
2. Phosphatase kit (Millipore, Cat# 14-405).
3. Cy3, and Cy5 dyes (GE Healthcare).
4. Complete® protease inhibitor.

2.5.2 Methods

1. Resuspend cells in lysis buffer.
2. Isolate and clean up proteins (Section "Sample Cleaning"), and determine protein concentration.
3. Take 30 μg of protein.
4. Split samples in two identical halves.
5. Label one of the fractions with Cy3 as explained under sample labeling (Section "Modified DIGE")—sample U (<u>U</u>naltered).
6. Label the other fraction with **Cy5**; sample A (<u>A</u>ltered).
7. Stop the reaction with Lysine (Section "Modified DIGE").
8. Clean up each sample individually using the cold methanol/ chloroform method (Section "Sample Cleaning").
9. Resuspend sample U in 200 μL phosphatase buffer included with the phosphatase kit; and resuspend sample A in 200 μL of phosphatase buffer supplemented with λ-PP included in the phosphatase kit.

   See **Note 4**.

10. Let the reaction of both samples proceed overnight at 30 °C with gentle agitation.

11. The next day clean up both samples (Section "Sample Cleaning").

12. Dissolve the final pellets in 20 μL lysis buffer supplemented with protease inhibitor "Complete®," added accordingly to manufacturer's instructions.

13. Find the concentration of samples U and A.

14. Mix equal amounts of samples U and A, shake them gently, add a volume of 2×-buffer (Section 2.6.2) equivalent to the total volume of the pooled U and A samples.

15. Run a 2D gel (Section 2.6.2).

16. Scan the gel as explained under Section "Gel imaging."

17. Use the images for differential analysis as explained in Section "DeCyder-Based Differential Protein Modification Analysis."
See **Note 5**.

2.6   Modified 2D-DIGE

2.6.1   Materials

1. Ethanol.

2. Glacial acetic acid.

3. Bindsilane.

4. MilliQ water (18 Ω).

5. Tris.

6. Glycine.

7. Urea.

8. Thiourea.

9. IPG strips.

10. IPG For.

11. Typhoon TRIO+.

12. Low-fluorescence glass plates.

13. Dalt 6.

14. NHS-Cy2, NHS-Cy3, NHS-Cy5, Cy3-, Cy5-maleimide (GE Healthcare).

15. Acrylamide/Bis Acrylamide.

16. Alkanox.

17. Bromophenol blue.

18. CHAPS.

19. Complete protease inhibitor.

20. DMF.

21. DTT.

22. Iodoacetamide.

23. IPG buffer (Ampholytes: pI range as desired).

24. Sodium Orthovanadate (NaVO$_4$).

25. TEMED.

26. DeCyder 2D v 7.0 software (GE Healthcare).

2.6.2 Methods

Sample Cleaning

Proteomics experiments require properly prepared samples. Before the samples are labeled, they should be properly cleaned up, and the concentration should be determined with the highest accuracy possible. There are multiple methods for sample cleaning; we use frequently a method based on 2D Clean-Up kit (GE Healthcare). Another efficient method based on chloroform/methanol is described next; this method is very efficient and the quality of the sample is high enough for most proteomics requirements.

1. Store methanol in -20 °C the day before the cleaning procedure.

2. Bring up the predetermined amount of protein extract to 100 µL with d$_i$H$_2$O.

3. Add 300 µL (3 volumes) of water.

4. Add 400 µL (4 volumes) of methanol.

5. Add 100 µL (1 volume) of chloroform.

6. Vortex vigorously and centrifuge for 15 min, at 4 °C, 20817.2 $\times$ g (protein precipitate should appear at the interface).

7. Add another 400 µL of methanol.

8. Remove the water/methanol mix from the top of the interface (be careful NOT to disturb the interface).

9. Vortex vigorously and centrifuge as in 6 (the protein precipitate should appear now as a pellet to the bottom of the tube).

10. Remove the supernatant and briefly dry the pellet.

11. Resuspend the pellet in Lysis buffer.

Modified DIGE

Buffers and Solutions

– Lysis buffer

   12 g Urea (final concentration 8 M)

   5.0 mL of 20 % (w/v) CHAPS (final concentration 4 %)

   0.75 mL of 1 M Tris, pH 8.5 (final concentration 30 mM)

   Bring up to 25 mL with d$_i$H$_2$O

   To 1 mL of Lysis buffer add 10 µL Complete protease inhibitor

– 2$\times$ Sample buffer

   0.48 g Urea

   200 µL of 20 % CHAPS

   20 mg DTT—prepare fresh

20 μL ampholyte (pH 4–7 or 3–10)

Up to 1 mL with d_iH_2O

– Rehydration buffer

0.48 g Urea

200 μL of 20 % CHAPS

2 mg DTT—prepare fresh

10 μL ampholyte (pH 4–7 or 3–10)

Up to 1 mL with d_iH_2O

– 1× Equilibration buffer

18.2 g Urea

1.68 mL 1.5 M Tris, pH 8.8

17.25 mL 87 % Glycerol

1.0 g SDS

Bring volume up to 50 mL with d_iH_2O

– **DTT equilibration solution (DTT-ES)**: Add 20 mL of equilibration buffer to 0.1 g of DTT

– **Iodoacetamide equilibration solution (IAA-ES)**: Add 20 mL of equilibration buffer to 0.9 g of Iodoacetamide

– 3× Anode buffer

90 g Tris

432 g Glycine

30 g SDS

Up to 10 L with d_iH_2O

– Cathode buffer

30 g Tris

144 g Glycine

10 g SDS

Up to 10 L with d_iH_2O

– 12% SDS resolving gel (per 15 mL)

5 mL 1.5 M Tris pH 8.8

6 mL 30 % acrylamide

3.9 mL d_iH_2O

75 μL 20 % SDS

75 μL 10 % APS

25 μL TEMED

– 5 % SDS stacking gel (per 5 mL)

0.620 mL 1.5 M Tris pH 8.8

0.833 mL 30 % acrylamide

3.87 mL d$_i$H$_2$O

25 μL 20 % SDS

50 μL 10 % APS

5 μL TEMED

– Coomassie G250 solution
450 mL d$_i$H$_2$O

100 mL glacial acetic acid

3 g Coomassie Brilliant Blue G250

Protein Labeling with Cy Dyes

1. Dilute Cy dyes according to the manufacturer's instructions.

2. The final concentration of the Cy dye stock solution should be 1 mM. Aliquot into light-protected small tubes each containing 2 μL.

3. Prepare working solution by adding 3 μL of DMF (final concentration 400 pmol/μL) to each tube containing 2 μL Cy dye.

4. Label the protein as indicated by the manufacturer maintaining the ratio 8 pmol of Cy dye for each 1.0 μg of protein.

First Dimension Separation

First dimension separation will require Immobilized pH gradient (IPG) strips and specialized electrophoretic equipment which are commercially available.

1. Combine all samples with the different fluorescent labels and record the total volume.

2. Add equal volume of 2× sample buffer.

3. Place samples on ice for 15 min.

4. Add rehydration buffer to reach a final volume of 450 μL if running a 24 cm IPG strip, or 250 μL if running a 13 cm IPG strip.

5. Load the IPG strip into the specialized electrophoretic system and rehydrate with the prepared sample.

6. Run first dimension as follows:

– For 13 cm strips:

S1	30 V	12 h
S2	500 V	1 h
S3	1000 V	1 h
S4	8000 V	28,000 Vh

– For 24 cm strips:

Rehydrate (Active) at 30 V up to 450 Vh

S1	Step to 500	1 h
S2	Step to 1000	1 h
S3	Step to 8000	Up to 68 kVh

Second Dimension Separation

Second dimension separation will require a denaturing polyacrylamide gel.
See **Note 7**.

1. Prepare an appropriate 12 % resolving gel in appropriate low-fluorescence glass.

2. Overlay the gel with d_iH_2O or *n*-butanol and allow it to polymerize 2–4 h.

3. Remove the overlay liquid and wash well with d_iH_2O.

4. Prepare a thin 5 % stacking gel above the resolving gel.

5. Overlay the gel with d_iH_2O or *n*-butanol and allow it to polymerize 1–2 h.

6. Remove the overlay liquid and wash well with d_iH_2O.

7. Prepare fresh equilibration buffer, followed by fresh DTT-ES and IAA-ES as required.

8. Prepare 1× running buffer for washing the IPG strips; usually 100–200 mL.

9. Remove the strip from the holder using forceps and rinse the strip in 1× running buffer.

10. Wash strips in 20 mL DTT-ES. Wrap tubes in aluminum foil and place them on an orbital shaker for 10 min.

11. Rinse strips in 1× running buffer, then transfer to fresh tubes containing 20 mL of IAA-ES. Cover tubes in aluminum foil and place them on an orbital shaker for 10 min.

12. Rinse each strip in 1× running buffer, and then load each strip onto the prepared gel.

13. Melt an aliquot of 1 % agarose containing bromophenol blue. Cover the strip by slowly adding the agarose, allow the agarose to cool.

14. Load the plate into the electrophoresis chamber and add the appropriate amount of 3× running buffer. Connect the chamber to a circulating water cooler set at 14 °C, and cover the electrophoresis unit with aluminum foil.

15. Use the following settings for 13 cm strips/gels: 600 V, 9 mA per gel, 1 W per gel for 16 h and for 24 cm strips/gels use 600 V, 25 mA per gel, 1 W per gel for 16 h.

16. Stop the running when the bromophenol blue reaches the bottom of the gel.

Gel Imaging

Scan the gels using Typhoon Trio+ scanner following the manufacturer's instructions.

– Image Cy3, and Cy3-maleimide with 520/590 nm wavelengths; and

– Image Cy5, and Cy5-maleimide with 620/680 nm wavelengths.

DeCyder-Based Differential Protein Modification Analysis

The description presented next is based on DeCyder 2D v 7.0 software (GE Healthcare).

1. In DIA (Differential In gel Analysis) select a "spot number" of 4500 to create spot maps.

2. Filter spot maps with the built-in algorithm using "max slope" of 1.0. Edit manually spot maps to eliminate signals from dust particles.

3. In BVA (Biological Variation Analysis) normalize and standardize spot maps for each sample.

4. Match spot maps for each gel (individual samples and IC).

5. Perform statistical comparison (*t*-test and ANOVA) between modified and unmodified samples for each spot.

6. Use EDA (Extended Data Analysis) to determine the PCA (Principal Component Analysis) and the partition cluster analysis.

See **Note 6**.

2.6.3 Pick Gel

1. Prepare enough sample for creating a "Pick gel" (usually between 400 and 600 μg of total protein).

See **Note 8**.

2. Use low fluorescent glass plates. One plate is covered with bindsilane that allows the polymerized gel to stick to the glass. The other plate is covered with repelsilane or alternatively left uncovered so that it can be removed from the gel.

3. Add bindsilane to the glass plates as follows:

– Wash glass plates carefully with alkanox, and rinse extensively with d$_i$H$_2$O.

– Prepare the bindsilane solution:

   8 mL ethanol

   200 μL glacial acetic acid

   10 μL bindsilane

   1.8 mL d$_i$H$_2$O

– Disperse evenly 3 mL of bindsilane solution on a glass plate.

– Damp a kimwipe with ethanol, and wash the plate evenly.

– Let the plate dry for 1–2 h.

4. Add two adhesive markers at the edge of the glass for orientation.

5. Add repelsilane to the other glass plate as follows:
   - Wash glass plate carefully with alkanox, and rinse extensively with d_iH_2O.
   - Prepare the repelsilane solution:

     4.5 mL octamethylcyclooctasilane

     200 μL dimethyldichlorosilane
   - Disperse evenly 3 mL of repelsilane solution on a glass plate.
   - Damp a kimwipe with ethanol, and wash the plate evenly.
   - Let the plate dry for 1–2 h.

6. Prepare and run the gel with the two specially coated plates as previously described in section "Second Dimension Separation."

7. Remove the pick gel and separate the repelsilane glass from the gel assembly. The gel will stick to the bindsilane coated plate.

8. Stain the gel with Coomassie G250 solution with gentle agitation for 24–72 h at room temperature in a covered vessel to prevent evaporation.

9. Destain the pick gel with fresh destain solution, utilizing large kimwipe towels to wick the Coomassie blue out of the destain solution.

10. Repeat **step 9** twice or until the gel is clarified and the spots are clearly visible.

11. Scan the gel for visible color (using the blue filter on the Typhoon Trio+ scanner).

2.6.4 Spot Picking for Protein ID

After PTMs have been discovered utilizing this straightforward methodology, the protein ID must be determined. Spots can be excised by hand or may be robotically cut using commercially available robotic spot picking systems. The orientation marks placed on the bindsilane plate will be critical for appropriate machine alignment when high-throughput spot picking is necessary.

In this chapter we demonstrated the general principles to apply differential fluorescent labeling to target a few common PTMs that are observed in the proteomics strategies utilized to conquer cellular needs. Our experiments with modified 2D-DIGE have suggested that any PTM can be targeted for analysis using the techniques delineated here. Even though the tool kit for PTM analysis has been greatly expanded, the possibilities for enhancements in this field remain strong. The biological significance of the over 2 million putative PTMs remains to be discovered.

3 Notes

1. The order in which the samples are labeled is up to each researcher—it has been customary in DIGE-based proteomics to label the control samples with Cy3, the target samples with Cy5, and to rotate the labels at least once out of three repeats to even-out the effects of unequal labeling.

2. To identify and further characterize proteins via mass spectrometry it is required, in most cases, to prepare a pick gel, and from this gel remove the protein spots that will be further analyzed. *See* Section 2.6.3.

3. Ubiquitination/De-ubiquitination is a complex process that needs to be analyzed carefully. The method presented here uses the inhibition of ubiquitinating enzymes to create proteins that accumulate in the cell by not having an ubiquitin moiety. Alternatively, ubiquitinated proteins can be altered and analyzed in a similar fashion. In the example presented here, the samples were extracted from postsynaptic densities purified by a sucrose gradient following well-established protocols [15].

4. Prepare λ-PP according to the manufacturer's protocols, and add amount of λ-PP according to those instructions. We have used ~15 μg total protein with 0.4 μL of λ-PP. Millipore recommends 3.0 μL for 40 μg of total protein.

5. In principle, all spots that have equivalent labeling of Cy dyes 3 and 5 are not phosphorylated (under normal circumstances these are the spots that look yellow); any other spot has some form of phosphorylation. In our experiments we have observed changes in both p*I* and Mw.

6. If protein IDs are available for DeCyder analysis, Gene Ontology can be interrogated via the Extended Data Analysis module using the gene ontology (http://www.geneontology.org) and the KEGG (http://www.genome.jp/kegg/) databases.

7. Second dimension separation will require a denaturing polyacrylamide gel. For best results, the gel should be prepared on the day of separation. Alternatively, they can be purchased readymade from commercial suppliers. It is extremely important to utilize low-iron glass with DIGE methodology in order to minimize background fluorescence. Low-iron glass can be readily purchased commercially or locally prepared with laser precision. If handled with care, and cleaned properly low-iron glass is reusable. Gels can be prepared in a range of 7.5–20 % acrylamide. Often 12 % is utilized for optimized separation of a broad range of protein molecular weights.

8. Total protein should be prepared from a pool of all the samples that want to be analyzed.

References

1. International Human Genome Sequencing Consortium (2004) Finishing the euchromatic sequence of the human genome. Nature 431 (7011):931–945. doi:10.1038/nature03001

2. International HapMap C, Altshuler DM, Gibbs RA, Peltonen L, Altshuler DM, Gibbs RA, Peltonen L, Dermitzakis E, Schaffner SF, Yu F, Peltonen L, Dermitzakis E, Bonnen PE, Altshuler DM, Gibbs RA, de Bakker PI, Deloukas P, Gabriel SB, Gwilliam R, Hunt S, Inouye M, Jia X, Palotie A, Parkin M, Whittaker P, Yu F, Chang K, Hawes A, Lewis LR, Ren Y, Wheeler D, Gibbs RA, Muzny DM, Barnes C, Darvishi K, Hurles M, Korn JM, Kristiansson K, Lee C, McCarrol SA, Nemesh J, Dermitzakis E, Keinan A, Montgomery SB, Pollack S, Price AL, Soranzo N, Bonnen PE, Gibbs RA, Gonzaga-Jauregui C, Keinan A, Price AL, Yu F, Anttila V, Brodeur W, Daly MJ, Leslie S, McVean G, Moutsianas L, Nguyen H, Schaffner SF, Zhang Q, Ghori MJ, McGinnis R, McLaren W, Pollack S, Price AL, Schaffner SF, Takeuchi F, Grossman SR, Shlyakhter I, Hostetter EB, Sabeti PC, Adebamowo CA, Foster MW, Gordon DR, Licinio J, Manca MC, Marshall PA, Matsuda I, Ngare D, Wang VO, Reddy D, Rotimi CN, Royal CD, Sharp RR, Zeng C, Brooks LD, McEwen JE (2010) Integrating common and rare genetic variation in diverse human populations. Nature 467 (7311):52–58. doi:10.1038/nature09298

3. Doniger SW, Kim HS, Swain D, Corcuera D, Williams M, Yang SP, Fay JC (2008) A catalog of neutral and deleterious polymorphism in yeast. PLoS Genet 4(8), e1000183. doi:10.1371/journal.pgen.1000183

4. Ayoubi TA, Van De Ven WJ (1996) Regulation of gene expression by alternative promoters. FASEB J 10(4):453–460

5. Matlin AJ, Clark F, Smith CW (2005) Understanding alternative splicing: towards a cellular code. Nat Rev Mol Cell Biol 6(5):386–398. doi:10.1038/nrm1645

6. Walsh C (2005) Posttranslational modification of proteins: expanding nature's inventor, 1st edn. Roberts and Company Publishers, Englewood, CO

7. Khoury GA, Baliban RC, Floudas CA (2011) Proteome-wide post-translational modification statistics: frequency analysis and curation of the swiss-prot database. Sci Rep 1. doi:10.1038/srep00090

8. Lu CT, Huang KY, Su MG, Lee TY, Bretana NA, Chang WC, Chen YJ, Chen YJ, Huang HD (2013) DbPTM 3.0: an informative resource for investigating substrate site specificity and functional association of protein post-translational modifications. Nucleic Acids Res 41(Database issue):D295–D305. doi:10.1093/nar/gks1229

9. Shannon DA, Weerapana E (2013) Orphan PTMs: rare, yet functionally important modifications of cysteine. Biopolymers. doi:10.1002/bip.22252

10. Diez R, Herbstreith M, Osorio C, Alzate O (2010) 2-D fluorescence difference gel electrophoresis (DIGE) in neuroproteomics. In: Alzate O (ed) Neuroproteomics. Frontiers in Neuroscience, Boca Raton, FL

11. DeKroon RM, Robinette JB, Osorio C, Jeong JS, Hamlett E, Mocanu M, Alzate O (2012) Analysis of protein posttranslational modifications using DIGE-based proteomics. Methods Mol Biol 854:129–143. doi:10.1007/978-1-61779-573-2_9

12. Drisdel RC, Green WN (2004) Labeling and quantifying sites of protein palmitoylation. Biotechniques 36(2):276–285

13. DeKroon RM, Osorio C, Robinette JB, Mocanu M, Winnik WM, Alzate O (2011) Simultaneous detection of changes in protein expression and oxidative modification as a function of age and APOE genotype. J Proteome Res 10(4):1632–1644. doi:10.1021/pr1009788

14. Rubel CE, Schisler JC, Hamlett ED, DeKroon RM, Gautel M, Alzate O, Patterson C (2013) Diggin' on u(biquitin): a novel method for the identification of physiological E3 ubiquitin ligase substrates. Cell Biochem Biophys 67 (1):127–138. doi:10.1007/s12013-013-9624-6

15. Ehlers MD (2003) Activity level controls postsynaptic composition and signaling via the ubiquitin-proteasome system. Nat Neurosci 6(3):231–242. doi:10.1038/nn1013

Neuromethods (2016) 114: 263–273
DOI 10.1007/7657_2015_98
© Springer Science+Business Media New York 2015
Published online: 08 November 2015

Mass Spectrometric Detection of Detyrosination and Polyglutamylation on the C-Terminal Region of Brain Tubulin

Yasuko Mori, Alu Konno, Mitsutoshi Setou, and Koji Ikegami

Abstract

Polyglutamylation, detyrosination/tyrosination cycle, and conversion to the Δ2 form are three major types of posttranslational modifications (PTMs) in neuronal tissues, occurring on the C-terminal region of tubulin, which lies on the surface of microtubule (MT). Polyglutamylation is the addition of glutamic acid chain to a specific glutamic acid residue near the tubulin C-terminus. The detyrosination/tyrosination cycle is a reversible removal and re-addition of the genomically encoded tyrosine residue of the α-tubulin C-terminus. The detyrosinated α-tubulin is further converted into the irreversible Δ2 form through the removal of the penultimate glutamate residue. These PTMs change the interaction between motor proteins or MT-associated proteins and MTs, involved in neuronal growth and development as well as the maintenance of neuronal function. For analysis of these PTMs, mass spectrometry (MS) is a powerful approach. In this chapter, we provide a convenient procedure specialized for polyglutamylation of β-tubulin as well as a conventional procedure for all three modifications of both α- and β-tubulin.

Keywords: Microtubule (MT), Tubulin, Posttranslational modification (PTM), Polyglutamylation, Detyrosination, Tyrosination, Mass spectrometry (MS)

1 Introduction

In neuronal cells, the microtubule (MT) plays great roles in the development, establishment, and maintenance of the complex neuronal structures and function, acting as a building frame of cell structure and a railway for the intracellular transport system. The MT has a tubelike structure, which consists of protofilaments built of heterodimers of α- and β-tubulin, with the tubulin carboxyl-terminus (C-terminus) exposed on the microtubule surface. The C-terminal region of tubulin undergoes unique posttranslational modifications; polyglutamylation, detyrosination/tyrosination cycle, and conversion to Δ2 form in neuronal cells [1, 2].

Polyglutamylation is a form of PTM, where polymers consisting of multiple glutamic acids are attached to specific glutamic acid residues near the tubulin C-terminus. In many cases, anywhere from several to a dozen glutamic acids are added. There appears to be an optimal range for the number of glutamic acids incorporated. A loss of α-tubulin polyglutamylation impairs transport of a motor protein, Kif1A, and affects synaptic transmission [3]. A loss of β-tubulin polyglutamylation results in the slowdown of neurite outgrowth [4]. Over-polyglutamylation is more deleterious for neuronal structure and function, causes Purkinje cell degeneration, and results in severe ataxia [5, 6].

Detyrosination is a form of PTM, where the tyrosine residue of α-tubulin C-terminus encoded by the genome is enzymatically removed. The detyrosinated form of α-tubulin is subjected to two different PTMs. On the one hand it may undergo (re-) tyrosination, in which a tyrosine residue is re-added to the C-terminus. Or, it may be converted to the Δ2 form, where the penultimate glutamic acid residue is irreversibly removed, which represents 40–50 % of α-tubulin in the brain. Recently, a new form of PTMs, Δ3 form, is proposed, which can be generated through the further removal of the third last glutamic acid residue of α-tubulin from the Δ2 form [6]. The detyrosination/tyrosination cycle is crucial for brain development [7] and for the axon determination at the early stage of neuronal development [8].

To analyze these PTMs, immunochemical techniques are most often used as convenient methods. Nowadays, a variety of highly reliable monoclonal antibodies are available: GT335 for glutamylated tubulin, B3 for di-glutamylated α-tubulin, 1A2 and YL1/2 for tyrosinated α-tubulin, and AA12 for detyrosinated α-tubulin. Antibodies provide simple information as to the presence or absence of PTMs. However, they have a major weakness with respect to PTM analysis, especially of polyglutamylation. Antibodies are not suitable for counting the exact number of glutamic acids attached on tubulin. Detecting the modification directly by mass spectrometry (MS) provides accurate information about the length of the glutamic acid chain attached to tubulin. In this chapter, we present procedures of analyzing these PTMs of the tubulin C-terminal region, aiming to analyze the state of PTMs in nervous tissues [3] as well as to evaluate activities of PTM-performing enzymes in vitro [9]. We also provide a new approach that enables a facile procedure of PTM analysis of β-tubulin (Fig. 1). The conventional procedure (Fig. 1; 1 to 6) or the rapid convenient procedure (Fig. 1; 7 to 9) can be chosen according to the experimental goal.

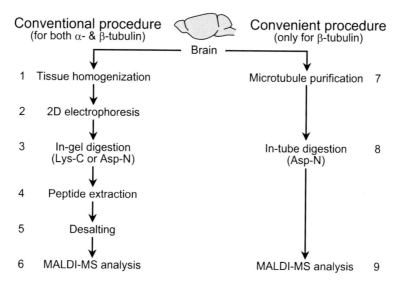

Fig. 1 A workflow of mass spectrometry of brain tubulin C-terminal PTMs. Two strategies can be chosen according to the aim. If purified tubulin is available and β-tubulin is the research target, the course of 7 through 9 is more convenient. Two enzymes, lysyl endopeptidase (Lyc-C) or endoprotenaise (Asp-N), can be chosen according to the research target, α-tubulin or β-tubulin

---

2 Materials

2.1 Tissue Homogenization

1. Mice: adult males.
2. CHAPS (Dojindo).
3. Urea (Sigma).
4. Thiourea (Sigma).
5. IPG buffer (GE Healthcare).
6. Dithiothreitol (DTT) (Sigma).
7. Protease inhibitor cocktail EDTA-free (Roche).
8. Extraction buffer: 7 M urea, 2 M thiourea, 2 % CHAPS, 2 % IPG buffer, 40 mM DTT, EDTA-free protease inhibitor cocktail (**Note 1**).
9. Glass Teflon homogenizer: 10 mL.

2.2 Two-Dimensional Electrophoresis

1. Multiphor II (GE Healthcare).
2. Power supply, EPS 3501 XL (GE Healthcare).
3. Water circulator, CCA-1110 (EYELA).
4. Glass plates and combs for slab gel electrophoresis (Biocraft).
5. Immobiline DryStrip (GE Healthcare).
6. Silicone oil (KF-96-L-1.5CS, Shin-Etsu Silicone; also available from GE Healthcare).

7. Equilibration buffer: 50 mM Tris–HCl, pH 6.8, 7 M urea, 20 % glycerol, 2 % SDS, 33 mM DTT, bromophenol blue (trace).

8. SDS, 95 % purity (L-5750, Sigma) (**Note 2**).

9. Coomassie brilliant blue staining solution (Wako).

2.3 In-Gel Digestion

1. Disposable scalpel.

2. Centrifugal concentrator, CC-105 (TOMY).

3. Ultrasonic cleaner, Model 2510 (Branson).

4. Lysyl endopeptidase (Lyc-C), mass spectrometry grade (Wako).

5. Endoproteinase Asp-N, sequencing grade (Roche).

6. Acetonitrile, HPLC grade (Wako).

7. Ammonium hydrogen carbonate (NH_4HCO_3) (Wako).

8. Iodoacetamide (Wako).

9. Trifluoroacetic acid (TFA) (Wako).

10. DTT solution: 10 mM DTT/100 mM NH_4HCO_3 (**Note 3**).

11. Alkylating solution: 40 mM iodoacetamide/100 mM NH_4HCO_3 (**Note 3**).

12. Lysyl endopeptidase solution: 2 μg/mL lysyl endopeptidase in 25 mM Tris–HCl, pH 9.0, 10 % acetonitrile.

13. Endoproteinase Asp-N solution: 20 μg/mL endoproteinase Asp-N in 25 mM NH_4HCO_3, 10 % acetonitrile.

2.4 Peptide Extraction

1. 0.1 % TFA/50 % acetonitrile.

2. 0.1 % TFA/80 % acetonitrile.

2.5 Desalting

1. NuTip (1–10 μL), Silica strong Anion (SAX) (Glygen).

2. Washing solution: 20 mM ammonium formate ($HCOONH_4$).

3. Elution solution: 1 M TFA.

2.6 Mass Spectrometry (MS)

1. Mass spectrometer, AXIMA-QIT (Shimadzu).

2. 2,5-Dihydroxybenzoic acid (DHB) (Bruker).

3. Matrix solution: 12.5 mg/mL DHB in 50 % acetonitrile/0.1 % TFA (**Note 3**).

*****

2.7 Microtubule Purification

1. Depolymerization buffer (DB): 50 mM MES-KOH, pH 6.6, 1 mM $CaCl_2$.

2. High-molarity PIPES buffer (HMPB): 1 M PIPES-KOH, pH 6.9, 10 mM $MgCl_2$, 20 mM EGTA.

3. Brinkley BR buffer 1980 (BRB80): 80 mM PIPES-KOH, pH 6.8, 1 mM $MgCl_2$, 1 mM EGTA.

4. PMSF (Sigma).

5. Leupeptin (Sigma).

6. Nucleotide solution: 100 mM ATP, 200 mM GTP.

7. Ultracentrifuge, Optima Max (Beckman).

8. Angle rotor, TLA100.2 or TLA100.3 (Beckman).

2.8 In-Tube Digestion

1. Purified tubulin (from Section 2.7).

2. Endoproteinase Asp-N solution: 40 μg/mL endoproteinase Asp-N in ultrapure water.

2.9 Mass Spectrometry (MS)

1. Mass spectrometer, Ultraflex II (Bruker).

2. Matrix solution: 20 mg/mL DHB in 70 % methanol/0.1 % TFA (**Note 3**).

3. Peptide calibration standard (Bruker).

4. ITO-coated glass slides (Bruker).

5. MTP Slide Adapter II (Bruker).

6. Airbrush.

3 Methods

3.1 Tissue Homogenization

1. Decapitate mouse head under anesthesia.

2. Dissect the brain immediately, and measure the weight.

3. Homogenize the brain in the 20-fold volume (v/w) of the extraction buffer with the glass Teflon homogenizer by ten strokes at 3000 rpm on ice.

4. Centrifuge the brain homogenate at $20,000 \times g$ at 4 °C for 10 min.

5. Collect the supernatant.

6. The supernatant is ready for being subjected to the two-dimensional electrophoresis (**Note 4**).

3.2 Two-Dimensional Electrophoresis

1. Rehydrate Immobiline DryStrip with the extraction buffer overnight in a 15 mL conical tube or sealed 25 mL disposable pipet (**Note 5**).

2. Place the rehydrated DryStrip on Multiphor II, and cover it by silicone oil.

3. Apply 100 μL of the sample into the sample cup.

4. Run the electrophoresis, i.e., the isoelectric focusing, in a gradient mode: fixed 500 V for 1 min; gradient increase of the voltage from 500 to 3500 V for 1.5 h; fixed 3500 V for 2–24 h (**Note 6**).

5. Equilibrate the DryStrip in the equilibration buffer for 30 min with gentle agitation.

6. Place the equilibrated DryStrip on the polyacrylamide gel, and pour a few hundred microliter of equilibration buffer between the DryStrip and the polyacrylamide gel.

7. Run the second electrophoresis, i.e., SDS-PAGE, according to general procedures (**Note 7**).

8. Fix the gel in 50 % methanol/10 % acetic acid for 20 min (**Note 8**).

9. Stain the fixed gel with the CBB solution.

3.3 In-Gel Digestion

1. Cut out tubulin spots from the gel with a disposable scalpel.

2. Mince the gel to 0.5–1 mm pieces (**Note 9**).

3. Wash the minced gel pieces three times with 0.2 mL of extra pure water briefly.

4. Destain the gel pieces by incubating them in 0.3 mL of 50 mM NH_4HCO_3/50 mM methanol at 40 °C for 3 min.

5. Repeat **step 4** at least three times (**Note 10**).

6. Immerse the gel pieces in 100 % acetonitrile with the volume enough to cover the gel (about 100 μL).

7. Dehydrate the gel pieces via vigorous mixing in 100 % acetonitrile with the ultrasonic cleaner.

8. Discard acetonitrile completely when the color of gel pieces gets white.

9. Completely dry up the rehydrated gel pieces in the centrifugal concentrator until the gel pieces become solid and white (**Note 11**).

10. Rehydrate the dried gel pieces in DTT solution at 50 °C for 1 h (**Note 12**).

11. Discard DTT solution, and dehydrate and dry up the gel pieces by repeating **steps 6** through **9** (**Note 11**).

12. Rehydrate the dried gel pieces in the alkylating solution at room temperature under the dark—wrapped by aluminum foil—for 30 min.

13. Add ultrapure water to cover the gel pieces, mix them by vortexing for a few seconds, and then discard the supernatant.

14. Incubate the gel pieces in 50 mM NH_4HCO_3/50 % methanol at 40 °C for 15 min, and then discard the supernatant.

15. Repeat **step 14** once.

16. Dehydrate and dry up the gel pieces by repeating **steps 6** through **9** (**Note 11**).

17. Immerse the dried gel pieces in 200 μL of the digestive enzyme solution (**Note 13**), and rehydrate them on ice for 10 min.

18. Vortex the gel pieces in the digestive enzyme solution briefly (**Note 14**), and then discard the solution (**Note 15**).

19. Incubate the gel pieces at 37 °C for more than 10 h.

3.4 Peptide Extraction

1. Soak the gel pieces in 100 µL of 0.1 % TFA/50 % acetonitrile, and mix them in the ultrasonic cleaner for 10 min to extract digested peptides.

2. Collect the supernatant into a new collecting plastic tube after a brief centrifugation (**Note 16**).

3. Add 50 µL of 0.1 % TFA/50 % acetonitrile onto the gel pieces, and mix them in ultrasonic cleaner for 10 min to extract the digested peptides.

4. Gather the supernatant into the collecting tube after a brief centrifugation (**Note 16**).

5. Repeat **steps 3** and **4** once.

6. Soak again the gel pieces in 50 µL of 0.1 % TFA/80 % acetonitrile, and mix them in ultrasonic cleaner for 10 min to further extract the digested peptides.

7. Gather the supernatant into the collecting tube after a brief centrifugation (**Note 16**).

8. Evaporate ~250 µL of the extracted peptides solution until the volume becomes 5–20 µL in the centrifugal concentrator (**Note 17**).

3.5 Desalting

1. Load a tip of Nutip (1–10 µL, SAX) to P10 pipette, and then wash the embedded media with 10 µL of the washing solution by three-time pipetting up and down (**Note 18**).

2. Repeat nine times **step 1**.

3. Adsorb the extracted peptides to the embedded media through the repeated pipetting up and down.

4. Wash the embedded media (and bound peptides) with 10 µL of the washing solution by three-time pipetting up and down.

5. Repeat nine times **step 4**.

6. Extract bound peptides with a small volume (several to 10 µL) of the elution solution.

7. The extracted peptides (enriched by acidic tubulin C-terminal peptides) are ready for mass spectrometry.

3.6 Mass Spectrometry (MS)

1. Mix the extracted peptides with an equal volume of the matrix solution.

2. Drop 0.5–1 µL of the mixture onto the sample plate, and then dry up the droplet (**Note 19**).

3. Analyze the dried sample in negative ion mode with a mass spectrometer, AXIMA-QIT, equipped with a 377 nm N_2 laser (**Notes 20** and **21**).

*****

3.7 Microtubule Purification

1. Decapitate mouse head under anesthesia.

2. Dissect the brain immediately, and measure the weight.

3. Homogenize the brain in the twofold volume (v/w) of ice-cold DB containing 1 mM PMSF and 2 μg/mL leupeptin with the glass Teflon homogenizer by ten strokes at 3000 rpm on ice (**Note 22**).

4. Centrifuge the homogenate at 5000 × g at 4 °C for 20 min.

5. Centrifuge the supernatant at 52,000 × g at 4 °C for 30 min by means of TLA100.2 or TLA100.3 rotor (**Note 23**).

6. Take the supernatant, and then add equal volumes of pre-warmed HMPB containing 1.5 mM ATP and 0.5 mM GTP, and pre-warmed pure glycerol (supernatant:HMPB:glycerol = 1:1:1).

7. Incubate the mixture at 37 °C for 45 min.

8. Centrifuge the mixture at 200,000 × g at 37 °C for 60 min (**Note 24**).

9. Discard the supernatant and rinse the pellet with pre-warmed DB.

10. Add 250 μL of ice-cold DB per 1 g of brain sample, and mix it gently by mildly pipetting up and down until the pellet is completely dissolved (**Note 25**).

11. Incubate the solution on ice for 20 min.

12. Centrifuge the solution at 80,000 × g at 4 °C for 30 min (**Note 23**).

13. Take the supernatant and then add pre-warmed HMPB containing 1.5 mM ATP and 0.5 mM GTP, and pre-warmed pure glycerol (supernatant:HMPB:glycerol = 1:1:1).

14. Incubate the mixture at 37 °C for 45 min.

15. Centrifuge the mixture at 160,000 × g at 37 °C for 35 min (**Note 24**).

16. Discard the supernatant and rinse the pellet with pre-warmed BRB80.

17. Add 50–100 μL of ice-cold BRB80 per the number of used brains, and mix it gently by mildly pipetting up and down until the pellet is completely dissolved (**Note 25**).

18. Incubate the solution on ice for 10 min.

19. Centrifuge the solution at 100,000 × g at 4 °C for 10 min (**Note 23**).

20. Take the supernatant and quantify the tubulin concentration by means of spectrometer (**Note 26**), and check the purity of tubulin by general SDS-PAGE.

21. The supernatant is ready for enzyme digestion (**Note 27**).

3.8 In-Tube Digestion

1. Add 10 µg of tubulin (0.2 µL of 50 mg/mL solution) and 10 ng of Asp-N (2.5 µL of 40 µg/mL solution) into 8 µL of 50 mM NH_4HCO_3.

2. Incubate the mixture at 37 °C for overnight.

3.9 Mass Spectrometry (MS)

1. Drop 1 µL of the enzyme-digested tubulin solution onto an ITO-coated glass slide, and dry up the droplet (**Note 19**).

2. Spray 1 mL of the matrix solution over the ITO-coated glass slide with an airbrush in a fume hood (**Note 28**).

3. Hold the glass slide on the slide adapter, and analyze the spot with a mass spectrometer, Ultraflex II, with the following parameters:

   - Detection mode: negative ion and reflector mode
   - Laser: Nd:YAG laser, 100 Hz, 200 shots
   - Laser energy and detector gain: optimal values are varied in each experiment (**Notes 20 and 21**)

4 Notes

1. Add DTT, IPG buffer, and protease inhibitors just before using the buffer.

2. Use the low-purity SDS to separate α- and β-tubulin vertically in the second electrophoresis, i.e., SDS-PAGE.

3. Prepare the solution freshly just before use.

4. The sample can be stored at −80 °C for future analyses.

5. The 15 mL conical tube is for DryStrip less than 11 cm in length, and the sealed 25 mL disposable pipet is for DryStrip more than 13 cm in length.

6. The duration of the last step depends on the length of DryStrip. Two-hour electrophoresis is enough for 7 cm DryStrip, while 24 cm DryStrip requires more than 20-h electrophoresis.

7. Make sure that the running buffer contains the low-purity SDS. This is essential for separating α- and β-tubulin.

8. Handle the gel with great care to prevent keratin from contaminating samples. All instruments (e.g., glass plates and buffer tanks) and reagents used after this step should be for exclusive

use to mass spectrometry. Wash instruments thoroughly with a detergent before every use to remove any protein deposits. In addition, wear fresh disposable gloves, and sleevelets, a mask, and a cap if available, after this step.

9. Handle the minced gel pieces with care to avoid losing them.

10. Repeat the step more, if the gel pieces still look blue.

11. Make sure that the gel has completely dried out. The condition of drying crucially affects the result of mass spectrometry.

12. Cover the dried gel pieces by DTT solution. The volume depends on the amount of gel pieces. A recommendation is ~100 μL.

13. Choose lysyl endopeptidase or endoproteinase Asp-N based on research interests. The former is suitable for α-tubulin, and the latter for β-tubulin.

14. Add more the digestive enzyme solution, if needed, to effectively vortex the sample.

15. It is very important to remove the extra enzyme solution. The remaining enzyme solution makes samples contaminated by peptides derived from the autolysis of enzymes.

16. Avoid taking small debris of gel into next step. It strongly interferes with the result of mass spectrometry.

17. The required time depends on the concentration of extracted peptide solution. It usually takes more time if the concentration of peptides is high.

18. Avoid drying the embedded media throughout the desalting after this step. Peptides are deposited tightly in the embedded media once the media is dried. It results in a severe loss of peptide yield.

19. Spot the calibration standard near the samples to achieve more accurate measurements.

20. Shoot laser at the edge of sample droplets. Droplets usually do not dry homogeneously and peptides tend to accumulate in the edge of the droplets.

21. Seek out the optimal parameters every time, because the optimal laser energy and detector gain are varied in each experiment.

22. Add PMSF and leupeptin just before use.

23. Cool the rotor well at 4 °C prior to use.

24. Warm the rotor well at 37 °C prior to use.

25. Avoid making air bubbles. Tubulin is easily degraded by air bubbles.

26. Extinction coefficient of tubulins at 280 nm is 1.15 $(mg/mL)^{-1}cm^{-1}$.

27. The purified tubulin can be stored at −80 °C as small aliquots after freezing them by liquid nitrogen for future analyses including in vitro analyses of enzyme activities.

28. Put some markers behind the ITO-coated glass slide to locate sample droplets. The sample droplets are sometimes almost transparent and difficult to be found under the camera of a mass spectrometer.

Acknowledgements

The chapter described is supported in part by a Grant-in-Aid for Challenging Exploratory Research 26670091 (to K.I.). The content is solely the responsibility of the authors. The authors declare no conflict of interest.

References

1. Ikegami K, Setou M (2010) Unique post-translational modifications in specialized microtubule architecture. Cell Struct Funct 35:15–22

2. Magiera MM, Janke C (2014) Post-translational modifications of tubulin. Curr Biol 24:R351–R354

3. Ikegami K et al (2007) Loss of alpha-tubulin polyglutamylation in ROSA22 mice is associated with abnormal targeting of KIF1A and modulated synaptic function. Proc Natl Acad Sci U S A 104:3213–3218

4. Ikegami K et al (2006) TTLL7 is a mammalian beta-tubulin polyglutamylase required for growth of MAP2-positive neurites. J Biol Chem 281:30707–30716

5. Rogowski K et al (2010) A family of protein-deglutamylating enzymes associated with neurodegeneration. Cell 143:564–578

6. Berezniuk I et al (2012) Cytosolic carboxypeptidase 1 is involved in processing α- and β-tubulin. J Biol Chem 287:6503–6517

7. Erck C et al (2005) A vital role of tubulin-tyrosine-ligase for neuronal organization. Proc Natl Acad Sci U S A 102:7853–7858

8. Konishi Y, Setou M (2009) Tubulin tyrosination navigates the kinesin-1 motor domain to axons. Nat Neurosci 12:559–567

9. Mukai M et al (2009) Recombinant mammalian tubulin polyglutamylase TTLL7 performs both initiation and elongation of polyglutamylation on beta-tubulin through a random sequential pathway. Biochemistry 48:1084–1093

Neuromethods (2016) 114: 275–287
DOI 10.1007/7657_2015_86
© Springer Science+Business Media New York 2015
Published online: 08 November 2015

Determination of Polyhydroxybutyrate (PHB) Posttranslational Modifications of Proteins Using Mass Spectrometry

Tong Liu, Wei Chen, Stacey Pan, Chuanlong Cui, Hong Li, and Eleonora Zakharian

Abstract

Posttranslational modifications (PTMs) of proteins are important determinants of their biological functions. Proteins undergo various PTMs throughout their life span. Some of these modifications are of a temporary nature and may control rapid on/off rates for activation or inactivation of particular proteins. Other types of PTMs are of a permanent nature. Those attachments take place upon protein synthesis and may substantially alter protein structures and function, and are removed only upon protein hydrolysis. In these modifications, the moieties of the modifier molecule are most likely bound covalently. Here we discuss such a type of PTM with a polyester, poly-(R)-3-hydroxybutyrate (PHB). PHB is a ubiquitous homopolymer that is present in all living organisms. In animals PHB was specifically found in the liver, kidney, heart, and brain. However, what role PHB plays in these tissues is not well understood. As a polymer-electrolyte, PHB has been recognized in mediating ion transport across the membrane, and thus it may be implicated in various signaling pathways carried in the central and peripheral nervous systems. In this chapter, we present a protocol for determination of PHB modification of the mammalian ion channel, TRPM8. The TRPM8 channel is the cold and menthol receptor in the peripheral nervous system, and an important mediator of pain stimuli. The procedures to determine the PHB moieties on the specific amino acids of TRPM8, including protein isolation, purification, digestion, mass spectrometry analysis, and database search, will be outlined here. Herein we also discuss the challenges to resolving PHBylated peptides that may arise due to the fragile chemical structure of the polyester and its disintegration during the experimental procedures and mass spectrometry.

Keywords: Protein purification, Poly-(R)-3-hydroxybutyrate (PHB), Transient receptor potential ion channel of melastatin subfamily member 8 (TRPM8), Posttranslational modification (PTM), Mass spectrometry (MS)

1 Introduction

PTMs of proteins are very important for modulating their activities and regulating cellular processes. The study of various PTMs will help biologists understand the mechanisms of cellular regulation. Mass spectrometry (MS) has become one of the cornerstone methodologies to identify PTMs. It can localize the PTM sites and provide their quantification information. However, using MS to analyze PTMs can prove very challenging due to the overall low stoichiometry of modified peptides, poor stability, and ionization efficiency of some PTMs in MS [1].

Poly-(R)-3-hydroxybutyrate (PHB) is a ubiquitous biological polymer present in all organisms. For a long period, PHB was thought to be present only in bacteria, where it plays the role of a carbon storage source. Later, this polyester was found widely distributed throughout higher eukaryotes [2]. However, its role in mammalian cells remained questionable, and it was not clear whether PHB is a "molecular rudiment" evolutionarily derived to more evolved organisms or has some important biological functions. A recent study localized PHB to the endoplasmic reticulum and plasma membranes of neurons, particularly those of the dorsal root ganglion [3], where the polymer may be involved in ion transport. The main obstacles in defining the role of PHB relate to the technical limitations of the current methodology for detecting this polymer. LC-MS/MS methods, as they improve, offer many advantages in the delineation of the localization and potential function of this PTM.

PHB synthesis is well studied in bacteria, where specific enzymes, PHB synthases, assemble the polymer from CoA esters of R-3-hydroxybutyrate. The synthesis occurs in three steps: (1) condensation of two molecules of acetyl-CoA to form acetoacetyl-CoA, (2) reduction of acetoacetyl-CoA by NADPH to form (R)-3-hydroxybutyryl-CoA, and (3) polymerization of (R)-3-hydroxybutyryl-CoA that results in the formation of PHB [4]. The enzymes and metabolic pathways for PHB synthesis in eukaryotes are yet unknown. It might be similar to that of cholesterol, as both PHB and cholesterol share a common intermediate, acetoacetyl-CoA, and both their syntheses are regulated by changes in intracellular concentrations of acetyl-CoA [5].

The PHB molecule is comprised of hydrophobic methyl groups alternating with hydrophilic ester groups and has a CoA-ester binding group at its C-terminal end (Fig. 1). PHB interacts with proteins by forming supramolecular complexes via covalent bonds and multiple hydrophobic interaction sites [2, 6, 7]. The molecular structure of the PHB polymer creates a highly flexible carbon backbone with a lipophilic outer surface [7]. Upon association with proteins, the high hydrophobicity of the polyester may substantially alter the physical properties and thus affect the function of their hosts. For instance, PHB modifications have significant impact on the temperature sensitivity of impacted proteins due to

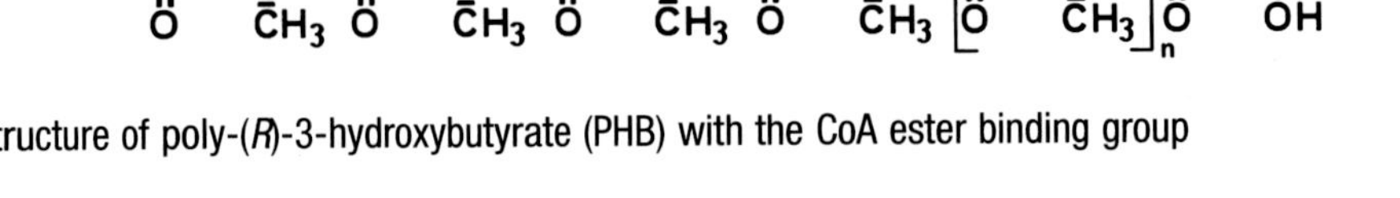

Fig. 1 Structure of poly-(R)-3-hydroxybutyrate (PHB) with the CoA ester binding group

the rapid conformational rearrangement of the polymer with temperature changes [3]. PHB is also important for protein folding and localization [8] as well as for ion permeation of some ion channels [9]. The high-energy C-terminal CoA-ester group, derived from PHB metabolic precursors, presumably acts as a cofactor for the enzymatic reaction in which a covalent bond to the protein is formed [10].

Analysis of PHB using mass spectrometry is very challenging due to the following reasons: (1) Rapid disintegration of the labile ester bonds of PHB during MS and MS/MS analysis makes it difficult to capture intact PHB-modified peptides and to assign the fragment ions for the localization of PHB modification sites; (2) PHB modification is not one single mass shift of an amino acid. It has multiple repeated units covalently linked to one amino acid. This adds great complexity during the database search, because it requires consideration of various number of PHB modification repeats in the precursor, as well as corresponding fragment ions; (3) Some of the masses of PHB-modified peptides are too large that it may affect the detection and peptide fragmentation in MS/MS analysis; (4) Some of the PHB-modified peptides are very hydrophobic, making them difficult to elute from RPLC during LC-MS/MS analysis.

In our recent study, we achieved elucidation of PHB as PTM of a mammalian ion channel protein from the transient receptor potential melastatin subfamily, member 8, TRPM8 [3]. Using MS/MS approaches, we have shown that PHB modifies various amino acids of the TRPM8 protein. The majority of these peptides reside on the intracellular N-terminus of the protein, and one modification localized on the extracellular side of the channel [3]. Furthermore, the extracellular PHB modification (Fig. 2) plays an

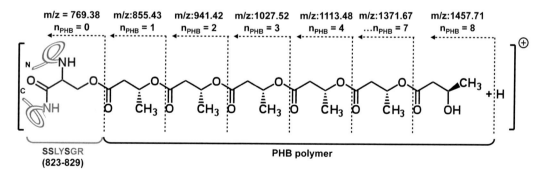

Fig. 2 Mass spectrometric analysis of the chloroform-extracted peptide SSLYSGR 823–829 of the TRPM8 protein derived from matrix-assisted laser desorption/ionization (MALDI) MS experiments: Molecular composition of PHBylated serine residue on the SSLYSGR (823–829) peptide with a number of PHB units attached via an ester bond; numbers indicate the PHB modification with a shift in the monoisotopic masses. Number of PHB units (*n*) varied in length from 1 to 26 units supposedly due to the breakage of the labile ester bonds of PHB under the MS beam

essential role in TRPM8 channel function: PHB-deficient mutants exhibit reduced temperature sensitivity of TRPM8 upon cold-induced activation and also show inhibited menthol- or icilin-evoked responses, suggesting that PHB mediates induced by the ligands conformational changes that result in channel openings [3]. This finding indicates that PHB is an important structural and functional component of this receptor protein, and therefore may be implicated in the networking of the peripheral nervous system.

The purpose of this chapter is to introduce the methods and detailed experimental procedures targeted for identifying PHB modifications using LC-MS/MS and also to describe technical difficulties and the possible solutions. This protocol is comprised of (1) TRPM8 protein extraction and purification from mammalian expression system, (2) protein digestion with trypsin, (3) chloroform isolation of PHBylated peptides, (4) LC-MS/MS experiments, and (5) bioinformatics analysis of PHBylated peptides.

2 Materials

2.1 Cell Culturing/Handling

Human embryonic kidney (HEK) 293 cells were cultured in Eagle's Minimum Essential Medium (EMEM) (ATCC, Manassas, VA) containing 10 % fetal bovine serum (FBS) in a 5 % CO_2 atmosphere.

1. Phosphate buffer saline (PBS): 1.058 mM potassium phosphate monobasic (KH_2PO_4), 155.17 mM sodium chloride (NaCl), 2.97 mM sodium phosphate dibasic ($Na_2HPO_4 \cdot 7H_2O$), pH 7.4, Mg^{2+}-free, Ca^{2+}-free (Invitrogen, Grand Island, NY).

2. Microcentrifuge 5402 R (Eppendorf, Hauppauge, NY).

3. Preferable centrifuge tubes are orange-cap tubes obtained from Corning Inc. (Corning, NY).

2.2 Protein Purification

1. The buffer used for protein extraction/purification is sodium chloride-based purification buffer (NCB) containing: 500 mM NaCl, 50 mM NaH_2PO_4, 20 mM Hepes, 2 mM Na-orthovanadate, 10 % Glycerol, pH 7.5 with addition of 1 mM of protease inhibitor PMSF, 5 mM β-Mercaptoethanol (Sigma-Aldrich, St. Louis, MO) before homogenizing the cells.

2. The membrane resuspension buffer: NCB with addition of a protease inhibitor cocktail (Roche, Indianapolis, IN), 20 μg/mL DNase, 20 μg/mL RNase (Fisher Scientific, Pittsburg, PA), 0.1 % Nonidet P40 (Roche, Germany), and 0.5 % dodecyl-maltoside (DDM) (CalBiochem, Darmstadt, Germany).

3. For protein purification/visualization: A/G protein magnetic beads (Pierce, Thermo Scientific, Rockford, IL); Myc-antibody

(Sigma-Aldrich, St. Louis, MO); 10 % TGX-ready gel, Precision Plus Protein Standards (All blue), Coomassie Brilliant Blue G-250 (Bio-Rad Labs. Inc., Hercules, CA).

2.3 Protein Estimation

1. BCA protein assay kit (Pierce, Thermo Scientific, Rockford, IL)
2. Protein standard (2.0 mg/mL bovine serum albumin)
3. Spectra-Max 190 microplate reader (Molecular Devices, Sunnyvale, CA).

2.4 Protein In-Gel Digestion

1. Wash solution: 50 mM ammonium bicarbonate in 30 % acetonitrile (ACN)
2. Reduction reagent: 10 mM DTT
3. Alkylation reagent: 100 mM iodoacetamide
4. 50 mM ammonium bicarbonate
5. Extraction buffer: 50 % (v/v) ACN and 0.1 % trifluoroacetic acid (TFA)
6. 0.1 % TFA solution
7. Trypsin solution: 20 ng/μL trypsin (Promega)
8. Eppendorf tubes, 0.5 and 1.5 mL

2.5 Partition of PHB-Modified Peptides

1. $CHCl_3$
2. 10 mL glass bottles

2.6 Mass Spectrometry

1. For matrix-assisted laser desorption ionization-time-of-flight MS (MALDI TOF MS): plates (AB Sciex), 4000 Series Explorer (AB Sciex), mass calibration standards kits (AB Sciex), MALDI matrix (7 mg α-cyano-4-hydroxycinnamic acid dissolved in 1 mL solvent containing 60 % ACN, 0.1 % TFA, 5 mM ammonium monobasicphosphate, 50 fmol/μL ach of glu-fibrinogen peptide (m/z 1570.677), and adrenocorticotropic hormone fragment 18–39 (m/z 2465.199))
2. For LC-MS/MS: C_4 reversed-phase column (50 μm × 250 mm, 5 μm, 300 Å, C_4, Dionex, Sunnyvale, CA, USA), Solvent A (2 % ACN, 0.1 % formic acid, FA), Solvent B (85 % ACN, 0.1 % FA), Xcalibur Software (Thermo Scientific), LTQ Velos ESI positive Ion Calibration Solution (Thermo Scientific).

2.7 Data Analysis Software

1. Mascot (Matrix Science Inc., Boston, MA, http://www.matrix-science.com)
2. Proteome Discoverer (Thermo Scientific)
3. Data Explorer (AB Sciex)
4. GPS Explorer (AB Sciex)
5. ExPASy FindMod Tool (Swiss Proteomics Bioinformatics Resources, http://web.expasy.org/findmod/)

2.8   Equipment and Materials

1. OptimaTM L-90K ultracentrifuge; used with Ti70 rotor (Beckman Coulter, Indianapolis, IN)

2. Microcentrifuge Z216MK (Hermle, Labnet, Woodbridge, NJ)

3. 4800 Plus MALDI TOF/TOF Analyzer (AB Sciex, Foster City, CA)

4. LTQ-Orbitrap Velos Mass Spectrometer (Thermo Scientific)

5. Ultimate™ 3000 nano HPLC (Thermo Scientific)

3   Methods

3.1   TRPM8 Protein Purification

1. For each purification, culture 18–20 dishes (10 cm) of HEK-293 cells stably expressing the TRPM8 protein with a Myc-tag (located on the N-terminus) to reach 70–80 % confluence.

2. Wash and remove the cells from dishes with PBS by pipetting the suspension up and down (*see* **Note 1**). Collect the cells in a 15 mL centrifuge tube.

3. Centrifuge the cells at 4000 × *g* for 5 min at 4 °C to obtain dense pellets, remove PBS completely, and store the pellet at −20 °C until use (*see* **Note 2**).

4. Resuspend the cell pellet in 10 mL of NCB-lysis buffer supplemented with 1 mM of protease inhibitor PMSF and 5 mM β-Mercaptoethanol.

5. Lyse the cells by the freeze-thawing method, repeat two times. For quick freezing purposes liquid nitrogen may be used, or alternatively, a mixture of ethanol and dry ice can be used (*see* **Note 3**).

6. Transfer the cell lysate into the ultracentrifuge tube. Spin the content at 40,000 × *g* for 2.5 h at 4 °C to pellet the membranes. Remove the supernatant thoroughly.

7. Resuspend the membranes in 3 mL NCB-resuspension buffer with addition of a protease inhibitor cocktail, 20 μg/mL DNase, 20 μg/mL RNase, 0.1 % Nonidet P40, and 0.5 % dodecyl-maltoside (DDM) (*see* **Note 4**), and place in three (1.5 mL) tubes. Close the cap tightly and cover tubes with Parafilm. Incubate the suspension overnight on the shaker with gentle rotation, at 4 °C.

8. On the same day, prepare the A/G protein magnetic beads. For each mL (about 15–20 mg/mL of protein) of the membrane prep, use 120 μL of the beads; total 45–60 mg of membrane protein could be extracted from 18–20 dishes (10 cm). Mix the beads intensely to achieve a homogenously distributed mixture, add 120 μL in new tubes for three samples, and place the tubes onto the magnetic stand. Incubate for about 3 min or until the

solution is clear. Remove the beads storage solution (20 % ethanol), and wash the beads twice with 500 μL of plain NCB buffer (*see* **Note 5**). After removing the last wash buffer, resuspend the magnetic beads in 500 μL NCB buffer and add 4 μL of anti-Myc-IgG (1/125 dilution). Close the cap tightly and cover tubes with Parafilm. Incubate the beads with the antibodies overnight on the rotating shaker with gentle rotation, at 4 °C.

9. On the next day, the protein suspension should look slightly opaque, which would indicate that the membranes are well solubilized in the detergent contained in NCB buffer, and that the TRPM8 protein is extracted from the membranes into the detergent micelles. No membrane clots should remain. Centrifuge the protein/membrane extracts for 1 h at 30,000 × *g*. After centrifugation, the TRPM8 protein/detergent complex will be in the aqueous supernatant.

10. Place the tubes with the beads onto the magnetic stand; allow about 3–5 min for the beads to settle; remove the NCB buffer containing anti-Myc-IgG. Resuspend the beads with fresh NCB buffer (500 μL). Wash the beads at least four to five times with 500 μL NCB to remove unbound antibody. Remove the last wash right before the membrane extracts are ready to add to the beads.

11. After centrifugation, add the supernatant with the TRPM8 protein/detergent complex onto the washed magnetic beads with conjugated α-Myc-antibodies. Mix gently but thoroughly, and incubate overnight on the shaker with gentle rotation, at 4 °C.

12. Next day, place the protein and beads containing tubes onto the magnetic stand. Allow them to stand for about 5–10 min (*see* **Note 6**).

13. Wash the beads four times with NCB buffer to remove impurities. Add 150 μL of 2× SDS-loading buffer into each sample (three tubes), mix the beads thoroughly, and place into the preheated 100 °C water bath on the magnetic stir to incubate for 10 min. Make sure to turn the magnetic stirring plate on to ensure rotation of the beads in the mixture. Note that magnetic beads are not visible in the SDS-loading buffer, and without rotation the beads will adhere to the magnet on the plate, which may interfere with protein release and result in a lesser yield.

14. Load the TRPM8 protein samples onto a 10 % TGX-ready gel by adding 50 μL from each sample (total of three samples will occupy nine wells), and 3 μL of the Protein Standards into the first well. Run the gel at 180 V for ~40 min, wash the resulting gel three times with milli-Q water, and stain with Coomassie

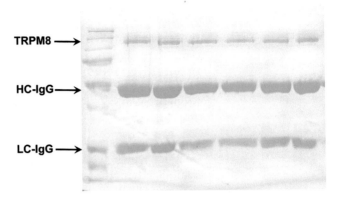

TRPM8 →

HC-IgG →

LC-IgG →

Fig. 3 The TRPM8 protein purified from HEK-293 cells. 10 % SDS-PAGE of TRPM8 stained with Coomassie stain. For mass spectrometry analysis, the TRPM8 protein was expressed in HEK-293 cells and purified by immunoprecipitation, using protein A/G magnetic beads conjugated with anti-Myc IgG. TRPM8 was eluted in SDS-loading buffer, and samples were boiled at 100 °C for 10 min. The *lanes* on the gel show the protein eluted from the beads. The TRPM8 bands were further excised and combined for the MS experiments

Brilliant Blue dye for 1 h by gentle agitation on shaker. Destain with milli-Q water (Fig. 3).

3.2 Protein In-Gel Digestion and Peptides Partition

1. Excise the protein bands with the molecular weight corresponding to TRPM8. Use a sharp scalpel to cut the gel into smaller pieces (~1 mm³)

2. Place the gel pieces into a 1.5 mL Eppendorf tube.

3. Add 200 μL of the Wash Solution (WS) and rinse the gel pieces for at least 30 min at RT.

4. Remove WS from the sample with a pipette. Repeat **steps 3** and **4** until Coomassie blue stain is washed off.

5. Add 200 μL of ACN and dehydrate the gel pieces for ~5 min at RT (*see* **Note 7**).

6. Remove the ACN from the sample with a pipette.

7. Add 30 μL of DTT solution into the gel pieces to reduce the proteins for 30 min at 37 °C.

8. Remove the excess DTT solution, and add 30 μL of iodoacetamide (IAA) solution for protein alkylation for another 30 min at 37 °C.

9. Remove excess IAA from the sample with a pipette.

10. Add 200 μL of ACN to dehydrate the gel pieces at RT.

11. Remove the ACN from the sample, and add 200 μL of 50 mM ammonium bicarbonate (AB) to rehydrate the gel pieces at RT for 10 min.

12. Remove AB from the sample and dehydrate the gel pieces with 200 μL of ACN.

13. Remove the ACN from the sample and air-dry the gel pieces at RT.

14. Add 30 μL of trypsin solution to the gel pieces and wait for ~10 min until the gel pieces are rehydrated by the trypsin solution. Remove the excess trypsin solution and then add an additional 5 μL of 50 mM AB. Incubate the sample at 37 °C for trypsin digestion overnight.

15. Extract the digested peptides by adding 30 μL of 0.1 % TFA with gentle vortex, and collect the supernatant into a 0.5 mL Eppendorf tube. Then add 30 μL of the extraction buffer.

16. Repeat **step 15** once and pool the peptides together.

17. The resulting peptides are partitioned between organic and aqueous phases against $CHCl_3$ in a 1:1 ratio at RT with slow rotation. The $CHCl_3$ fraction should be carefully separated from the hydrophilic substances. Avoid the contamination from interphase region. The $CHCl_3$ fraction is concentrated in a centrifugal evaporator (SpeedVac®, Savant, Asheville, NC) and subjected to analysis on either a 4800 MALDI TOF/TOF instrument (AB Sciex) or an Orbitrap Velos tandem mass spectrometry instrument (Thermo Scientific) coupled with an Ultimate® 3000 Nano HPLC (Dionex, Thermo Fisher, Bannockburn, IL) (*see* **Note 8**).

3.3 Analyze PHB-Modified Peptides Using MALDI TOF/TOF MS

1. The sensitivity and resolution of MALDI TOF/TOF MS is tuned and optimized using the mass standard mixture kit prior to the sample analysis.

2. Calibrate the mass accuracy of the mass spectrometer in MS reflector mode using six peptide masses in the mass standard mixture. Select "plate mode & default calibration" in the "cal type" to update the Detector offset, TOF offset, and B-factor in reflector mode. Update the MS/MS default calibration using the fragments from GluFib (m/z 1570.677) (*see* **Note 9**).

3. Peptides from the $CHCl_3$ fraction are dissolved in 60 % ACN containing 0.1 % TFA and mixed with MALDI matrix in a 1:1 ratio and spotted on a MALDI plate.

4. Create a new spot set in 4000 series Explorer software. Set up the acquisition and processing method. For MS analysis, m/z 700–5000 is set for mass range; 750 laser shots are accumulated for MS spectra at a laser intensity of 3500 using positive ion reflector mode. GFP (m/z 1570.677) and ACTH 18–39 (m/z 2465.199) are used for internal calibration. For MS/MS analysis, the precursor mass corresponding to the theoretical m/z of PHB-modified peptides is selected for CID

fragmentation with 1 kV collision energy. Fifteen hundred laser shots are accumulated for each MS/MS spectra at a laser intensity of 4200. The precursor mass window is set at a relative 200 resolution (FWHM) (*see* **Note 10**).

3.4 Analyze PHB-Modified Peptides Using LC-MS/MS

1. LTQ-Orbitrap Velos MS instrument was used for LC-MS/MS analysis. Tune and calibrate the sensitivity, resolution, and mass accuracy of the instrument using the LTQ Velos ESI positive Ion Calibration Solution (Pierce).

2. The peptides from $CHCl_3$ partition are first concentrated by a SpeedVac, and subjected to nano-C_4 reversed-phase liquid chromatography separation with a 180-min binary gradient of Solvent A (2 % ACN, 0.1 % FA) and Solvent B (85 % ACN, 0.1 % FA) (*see* **Note 11**).

Time (min)	Solvent A (%)	Solvent B (%)
0	98	2
6	98	2
6.5	95	5
120	50	50
165	5	95
180	5	95

3. The eluted peptides are directly introduced into a nanoelectrospray ionization source on the LTQ-Orbitrap Velos MS system with a spray voltage of 2.15 kV and a capillary temperature of 275 °C. MS spectra are acquired in the positive ion mode with a mass scanning range of m/z 350–2000; 10 most abundant ions are selected for collision-induced dissociation (CID) fragmentation. The precursor isolation width is set to 3 m/z, and a minimum ion threshold count is set to 3000. The lock mass feature was used for accurate mass measurements.

3.5 Database Search, Peptide, and Their PHB Modification Site Identification

1. The MS/MS spectra are searched against a Swissprot rat database (7631 sequences) using MASCOT (V. 2.3) search engine. Since there is no predefined PHB modification available in Mascot search engine, we manually added a new modification with 1–20 units of PHB ($C_4H_6O_2$) on serine residues into Mascot Configuration. For example, 86.0368 Da, 172.0736 Da, or 258.1103 Da were added for 1, 2, or 3 units of PHB modification, respectively.

2. The following parameters are used for peptide identification: MS error window is set as 10 ppm and an MS/MS error window is 0.5 Da. Methionine oxidation, cysteine

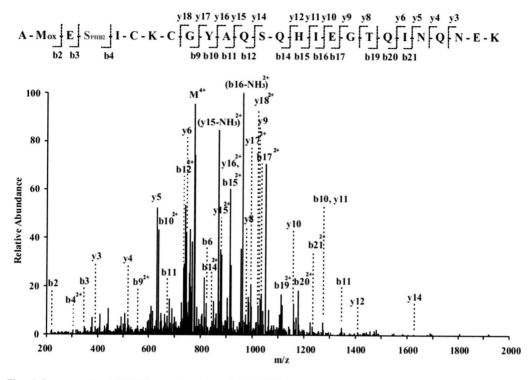

Fig. 4 An example of MS/MS spectrum from LC-MS/MS analysis. A quadruply charged ion (*m/z* 775.354) corresponding to peptide [63]AMESICKCGYAQSQHIEGTQINQNEK[88] with a Met[64] oxidation and 2 units of PHB modification on the Ser66 residue (with a Mascot score of 34) is presented here as an example. The observed y- and b-ion series confirmed the peptide sequence. Mass differences between b3 and b4 suggested the localization of the addition of two PHB units on Ser[66] residue

carbamidomethyl (IAM) modification, and up to 20 units of PHB modification of serine were set as variable modifications. Trypsin digestion allowing up to two missed cleavages is specified and peptide charge states are selected as 2^+, 3^+, and 4^+.

3. Only peptides identified with confidence interval (C. I.) values of 95 % and better are considered confidently identified. The PHB modification site is also manually evaluated based on the presence of continuous b- and/or y-series of ions (Fig. 4).

---

4   Notes

1. We prefer to collect the cells expressing TRPM8 with PBS at RT. HEK cells are easy to remove from dishes by applying large volumes (~7 mL) of PBS and pipetting the cells suspension up and down until all the cells are detached. Add more PBS to rinse the dish and combine the suspensions.

2. Store the cell pellet at −20 °C. In cases when the pellets are used immediately, we still recommend placing the cells into the

−20 °C freezer for ~30 min prior to protein extraction. Frozen cells are more easily lysed, and it helps to obtain more homogenous cell lysis suspension.

3. The choice of the tubes is important for the freezing step. Check the cryo-resistance of the tubes to prevent crack formation. We prefer the polypropylene tubes, purchased from Corning (*see* Section 2).

4. DDM should be weighed and always freshly added to the NCB-resuspension buffer. Use sterile spatula to remove the membrane pellets from the centrifuge tube as they are usually very sticky and require mechanical force to break the dense pellet apart. Make sure not to leave large clots of the pellet, as this will prevent freeing and extracting the protein from the membranes into the detergent micelles. This step is very important to maximize the yield of the purified proteins.

5. Make sure to allow enough time for beads to precipitate on the magnet by keeping the beads on the magnetic platform for at least 3–5 min. Never centrifuge the beads at high speed as it may damage the resin.

6. After incubating the beads with protein samples, the suspension may be dense and viscous, thus allow the beads to precipitate for a longer time than during the regular washing steps without the protein sample. Usually, beads in the presence of protein supernatant adhere to the magnetic base in about 5–10 min. Check the solution's clarity before removing the buffer. Allow for a longer time if needed.

7. After adding ACN, the gel pieces will change to a white color, and the size of the gel will become much smaller.

8. The chloroform partition is used for separating the peptides into a hydrophilic group and a hydrophobic group. Based on the previous publications [11–13], the association with PHB may render amphiphilic or even hydrophilic peptides soluble in chloroform. In our study, we found that the majority of the peptides that had been identified as PHB-modified before partition were extracted into the chloroform layer.

9. To ensure optimized mass accuracy, the high voltage for MALDI source needs to be turned on for at least 30 min before updating the default calibration. This step would minimize the fluctuation in accelerating voltages to provide more accurate calibration.

10. The composition of each PHB unit is $C_4H_6O_2$. The modification is at serine residue with the addition mass of 86.0368 Da. We added 1–20 units of PHB modification in the configuration of local Mascot search engine and set the PHB modification as variable modification during the data search.

11. Some of the peptides in CHCl$_3$ fraction are very hydrophobic, so we choose to use C$_4$ trap column and nano column for peptides separation.

Acknowledgements

This work was supported by the American Heart Association SDG-2640223 grant to E. Z., and the National Institutes of Health grant R01GM098052 to E. Z. The project described was supported by a grant from the Foundation of UMDNJ and a grant (P30NS046593) from the National Institute of Neurological Disorders and Stroke. The content is solely the responsibility of the authors and does not necessarily represent the official views of the National Institute of Neurological Disorders and Stroke or the National Institutes of Health. The authors report no conflict of interest.

References

1. Parker CE, Mocanu V, Mocanu M, Dicheva N, Warren MR (2010) Mass spectrometry for post-translational modifications, Chapter 6. In: Alzate O (ed) Neuroproteomics. CRC Press, Boca Raton, FL

2. Seebach D, Brunner A, Burger HM, Schneider J, Reusch RN (1994) Isolation and 1H-NMR spectroscopic identification of poly(3-hydroxybutanoate) from prokaryotic and eukaryotic organisms. Determination of the absolute configuration (R) of the monomeric unit 3-hydroxybutanoic acid from Escherichia coli and spinach. Eur J Biochem 224:317–328

3. Cao C, Yudin Y, Bikard Y, Chen W, Liu T, Li H, Jendrossek D, Cohen A, Pavlov E, Rohacs T, Zakharian E (2013) Polyester modification of the mammalian TRPM8 channel protein: implications for structure and function. Cell Rep 4:302–315

4. Reusch RN (2012) Physiological importance of poly-(R)-3-hydroxybutyrates. Chem Biodivers 9:2343–2366

5. Norris V, Bresson-Dumont H, Gardea E, Reusch RN, Gruber D (2009) Hypothesis: poly-(R)-3-hydroxybutyrate is a major factor in intraocular pressure. Med Hypotheses 73:398–401

6. Reusch RN (1989) Poly-beta-hydroxybutyrate/calcium polyphosphate complexes in eukaryotic membranes. Proc Soc Exp Biol Med 191:377–381

7. Cornibert J, Marchessault RH (1972) Physical properties of poly-hydroxybutyrate. IV. Conformational analysis and crystalline structure. J Mol Biol 71:735–756

8. Negoda A, Negoda E, Reusch RN (2010) Oligo-(R)-3-hydroxybutyrate modification of sorting signal enables pore formation by Escherichia coli OmpA. Biochim Biophys Acta 1798:1480–1484

9. Negoda A, Negoda E, Reusch RN (2010) Importance of oligo-R-3-hydroxybutyrates to S. lividans KcsA channel structure and function. Mol Biosyst 6:2249–2255

10. Zhang Z, Tan M, Xie Z, Dai L, Chen Y, Zhao Y (2011) Identification of lysine succinylation as a new post-translational modification. Nat Chem Biol 7:58–63

11. Castuma CE, Huang R, Kornberg A, Reusch RN (1995) Inorganic polyphosphates in the acquisition of competence in Escherichia coli. J Biol Chem 270:12980–12983

12. Pavlov E, Zakharian E, Bladen C, Diao CT, Grimbly C, Reusch RN, French RJ (2005) A large, voltage-dependent channel, isolated from mitochondria by water-free chloroform extraction. Biophys J 88:2614–2625

13. Seebach D, Fritz MG (1999) Detection, synthesis, structure, and function of oligo(3-hydroxyalkanoates): contributions by synthetic organic chemists. Int J Biol Macromol 25:217–236

Neuromethods (2016) 114: 289–290
DOI 10.1007/978-1-4939-3472-0
© Springer Science+Business Media New York 2016

INDEX

Printed in the United States
By Bookmasters